Random Operators

Disorder Effects on Quantum Spectra and Dynamics

Random Operators

Disorder Effects on Quantum Spectra and Dynamics

Michael Aizenman
Simone Warzel

Graduate Studies in Mathematics
Volume 168

American Mathematical Society
Providence, Rhode Island

2010 *Mathematics Subject Classification.* Primary 82B44, 60H25, 47B80, 81Q10, 81Q35, 82D30, 46N50.

For additional information and updates on this book, visit
www.ams.org/bookpages/gsm-168

Library of Congress Cataloging-in-Publication Data

Aizenman, Michael.
Random operators : disorder effects on quantum spectra and dynamics / Michael Aizenman, Simone Warzel.
pages cm. — (Graduate studies in mathematics ; volume 168)
Includes bibliographical references and index.
ISBN 978-1-4704-1913-4 (alk. paper)
1. Random operators. 2. Stochastic analysis. 3. Order-disorder models. I. Warzel, Simone, 1973– II. Title.

QA274.28.A39 2015
535′.150151923—dc23

2015025474

∞ The paper used in this book is acid-free and falls within the guidelines
established to ensure permanence and durability.
Visit the AMS home page at `http://www.ams.org/`

10 9 8 7 6 5 4 3 2 1 20 19 18 17 16 15

Dedicated to Marta by Michael

and to Erna and Horst by Simone

Contents

Preface

Disorder effects on quantum spectra and dynamics have drawn the attention of both physicists and mathematicians. In this introduction to the subject we aim to present some of the relevant mathematics, paying heed also to the physics perspective.

The techniques presented here combine elements of analysis and probability, and the mathematical discussion is accompanied by comments with a relevant physics perspective. The seeds of the subject were initially planted by theoretical and experimental physicists. The mathematical analysis was, however, enabled not by filling the gaps in the theoretical physics arguments, but through paths which proceed on different tracks. As in other areas of mathematical physics, a mathematical formulation of the theory is expected both to be of intrinsic interest and to potentially also facilitate further propagation of insights which originated in physics.

The text is based on notes from courses that were presented at our respective institutions and attended by graduate students and postdoctoral researchers. Some of the lectures were delivered by course participants, and for that purpose we found the availability of organized material to be of great value.

The chapters in the book were originally intended to provide reading material for, roughly, a week each; but it is clear that for such a pace omissions should be made and some of the material left for discretionary reading. The book starts with some of the core topics of random operator theory, which are also covered in other texts (e.g., [**105, 82, 324, 228, 230, 367**]). From Chapter 5 on, the discussion also includes material which has so far been presented in research papers and not so much in monographs on the subject. The mark * next to a section number indicates material which the reader is advised to skip at first reading but which may later be found useful. The selection presented in the book is not exhaustive, and for some topics and methods the reader is referred to other resources.

During the work on this book we have been encouraged by family and many colleagues. In particular we wish to thank Yosi Avron, Marek Biskup, Joseph Imry, Vojkan Jaksic, Werner Kirsch, Hajo Leschke, Elliott Lieb, Peter Müller, Barry Simon, Uzy Smilansky, Sasha Sodin, and Philippe Sosoe for constructive suggestions. Above all Michael would like to thank his wife, Marta, for her support, patience, and wise advice.

The editorial and production team at AMS and in particular Ina Mette and Arlene O'Sean are thanked for their support, patience, and thoroughness. We also would like to acknowledge the valuable support which this project received through NSF research grants, a Sloan Fellowship (to Simone), and a Simons Fellowship (to Michael). Our collaboration was facilitated through Michael's invitation as J. von Neumann Visiting Professor at TU München and Simone's invitation as Visiting Research Collaborator at Princeton University. Some of the writing was carried out during visits to CIRM (Luminy) and to the Weizmann Institute of Science (Rehovot). We are grateful to all who enabled this project and helped to make it enjoyable.

Michael Aizenman, Princeton and Rehovot
Simone Warzel, Munich
2015

Chapter 1

Introduction

Quantum dynamics is famously described by a unitary evolution in the Hilbert space of states. Among the solvable classes of models, for which a qualitatively complete theory could be obtained, is the the Bloch-Floquet theory of periodic operators. The structure of periodic operators' eigenfunctions has long provided the basic reference point for condensed matter theory. However, our luck with the availability of explicitly analyzable models starts to run out once disorder (and particle interactions) are incorporated.

As it turns out, a certain amount of disorder in condensed matter is hard to avoid and for some purposes is also advantageous. The spectral and dynamical effects of disorder have attracted a great deal of attention among physicists, mathematicians, and those who enjoy working at the fertile interface of the two subjects. Along with a rich collection of results, their research has yielded a number of basic principles, expressing physics-style insights and mathematically interesting theory in which are interwoven elements from probability theory, functional analysis, dynamical systems, topology, and harmonic analysis (not all of which are fully covered in this book). Yet deep challenges remain, and fresh inroads into this territory are still being made.

The topics presented in this book are organized into interlinked chapters whose themes can be read from their titles. The goal of this introduction is to sketch the central mathematical challenge concerning the effects of disorder on quantum spectra and dynamics and to mention some of the concepts which play an essential role in the theory which is laid down here.

Admittedly, at first reading the concepts mentioned below may not be clear to readers who are new to the subject. In that case the reader is encouraged to skip the text and return to it after gaining some familiarity with the relevant sections in the book.

1.1. The random Schrödinger operator

The quantum state of a particle moving in d-dimensional space is described by a wave function $\psi \in L^2(\mathbb{R}^d)$. It evolves in time under the unitary group of operators $\exp(-itH/\hbar)$ generated by the Schrödinger operator

$$H = -\frac{\hbar^2}{2m}\Delta + V(x) \tag{1.1}$$

with Δ the Laplacian and $V : \mathbb{R}^d \mapsto \mathbb{R}$ the external potential.

Disorder may be incorporated into quantum models through the addition of random terms in the potential, possibly as an addition to a periodic potential which represents an underlying lattice structure. Models incorporating such terms have appeared in the discussions of substitutional alloys, of metals with impurities, and also in the theory of normal modes of large structures.

Somewhat similarly, the positions of electrons in a metal are described in terms of lattice sites which represent the Wigner-Seitz cells. Simplifying this further by restricting to one quantum state per cell (the tight-binding approximation), allowing as elementary moves only nearest-neighbor hopping, and pretending that the electron-electron interaction is sufficiently represented by an effective one-particle potential, one is led to a one-particle Hamiltonian for the system in the form of a discrete random Schrödinger operator

$$\boxed{H = -\Delta + \lambda V} \qquad \text{on } \ell^2(\mathbb{Z}^d). \tag{1.2}$$

Here Δ is the second difference operator (defined in (2.3) below). As will be done subsequently, the physical constants which appear in (1.1) are dropped; their value in this context being a matter of phenomenology. The operator V acts as multiplication by random variables $(\omega(x))$, which are often taken to be independent and identically distributed (iid). We shall not discuss the validity of the approximations which were made in formulating this model but rather focus on their implications.

In other examples of Schrödinger operators with random potential the operator is of the form

$$H = -\Delta + V_0(x) + \lambda \sum_{\alpha} \omega_\alpha \, u(x - x_\alpha) \tag{1.3}$$

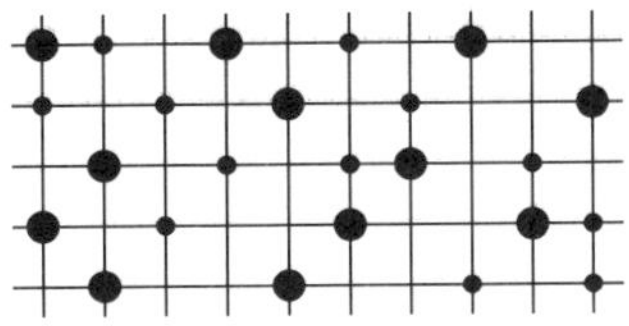

Figure 1.1. A disordered lattice system; the dots representing the on-site potential.

where Δ is the regular Laplacian on $\mathbb{R}^d$, V_0 is a periodic potential, and the sites x_α may range over either the lattice $\mathbb{Z}^d$ or a random discrete subset of $\mathbb{R}^d$ generated through a Poisson process of constant intensity. In the lattice case, the disorder is incorporated by taking the coefficients ω_α to be iid random variables. In the second case, disorder is already present in the location of the scatterers, but a further modeling choice can be made with ω_α either *iid* or constant. The formulation of the model over the continuum is not expected to make an essential change in the basic phenomena discussed here. These concern the long scale behavior of the eigenfunctions and of the dynamics. Yet the analysis would require addressing a number of issues related to the unboundedness (at short distances) of the kinetic term $(-\Delta)$. Omitting randomness in (ω_α) one would also give up random parameters on which the dependence of H is monotone. The monotonicity is a convenient feature which we shall adapt for this presentation.

To summarize the point let us restate that our main goal here is not to cover all the variants of random Schrödinger operators but rather to focus on the qualitative spectral and dynamical implications of disorder in the context of the relatively simpler versions of such random operators. And since in the discrete version of (1.2) one avoids a layer of difficulties which may be skipped in the first presentation of the main issues discussed here, we will restrict the discussion to the discrete models.

1.2. The Anderson localization-delocalization transition

It is instructive to note that the operator in (1.2) is a sum of two terms with drastically opposed spectral properties (terms which are explained in Chapter 2 and Appendix A).

The kinetic term $-\Delta$: It is of *absolutely continuous spectrum*. The plane waves $(e^{ik\cdot x})$ provide for it a spanning collection of generalized eigenfunctions which are obviously *extended*, and the evolution it generates is *ballistic* in the sense that for a generic initial state

$\psi_0 \in \ell^2(\mathbb{Z}^d)$ as $t \to \infty$,

$$\langle \psi_0, \, e^{-it\Delta} \, |x|^2 \, e^{it\Delta} \, \psi_0 \rangle \;\approx\; 2d \, \|\psi_0\|^2 \, t^{2\nu} \tag{1.4}$$

with $\nu = 1$.

The potential V**:** It acts as a multiplication operator on $\ell^2(\mathbb{Z}^d)$ and has a *pure-point spectrum* which consists of the countably infinite collection of its values which *densely* cover the support of the probability distribution of the single-site potential. Its eigenfunctions are the *localized* delta functions (δ_x) and the corresponding dynamics exhibit an extreme form of localization (in particular (1.4) holds with $\nu = 0$).

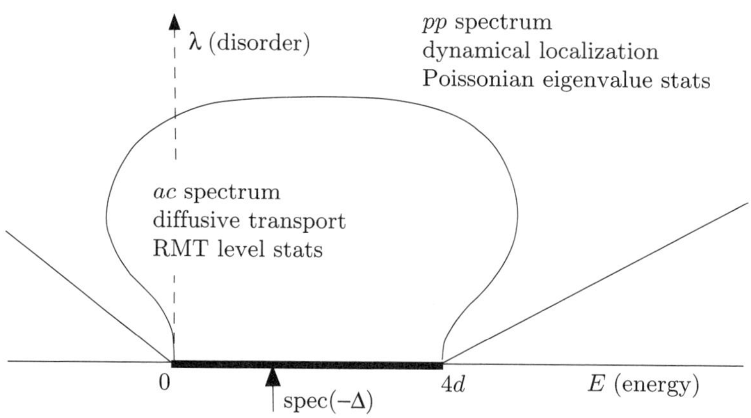

Figure 1.2. The predicted shape of the phase diagram of the Anderson model (1.2) in dimensions $d > 2$ for site potentials given by bounded iid random variables with a distribution similar to (1.5).

In his seminal work P. W. Anderson posited [**27**] that under random potential there would be a transition in the transport properties of the model which heavily depend on the dimension d of the underlying lattice, the strength $\lambda \in \mathbb{R}$ of the disorder, and the energy. The term **mobility edge** was coined for the boundary of the regime at which conduction starts.

Subsequent works [**132, 1**] have led to the current widely held, but not proven, conjecture that such phase transitions would be seen in dimensions $d > 2$, where operators such as (1.2) may have phase diagrams as depicted in Figure 1.2, which for the sake of concreteness is sketched having in mind the iid random variables with the uniform distribution in the unit interval:

$$\rho(dv) \;=\; 1[|v| < 1/2] \, dv \, . \tag{1.5}$$

The diagram's essential features are [**27, 307**]:

1. For $\lambda > 0$ the particle's energy ranges over the *set sum* of the spectra of the two terms in H, which here is the interval $\left[-\frac{1}{2}\lambda, \frac{1}{2}\lambda + 4d\right]$.
2. For small values of $\lambda > 0$, a pair of mobility edges separates the outer regimes of *localized states* from the intermediate energy regime which is conjectured to correspond to **diffusive transport** ($\nu = 1/2$).
3. At **high disorder**, i.e., for all λ exceeding some critical value, the spectrum is **completely localized**.

Disorder has a particularly drastic effect in one dimension, where it produces **complete localization** at any strength $\lambda \neq 0$, as was first pointed out by N. F. Mott and W. D. Twose [**306**]. The localization theory for $d = 1$ was further expanded by R. E. Borland [**64**] and others and established rigorously in the works by K. Ishii [**200**] (absence of *absolutely continuous* spectrum) and I. Goldsheid, S. Molchanov, and L. Pastur [**176**] (proof that the spectrum is *pure point*).

One-dimensional systems can also be regarded as *quantum wires* and from this perspective it is natural to approach the conductive properties through **reflection** and **transmission** coefficients. Such an approach was championed by R. Landauer [**280**]. The two approaches, through spectral characteristics and/or reflection coefficients, are nicely tied together in the Kotani theory [**263, 349**] (which seems to be largely unknown among physicists). It yields the general statement that for one-dimensional Schrödinger operators with shift-invariant distribution absolutely continuous spectrum is possible only for potentials which are **deterministic** under shifts, and it occurs only if the wire is **reflectionless**. The discrete version of the Kotani theory, which was formulated by B. Simon [**349**], is presented in Chapter 12 and used there as the lynchpin for the proof of complete localization for one-dimensional random Schrödinger operators. (Our presentation differs in this respect from the more frequently seen approaches to the one-dimensional case.)

The first rigorous proofs of Anderson (spectral) localization for $d > 1$ relied on the **multi-scale method** of J. Fröhlich and T. Spencer [**165**]. The method drew some inspiration from the Kolmogorov-Arnold-Moser (KAM) technique for the control of resonances and proofs of the persistence of integrability. It is of relevance and use also for quasi-periodic systems such as quasi-crystals. The **fractional moment method**, which arrived a bit later [**8**], was more specifically designed for random systems. It allows an elementary proof of localization, which we present in Chapter 6 and in more detail in Chapters 10 and 11. Through the relations which are derived in

Chapter 7, it also yields estimates on the **eigenfunction correlator** and hence allows to prove also **dynamical localization** with simple exponential bounds [**5**].

The localization-delocalization transition has been compared to phase transitions in statistical mechanics. The analogy has inspired the renormalization group picture suggested in [**1**], which is one of the arguments quoted in support of the above dimension dependence. Other helpful analogies are found in Chapter 9, where we present methods for establishing exponential decay of two-point functions and finite-volume criteria, which have also played a role in the analysis of the phase transitions in percolation and Ising systems.

The differences in the nature of the eigenfunctions in the regimes described above are also manifested in the different degrees of level repulsion and thus in differences in the **spectral statistics** on the scale of the typical level spacing in finite-volume versions of the model.

As will be illustrated in Chapter 17, in the pure-point regime of localized eigenstates the level repulsion is off, and the level statistics is that of a **Poisson process** of the appropriate density. This was first proven for one-dimensional systems by S. Molchanov [**303**] and for multi-dimensional discrete systems by N. Minami [**301**] (under the assumption of rapid decay of the Green function's fractional moments).

An intriguing conjecture is that in the regime of extended states the statistics may be close to those of the **random matrix ensemble**. Since the randomness is limited in Schrödinger operators to just the diagonal part, such a result does not yet follow from the recent results on classical matrices ensembles [**143, 374, 375, 144, 145**] and this challenge remains open.

In Chapters 13 and 14 we discuss some of the implications of Anderson localization for condensed matter physics concerning the **conduction** properties and the **integer quantum Hall effect** (IQHE). Disorder was found to serve as an enhancing factor in the IQHE. The latter provides an example of exquisite physics (allowing to determine e^2/h experimentally to precision 10^{-9} [**253**]) intertwined with mathematical notions of operator theory, topology, and probability [**282, 386, 38, 48, 41, 49**].

1.3. Interference, path expansions, and the Green function

Localization in quantum systems is ultimately an expression of destructive phase interference. However, the extraction of localization bounds through estimates relating directly to path interference is beyond the reach of available methods (though a certain success has been chalked up in [**339**]). Instead, typically the analysis proceeds through the study of the Green function:

$$G(x,y;z) := \left\langle \delta_x, \frac{1}{H-z}\,\delta_y \right\rangle = -i \int_0^\infty \langle \delta_x, e^{-it(H-z)}\,\delta_y\rangle\, dt\,. \tag{1.6}$$

Like the unitary operator to which it is related, the Green function's value is affected by the interference of path-dependent amplitudes, though in this case over paths of varying time duration. While this point of view is good to keep in mind, in practice the analysis is most often carried through methods which are enabled by the Green function's algebraic and analytic properties. Altogether, the Green function provides a remarkable tool for a number of reasons:

Informative: There is an easy passage from *bounds* and other qualitative information on $G(x,y;E+i0)$ to a host of quantities of interest about the model: the operator's spectrum, the nature of its *eigenfunctions* (Chapter 5), *time evolution* (Chapter 2), *conductance* (Chapter 13), the kernel of the *spectral projection* and hence also the ground state's n point functions for the related many-particle system of free fermions (Chapter 13).

Algebraic relations: The Green function's analysis is facilitated by various relations that are implied by elementary linear algebra. Among these are the resolvent identity, rank-one perturbation formula, Schur complement or Krein-Feshbach-Schur projection formulas (Chapter 5), and geometric decoupling relations (Chapter 11).

Path expansions: Resolvent expansions, an example of which can be obtained by treating the hopping term in H as a perturbation on the local potential, allows us to express $G(x,y;E+i0)$ in terms of a sum of path amplitudes, over paths linking the sites x and y. Partial resummation of the terms, organized into loop-erased paths, yields the very useful Feenberg expansion (Chapter 6). The expansion was applied to the localization problem in Anderson's original paper [**27**], and it remains a source of much insight on the Green function's structure.

Locality: Underlying some of the relations discussed below is the fact that the Green function is associated with a local operator. In this regard, the two-point function $G(x,y;E+i0)$ resembles the connectivity function of percolation models and the correlation function of Ising spin systems. This analogy has led to some useful tools for the analysis of the localization regime such as finite-volume criteria (Chapter 9).

Herglotz property: In its dependence on the energy parameter z, for any given ψ (element of the relevant Hilbert space) $\langle\psi, (H-z)^{-1}\psi\rangle$ is a holomorphic function taking the upper half-plane into itself and thus a function in the Herglotz-Pick class (Appendix B). Some of the general properties of functions in this class are behind the success and relevance of the *fractional moment method* (Chapter 8) which has yielded an effective tool for establishing Anderson localization and studying its dynamical implications.

1.4. Eigenfunction correlator and fractional moment bounds

A good example of the utility of the Green function is its relation to the **eigenfunction correlator** $Q(x, y; I)$:

$$Q(x,y;I) \;=\; \sum_{E\in\sigma(H)\cap I} |\psi_E(x)|\,|\psi_E(y)|\,. \tag{1.7}$$

written here for a matrix with simple spectrum, with the sum extending over the normalized eigenfunctions of energies in the specified interval $I \subset \mathbb{R}$. A natural generalization of this kernel is presented in Chapter 7. One can learn from it both about the dynamics and the structure of the eigenfunctions. Its average $\mathbb{E}[\cdot]$ over the random potential obeys for all $s \in (0,1)$

$$\boxed{\mathbb{E}\left[Q(x,y;I)\right] \;\le\; C_s(\rho)\; \liminf_{\eta\downarrow 0} \int_I \mathbb{E}\left[|G(x,y;E+i\eta)|^s\right] dE}\,, \tag{1.8}$$

with $C_s(\rho) < \infty$ for a broad class of distributions. A technically convenient expression of localization is in bounds on the two-point function

$$\tau(x,y;E) \;:=\; \mathbb{E}\left[|G(x,y;E+i0)|^s\right]\,, \tag{1.9}$$

at some $s \in (0,1)$, e.g., an exponential bound of the form

$$\sum_{y:\mathrm{dist}(x,y)\ge R} \tau(x,y;E) \;\le\; A\, e^{-R/\xi} \tag{1.10}$$

at $\xi < \infty$ which depends on the energy and the distribution of the random variables. Since for any spectral projection $P_I(H)$ (to the subspace of energies in $I \subset \mathbb{R}$) the time evolution operator satisfies

$$\left|\langle\delta_x, P_I(H)\, e^{-itH}\delta_y\rangle\right| \;\le\; Q(x,y;I)\,, \tag{1.11}$$

the fractional moments bound (1.10) if holding at some $s \in (0,1)$ implies, through (1.8), exponential *dynamical localization. Spectral localization*, in the sense of exponential localization of all the eigenfunctions for *almost every* realization of the random operator, can then be deduced using other standard tools which are discussed in Chapters 2 and 7. Fractional moment techniques yield proofs of Anderson localization in various regimes in the (E, λ)-phase space, starting with the high disorder regime for (regular) graphs of a specified degree (Chapter 11).

1.5. Persistence of extended states versus resonant delocalization

With the localization being now somewhat understood (though not completely, in particular in reference to two dimensions) the persistence of extended states, or delocalization, for random Schrödinger operators continues to offer an outstanding challenge. The main case for which it has been established rigorously is that of regular tree graphs, which are discussed in Chapter 16. A lesson which can be drawn from the analysis of the Anderson model in that case is that there may be two different mechanisms for extended states in the presence of disorder:

Continuity: For tree graphs, and some graphs close to those, there exist continuity arguments which allow us to prove the persistence of absolutely continuous spectrum at weak disorder, at least perturbatively close to the disorder-free operator's spectrum [**246, 14, 161**].

Resonant delocalization: On graphs with rapid growth of the volume, as function of the distance, localization may be unstable to the formation of extended states through rare resonances among local quasi-modes. An argument based on this observation yields for random Schrödinger operators on tree graphs a delocalization criterion whose reach appears to be *complementary* to that of the fractional moment localization criterion. And in case the random potential is unbounded it implies absolutely continuous spectrum even at weak disorder and well away from the ℓ^2-spectrum of the free operator (i.e., the graph Laplacian) [**21**]. The consequences are no less striking for the Anderson model on tree graphs with *bounded potential*, for which it was proven that a minimal **disorder threshold** needs to be met for there to be a mobility edge beyond which localization sets in [**19**].

Further implications of the second mechanism are still being explored. Among the interesting questions are

1. its possible manifestation in many-particle systems, with implications for conductance (regimes of "bad metallic conductivity") [**25, 44**],
2. the nature of eigenstates, which may be delocalized in the sense of geometric spread yet also non-ergodic in the sense that they violate a heuristic version of the equidistribution principle [**318**],

3. spectral statistics (intermediate phase which neither shows Poisson nor random matrix statistics [**58**]).

1.6. The book's organization and topics not covered

Included in a number of chapters are methods which are of relevance beyond the specific context of random Schrödinger operators. As can be seen in the table of contents, the first four chapters present some of the core material on the subject. These topics are also covered in other textbooks and extended reviews on random operators, such as [**228, 105, 82, 324, 367, 230**]. The discussion in the remaining chapters centers on methods and results which have so far been presented mainly in research papers and not much in monographs on the subject. Included there are also some recent results and comments on work in progress.

Let us conclude by noting that localization by disorder is a phenomenon of relevance in the broad range of systems governed by wave equations. That includes, beyond the Schrödinger equation, sound waves and *normal modes in vibrating systems* and also light propagation in disordered medium; see [**85, 113, 152, 153, 24**] and the references therein. In fact, since photons even in non-linear optical media do not interact as strongly as electrons do, direct observations of Anderson localization were purportedly first realized in *photonics systems*; see [**344**] and also the overview [**276**] (which is regrettably short on mathematical references to the subject).

This book is far from being exhaustive in terms of the subjects and methods covered. For that, one may need to add a rich collection of topics, including *quasi-periodic* operators [**324, 68**], the *multi-scale method* for establishing localization [**230, 367, 169, 170**], the *transfer-matrix* approach to localization in one dimension [**66, 82**], *quantum graphs* [**266, 339, 15, 146, 262**], *random network models* and *random quantum walks* [**31, 32, 212, 187, 213, 188, 214, 189**], *supersymmetric models* of Wegner and their relatives [**120, 121, 119**], *random-matrix* models of disordered systems [**133**], and then of course disorder effects in *non-linear dynamics*, such as the non-linear Schrödinger evolution, and the *quantum kicked rotator* [**157, 68**]. Also not discussed here are currently emerging questions and observations concerning multi-particle systems [**17, 88, 89, 90, 147**] and many-particle localization [**198**]. Further references to the above and to other topics are made in remarks and in Notes which are included in many chapters. More on the relevant *physics concepts* can be found in [**289, 399, 61, 199**].

Chapter 2

General Relations Between Spectra and Dynamics

The quantum time evolution of a system whose state vectors ψ range over a Hilbert space $\mathcal{H}$ is

$$\psi(t) := e^{-itH}\psi\,, \quad t \in \mathbb{R}\,, \tag{2.1}$$

with e^{-itH} a unitary group of operators whose generator H is a self-adjoint operator which is referred to as the quantum Hamiltonian. The **dynamical properties** of such evolution are closely related to the **spectral characteristics** of its generator. In this chapter we shall summarize some of the highlights of this relation.

The prototypical example of the operators which will be considered below is the Schrödinger operator

$$H \;=\; -\Delta \;+\; V\,, \tag{2.2}$$

acting in $\mathcal{H} = L^2(X)$, with X a somewhat generic symbol for the position or the configuration space, Δ denoting the corresponding Laplacian, and V a multiplication operator corresponding to a real-valued function on X satisfying some suitable regularity assumptions ([**105, 328**]). This setup includes the following:

1. A particle moving over the continuum, $X = \mathbb{R}^d$, or confined there to a finite box $\Lambda = [-L, L]^d \subset \mathbb{R}^d$. In the latter case, the self-adjointness

of the Laplacian requires us to limit its domain to functions satisfying suitable boundary conditions (cf. [**328, 329**]).

2. A system of N particles, with the configuration space $X = \Lambda^N$.
3. A quantum particle hopping on the vertices of a graph $X = \mathbb{G}$, in which case the Hilbert space is $\ell^2(\mathbb{G})$ and Δ is the graph Laplacian:

$$(\Delta\psi)(x) := \sum_{y\in\mathbb{G}\,:\,d(x,y)=1} [\psi(y) - \psi(x)] \tag{2.3}$$

with d denoting the metric on $\mathbb{G}$. This covers the tight-binding model for single electrons on a subset of the d-dimensional square lattice $\mathbb{Z}^d$.

The operators can be modified through the incorporation of a static electromagnetic field, a possibility which is of interest for the discussion in Chapter 13. Further structure can also be added through suitable choices of the potential, e.g., periodic background modified by random terms, as presented in Chapter 3. The principles discussed in the present chapter apply in equal measure to all such examples.

2.1. Infinite systems and their spectral decomposition

One may start from the simple, but perhaps somewhat astounding, observation that for any finite-dimensional quantum system the quantum time evolution is *quasi-periodic* in time. That might sound like a contradiction to the commonplace observations of *relaxation* in physical systems. Nevertheless a mathematical understanding of relaxation phenomena is attainable by paying closer attention to time scales and to the order of limits (first $L \to \infty$ and then $t \to \infty$) when the time evolution is considered for a large finite system of size L.

More explicitly, quantum observables are associated with self-adjoint operators, A, with the collection of possible outcomes of the measurement of each given by the operator's spectrum. For a system which at time $t = 0$ is in the state $\psi \in \mathcal{H}$ and for which an observable A is measured at a time $t > 0$, the probability of finding its value falling in the range $I \subset \mathbb{R}$ is given by

$$\mathrm{Prob}_{\psi(t)}(A \in I) := \langle\psi(t), P_I(A)\,\psi(t)\rangle = \|P_I(A)\,\psi(t)\|^2 , \tag{2.4}$$

where $P_I(A)$ is the spectral projection operator. Denoting by (E_n) the eigenvalues of H and by (ψ_n) the corresponding normalized eigenvectors (in case $\dim\mathcal{H} < \infty$),

$$\mathrm{Prob}_{\psi(t)}(A \in I) = \sum_{n,m} e^{it(E_n-E_m)}\,\langle\psi\,,\psi_n\rangle\,\langle\psi_n, P_I(A)\,\psi_m\rangle\langle\psi_m,\psi\rangle .$$

The fact that as $t \to \infty$ this function does not converge to a limit and that it repeatedly assumes values arbitrarily close to $\|P_I(A)\,\psi\|^2$ is reminiscent of the Poincaré recurrence phenomenon of classical mechanics of finite systems. There, the time evolution is described by symplectic and thus measure-preserving flows in the classical phase space. If taken literally, the recurrence has the implication that if gas is released into a room from a bottle, then with probability one there will be a moment when all the gas will be found back in the flask.

The paradox is resolved by the observation that the recurrence time for a macroscopic system is so long that well before the rare event happens the door will be opened, rendering this model of gas in the room insufficient. (The number of degrees of freedom in this experiment can be estimated through Avogadro's number: $N_A \approx 6 \cdot 10^{23}$ particles per mol). Actually, by the recurrence time for such an event far more grievous deviations from the idealized description will occur and most likely even the lab will no longer be there.

It is therefore physically relevant and mathematically convenient to consider the time evolution in the infinite-volume limit, with the understanding that this aims at capturing the way the system appears on the scale of the interatomic separation. For one particle, or any other fixed number N of particles, the corresponding state space is described by the Hilbert space $\mathcal{H} = L^2(X)$ with $X = \mathbb{R}^{Nd}$ or $\mathbb{Z}^{Nd}$, as appropriate to the model.

For infinite systems the **spectral representation** for the mean values of bounded and continuous functions $f \in C_b(\mathbb{R})$ of a self-adjoint operator H, in a normalized state $\psi \in \mathcal{H}$, takes the form

$$\boxed{\langle \psi \,, f(H)\psi \rangle \;=\; \int_{\mathbb{R}} f(E)\, \mu_\psi(dE)} \tag{2.5}$$

which generalizes the finite-dimensional eigenvalue expansion:

$$\langle \psi \,, f(H)\psi \rangle \;=\; \sum_n f(E_n)\; |\langle \psi_n, \psi \rangle|^2 \,. \tag{2.6}$$

The probability measure $\mu_\psi(dE)$ in (2.5) describes the spectral distribution of H in the state ψ. The **spectrum** $\sigma(H)$ of H is the minimal closed set which includes the support of these **spectral measures** for an $\mathcal{H}$-spanning set of vectors ψ.

Spectral measures $\mu_\psi(dE)$ admit the Lebesgue decomposition into their pure-point (*pp*), absolutely continuous (*ac*), and singular continuous (*sc*) components:

$$\boxed{\mu_\psi(dE) \;=\; \mu_\psi^{\mathrm{pp}}(dE) + \mu_\psi^{\mathrm{ac}}(dE) + \mu_\psi^{\mathrm{sc}}(dE)} \,. \tag{2.7}$$

The first term on the right side corresponds to a sum of point measures over proper (square summable or square integrable) eigenfunctions:

(2.8) $$\boxed{\mu_\psi^{\text{pp}} = \sum_n |\langle \psi, \psi_n \rangle|^2 \, \delta_{E_n}}\,.$$

The second term on the right side in (2.7) is an absolutely continuous measure (with respect to the Lebesgue measure dE), i.e., one of the form

$$\mu_\psi^{\text{ac}}(dE) = g_\psi(E)\, dE$$

with $g_\psi \in L^1(\mathbb{R})$ for which a more explicit expression is given in (2.11) below. The last term in (2.7) is a singular continuous measure (i.e., one which is supported on a set of zero Lebesgue measure, yet which does not charge single points).

Corresponding to the decomposition of measures is the orthogonal decomposition of $\mathcal{H}$ into the closed subspaces

(2.9) $$\mathcal{H}^\# := \left\{ \psi \in \mathcal{H} \,|\, \mu_\psi = \mu_\psi^\# \right\} \quad \text{with } \# = \text{pp, sc, ac}.$$

Some of the elements of the theory of spectral representation are summarized in Appendix A, and further discussion can be found in textbooks such as [**331, 328, 329, 118, 380**]. Among the notable general results are:

1. The spectral measure of a vector $\psi \in \mathcal{H}$ is encoded in properties of the corresponding diagonal element of the operator's resolvent:

   (2.10) $$\langle \psi, (H-z)^{-1} \psi \rangle = \int (E-z)^{-1} \mu_\psi(dE)\,.$$

   As a function of the complex parameter $z \in \mathbb{C}$, (2.10) is analytic away from the spectrum $\sigma(H)$ and over the upper half-plane it has the Herglotz property (Appendix A and Appendix B).

2. The absolutely continuous component of the spectral measure is given by

   (2.11) $$\boxed{\mu_\psi^{\text{ac}}(dE) = \frac{1}{\pi} \operatorname{Im}\langle \psi, (H-E-i0)^{-1} \psi \rangle \, dE}\,,$$

   where $\langle \psi, (H-E-i0)^{-1} \psi \rangle := \lim_{\eta \downarrow 0} \langle \psi, (H-E-i\eta)^{-1} \psi \rangle$, the limiting known to exist at almost every E, assuming only that H is self-adjoint.

3. For operators H which in a natural sense are limits of finite-volume approximants, H_L, it is convenient to define all as acting on the same Hilbert

space. Then, under mild assumptions, the measures $\mu_\psi(dE)$ (at specified $\psi \in \mathcal{H}$) are the limits, in the weak sense of probability measures on $\mathbb{R}$, of the point spectral measures corresponding to the finite-volume restrictions of the operators. The relevant technical assumption is strong-resolvent convergence of H_L to H. For Schrödinger operators this condition holds quite broadly, even allowing a variety of finite-volume boundary conditions. Their effects are not seen in the infinite-volume limit (with $L \to \infty$ taken first before $t \to \infty$) even though the choice of the boundary conditions affects the finite-volume spectra.

4. In an extension of (2.5) to the off-diagonal terms, one may write for all bounded and continuous $f \in C_b(\mathbb{R})$ and all $\phi, \psi \in \mathcal{H}$

$$\langle \phi\,,\, f(H)\,\psi\rangle \;=\; \int f(E)\,\mu_{\phi,\psi}(dE)\,, \tag{2.12}$$

with $\mu_{\phi,\psi}(dE)$ complex-valued Borel measures. These measures can be recovered from the diagonal ones through the polarization identity: $\mu_{\phi,\psi} := \frac{1}{4}(\mu_{\phi+\psi} - \mu_{\phi-\psi} - i\,\mu_{\phi+i\psi} + i\,\mu_{\phi-i\psi})$.

We next turn to general relations between the spectral properties of H and the dissipative properties of the dynamics it generates, which are related through the Wiener theorem and its extensions.

2.2. Characterization of spectra through recurrence rates

If a quantum system is initiated in a state $\psi \in \mathcal{H}$, the probability of finding it back in this state at time $t > 0$, if that is experimentally tested for, is

$$\left|\langle \psi, e^{-itH}\psi\rangle\right|^2 \;=\; |\hat{\mu}_\psi(t)|^2 \tag{2.13}$$

with

$$\hat{\mu}_\psi(t) \;:=\; \int e^{-itE}\mu_\psi(dE) \;=\; \langle \psi, \psi(t)\rangle\,, \tag{2.14}$$

which is the **Fourier transform** of the spectral measure of H associated with ψ. As was commented above, finite systems are recurrent: not only does the above return probability not decay to zero, but it will repeatedly assume values arbitrarily close to 1. However, recurrence is avoidable in models with an infinite-dimensional Hilbert space. In particular, the **Riemann-Lebesgue lemma** implies that for any ψ with absolutely continuous (ac) spectrum

$$\lim_{t\to\infty} \hat{\mu}_\psi^{\mathrm{ac}}(t) = 0\,. \tag{2.15}$$

By the following celebrated observation of N. Wiener a slightly weaker form of relaxation, with (2.15) replaced by decay of the Cesàro time-average, is actually equivalent to the continuity of the spectral measure μ_ψ, i.e., to the vanishing of the pure-point part μ_ψ^{pp}.

Theorem 2.1 (Wiener). *Let μ be a finite complex Borel measure on $\mathbb{R}$. Then*

$$\boxed{\lim_{T\to\infty}\frac{1}{T}\int_0^T |\hat{\mu}(t)|^2\,dt = \sum_{E\in\operatorname{supp}\mu^{\mathrm{pp}}} |\mu(\{E\})|^2}\,. \tag{2.16}$$

Proof. By Fubini's theorem,

$$\frac{1}{T}\int_0^T |\hat{\mu}(t)|^2\,dt \;=\; \int\!\!\int \left(\frac{1}{T}\int_0^T e^{i(E-E')t}\,dt\right)\mu(dE')\,\overline{\mu(dE)}\,. \tag{2.17}$$

The innermost integral is uniformly bounded by one and converges pointwise to Kronecker's delta $\delta_{E,E'}$. The dominated convergence theorem then implies

$$\begin{aligned}\lim_{T\to\infty}\frac{1}{T}\int_0^T |\hat{\mu}(t)|^2\,dt &= \int\!\!\int \delta_{E,E'}\,\mu(dE')\,\overline{\mu(dE)}\\ &= \int \mu(\{E\})\,\overline{\mu(dE)} = \sum_{E\in\operatorname{supp}\mu^{\mathrm{pp}}} |\mu(\{E\})|^2\,. \quad \square\end{aligned} \tag{2.18}$$

The next result goes beyond Wiener's theorem by providing an explicit bound on the decay rate of the recurrence frequency. Loosely speaking, the bound depends on a parameter which reflects on the spectral measure's fractal dimension.

Definition 2.2. Let μ be a finite Borel measure on $\mathbb{R}$ and let $\alpha \in [0,1]$. Then μ is called **uniformly α-Hölder continuous (UαH)** if there is some $C < \infty$ such that for all intervals I with $|I| < 1$ one has

$$\mu(I) \le C\,|I|^\alpha\,. \tag{2.19}$$

At $\alpha = 0$ the condition (2.19) provides no information and is satisfied even in the case of pure-point spectrum. At $\alpha = 1$, (2.19) requires the measure to be absolutely continuous with respect to the Lebesgue measure with a bounded density. Operators of relevance for physics for which the condition is found to hold at the less standard values $\alpha \in (0,1)$ occur in relation to models with quasi-period potentials. Prominent examples are the Harper Hamiltonian at criticality and the Fibonacci Hamiltonian. For such situations one has the following refinement of (2.16).

Theorem 2.3 (Strichartz-Last). *Let H be a self-adjoint operator on some Hilbert space $\mathcal{H}$ and assume the spectral measure of ψ with respect to H is $U\alpha H$ for some α. Then there is $C_\psi < \infty$ such that for all $\phi \in \mathcal{H}$ and all $T > 0$*

$$\frac{1}{T}\int_0^T \left|\langle \phi, e^{-itH}\psi\rangle\right|^2 dt \;\leq\; \frac{C_\psi \, \|\phi\|^2}{T^\alpha}\,. \tag{2.20}$$

Proof. The spectral theorem, in its unitary mapping version (Proposition A.2), yields the representation

$$\langle \psi, e^{-itH}\phi\rangle \;=\; \int e^{-itE} f(E)\mu_\psi(dE) \;=:\; \widehat{f\mu_\psi}(t)\,, \tag{2.21}$$

where $\mu_\psi(dE)$ is the spectral measure corresponding to ψ (the function which in $L^2(\mathbb{R},\mu_\psi)$ is represented by one) and f is the function which represents ϕ in $L^2(\mathbb{R},\mu_\psi)$, or more precisely which represents the projection of ϕ onto the space for which ψ serves as a cyclic vector under the action of H.

In the above terminology, the claim is that if a positive measure μ is UαH for some α, then there exists some $C < \infty$ such that for all $f \in L^2(\mathbb{R},\mu)$

$$\frac{1}{T}\int_0^T \left|\widehat{f\mu}(t)\right|^2 dt \leq \frac{C}{T^\alpha}\int |f(E)|^2\mu(dE)\,. \tag{2.22}$$

For the proof it helps to bound the indicator function of $[0,T]$ by a Gaussian and compute the Fourier transform:

$$\begin{aligned}
\frac{1}{T}\int_0^T \left|\widehat{f\mu}(t)\right|^2 dt &\leq \frac{e}{T}\int_{\mathbb{R}} e^{-\frac{t^2}{T^2}} \left|\widehat{f\mu}(t)\right|^2 dt \\
&\leq \frac{e}{T}\int |f(E)|\,|f(E')| \left(\int e^{-\frac{t^2}{T^2}}\, e^{it(E-E')}dt\right) \mu(dE)\mu(dE') \\
&\leq e\sqrt{\pi}\int |f(E)|^2 \left[\int e^{-\frac{T^2}{4}(E-E')^2}\mu(dE')\right] \mu(dE)\,,
\end{aligned} \tag{2.23}$$

where the last step results from an application of the Cauchy-Schwarz inequality. For $T > 1$, the integral in brackets is estimated using UαH of μ_ψ:

$$\sum_{n=0}^{\infty}\int_{\frac{n}{T}\leq|E-E'|<\frac{n+1}{T}} e^{-\frac{T^2}{4}(E-E')^2}\,\mu(dE') \leq \frac{2C}{T^\alpha}\sum_{n=0}^{\infty} e^{-\frac{n^2}{4}}\,.$$

This completes the proof since in case $0 < T \leq 1$, one has $\sup_{t\in\mathbb{R}}\left|\widehat{f\mu}(t)\right|^2 \leq \int |f(E)|^2\mu(dE)$, thanks to the Cauchy-Schwarz inequality. $\square$

2.3. Recurrence probabilities and the resolvent

The decay rate of the time-averaged recurrence probabilities, which the above general results link to spectral information, can also be expressed in terms of the behavior of the resolvent:

Theorem 2.4. *For any $\alpha \leq 1$ if one of the following limits exists, then so does the other and the equality holds:*

$$\begin{aligned}
&\lim_{T\to\infty} T^{\alpha} \int_0^T \left|\langle \phi, e^{-itH}\psi\rangle\right|^2 \frac{dt}{T} \\
(2.24) \qquad &= \lim_{\eta\downarrow 0} \frac{\eta^{1-\alpha}}{\pi\, 2^{\alpha}\,\Gamma(2-\alpha)} \int_{\mathbb{R}} \left|\langle \varphi\,, (H-E-i\eta)^{-1}\,\psi\rangle\right|^2 dE\,,
\end{aligned}$$

where $\Gamma(\lambda) := \int_0^\infty e^{-x} x^{\lambda-1} dx$.

Proof. For $\eta > 0$, the function $F(t) := e^{-\eta t}\langle \varphi\,, e^{-itH}\psi\rangle\, 1_{[0,\infty)}(t)$ is both integrable and square integrable. Its Fourier transform is given by

$$\begin{aligned}
\widehat{F}(-E) &= \int e^{itE} F(t)\, dt = \int_0^\infty e^{-t(\eta - iE)} \int e^{-itu} \mu_{\varphi,\psi}(du)\, dt \\
(2.25) \qquad &= -i\,\langle \varphi\,, (H-E-i\eta)^{-1}\,\psi\rangle\,,
\end{aligned}$$

where use was made of Fubini's theorem and the representation (2.12) in terms of the spectral measure associated with H and φ, ψ. Thus, by Parseval's theorem on the Fourier transform,

(2.26)

$$2\eta \int_0^\infty e^{-2\eta t} \left|\langle \varphi\,, e^{-itH}\psi\rangle\right|^2 dt = \frac{\eta}{\pi} \int \left|\langle \varphi\,, (H-E-i\eta)^{-1}\,\psi\rangle\right|^2 dE\,.$$

The claimed statement then follows from the following version of the Abelian-Tauberian principle on the relation of the Cesàro averages to the Abelian mean. □

The proof used the following relation of the Cesàro and the Abelian average known as Abelian-Tauberian theorem.

Proposition 2.5 (Abelian-Tauberian relation). *For any non-negative function $f : [0,\infty) \to [0,\infty)$ and constant $\alpha \leq 1$, if one of the following limits exists, then so does the other and the equality holds:*

$$(2.27) \qquad \lim_{T\to\infty} T^{\alpha} \int_0^T f(t) \frac{dt}{T} = \lim_{\eta\downarrow 0} \frac{\eta^{1-\alpha}}{\Gamma(2-\alpha)} \int_0^\infty e^{-\eta t} f(t)\, dt\,.$$

The direction conditioned on the existence of the Cesàro convergence, which is the easier one, forms the content of the Abelian theorem. The converse direction is the Tauberian theorem (in a version of J. Kamarata [**215**]) for which our favorite proof can be found in [**357**].

2.4. The RAGE theorem

Building on Wiener's theorem is the following dynamical characterization of the subspaces of $\mathcal{H}$ associated with the continuous part $\mathcal{H}^{\mathrm{c}} := \mathcal{H}^{\mathrm{ac}} \oplus \mathcal{H}^{\mathrm{sc}}$ and pure-point part $\mathcal{H}^{\mathrm{pp}}$ of the time evolution generator H. The result is named after its contributors D. Ruelle [**337**], W. Amrein and V. Georgescu [**26**] and V. Enss [**138**].

Theorem 2.6 (RAGE). *Let H be a self-adjoint operator on some Hilbert space $\mathcal{H}$ and let A_L be a sequence of compact operators which converge strongly to the identity. Then*

$$\mathcal{H}^{\mathrm{c}} = \left\{ \psi \in \mathcal{H} \mid \lim_{L\to\infty} \lim_{T\to\infty} \frac{1}{T} \int_0^T \left\| A_L e^{-itH} \psi \right\|^2 dt = 0 \right\}, \tag{2.28}$$

$$\mathcal{H}^{\mathrm{pp}} = \left\{ \psi \in \mathcal{H} \mid \lim_{L\to\infty} \sup_{t\in\mathbb{R}} \left\| (1 - A_L) e^{-itH} \psi \right\| = 0 \right\}. \tag{2.29}$$

In the proof given below, use is made of the following implication of the Wiener theorem.

Lemma 2.7. *Let H be a self-adjoint and let A be a compact operator on some Hilbert space $\mathcal{H}$. Then*

$$\boxed{\lim_{T\to\infty} \frac{1}{T} \int_0^T \left\| A\, e^{-itH} \psi \right\|^2 dt = 0} \tag{2.30}$$

for all $\psi \in \mathcal{H}^{\mathrm{c}}$, i.e., initial states for which the spectral measure of H is continuous.

Proof. For any rank-one operator $A = |\rho\rangle\langle\phi|$ the claim follows from (2.16) applied to the complex-valued spectral measure $\mu_{\phi,\psi}$. From rank-one operators (2.30) directly extends to the finite-rank case. Any compact operator A may in turn be approximated by finite-rank operators. More precisely, for every $\varepsilon > 0$ there exists A_ε of finite rank such that $\|A - A_\varepsilon\| < \varepsilon$ and hence

$$\|A\, \psi(t)\| \leq \|A_\varepsilon\, \psi(t)\| + \|(A - A_\varepsilon)\, \psi(t)\| \leq \|A_\varepsilon\, \psi(t)\| + \varepsilon\, \|\psi\| \,. \tag{2.31}$$

The Cesàro average of the first term tends to zero in the long-time limit. This completes the proof since $\varepsilon > 0$ was arbitrary. $\square$

Proof of Theorem 2.6. If $\psi \in \mathcal{H}^{\mathrm{c}}$, then, by the previous lemma, for every L,

$$\lim_{T\to\infty} \frac{1}{T} \int_0^T \|A_L \psi(t)\|^2 \, dt = 0. \tag{2.32}$$

We now claim that for any $\psi \in \mathcal{H}^{\mathrm{pp}}$

$$\lim_{L\to\infty} \sup_{t\in\mathbb{R}} \|(1 - A_L)\psi(t)\| = 0 \,. \tag{2.33}$$

For the proof, let us note that by the uniform boundedness principle the assumed strong convergence A_L to one implies $\sup_L \|A_L\| < \infty$. We then expand ψ into proper eigenfunctions (ψ_n) of H. For every $\varepsilon > 0$ the sum may be split into a finite collection of N terms and a remainder with norm less than $\varepsilon > 0$:

$$\psi = \sum_{k=1}^{N} \langle \psi_k, \psi \rangle \, \psi_k + \phi_N \,, \qquad \|\phi_N\| < \varepsilon \,. \tag{2.34}$$

Since $\lim_{L\to\infty} \|(1 - A_L)\psi_k\| = 0$ for every k, the first term contributes zero to the limit in (2.33), while the rest contributes at most $(1+\sup_L \|A_L\|)\,\varepsilon$. This implies (2.33) as $\varepsilon > 0$ may be chosen arbitrarily small.

Every ψ may be uniquely decomposed into a component $\psi^{\mathrm{c}} \in \mathcal{H}^{\mathrm{c}}$ and an orthogonal one $\psi^{\mathrm{pp}} \in \mathcal{H}^{\mathrm{pp}}$. To complete the proof it therefore remains to show that (i) the limit in (2.28) does not tend to zero for $\psi^{\mathrm{pp}} \neq 0$ and (ii) the limit in (2.29) does not tend to zero for $\psi^{\mathrm{c}} \neq 0$.

The proof of the first assertion relies on (2.33) which, through the triangle inequality, implies that

$$\|A_L \psi^{\mathrm{pp}}(t)\| \geq \big|\|\psi^{\mathrm{pp}}\| - \|(1 - A_L)\psi^{\mathrm{pp}}(t)\|\big| \,, \tag{2.35}$$

which is strictly positive for large enough L.

The second assertion follows by contradiction. Suppose (2.33) applies to ψ^{c}. Then Lemma 2.7 implies again, by the triangle inequality,

$$\begin{aligned} 0 \;&=\; \left(\lim_{L\to\infty} \lim_{T\to\infty} \frac{1}{T} \int_0^T \|(1 - A_L)\psi^{\mathrm{c}}(t)\|^2 \, dt \right)^{\frac{1}{2}} \\ &\geq\; \|\psi^{\mathrm{c}}\| - \left(\lim_{L\to\infty} \lim_{T\to\infty} \frac{1}{T} \int_0^T \|A_L \psi^{\mathrm{c}}(t)\|^2 \, dt \right)^{\frac{1}{2}} \;=\; \|\psi^{\mathrm{c}}\|, \end{aligned} \tag{2.36}$$

which contradicts $\psi^{\mathrm{c}} \neq 0$. □

An illustrative and important example of the RAGE theorem is the case $\mathcal{H} = \ell^2(\mathbb{G})$ with $\mathbb{G}$ the vertex set of a graph with a metric d and A_L the projections to $B_L := \{x \,|\, d(x,0) \leq L\}$, i.e., to concentric balls of increasing radius about a fixed vertex $0 \in \mathbb{G}$. Equations (2.15), (2.28), and (2.29) imply the following properties of the position probabilities,

$$\mathrm{Prob}_{\psi(t)}\,(x \in B) = \sum_{x\in B} |\psi(x;t)|^2 \,,$$

and characterizations of the different spectra:

1. States $\psi^{\mathrm{c}} \in \mathcal{H}^{\mathrm{c}}$ in the subspace of **continuous** spectrum of H are those which under the time evolution spend an asymptotically vanishing fraction of time in any finite region; that is,

$$\lim_{L\to\infty} \lim_{T\to\infty} \frac{1}{T} \int_0^T \mathrm{Prob}_{\psi^{\mathrm{c}}(t)}\left(x \in B_L\right) dt = 0 \tag{2.37}$$

(nevertheless, recurrence over a vanishing fraction of time is not ruled out).

2. States $\psi^{\mathrm{ac}} \in \mathcal{H}^{\mathrm{ac}}$ in the range of the **absolutely continuous** spectral projection of H asymptotically leave any finite region:

$$\mathrm{Prob}_{\psi^{\mathrm{ac}}(t)}\left(x \in B_L\right) \underset{t\to\infty}{\longrightarrow} 0 \tag{2.38}$$

(though this is not a sufficient condition for $\psi \in \mathcal{H}^{\mathrm{ac}}$).

3. The **pure-point** spectrum corresponds to the bound states, $\psi^{\mathrm{pp}} \in \mathcal{H}^{\mathrm{pp}}$, which are forever confined, up to an arbitrarily small term, to a sufficiently large ball

$$\sup_{t\in\mathbb{R}} \mathrm{Prob}_{\psi^{\mathrm{pp}}(t)}\left(x \notin B_L\right) \underset{L\to\infty}{\longrightarrow} 0\,. \tag{2.39}$$

Similar interpretations exist for quantum systems on a continuous manifold, e.g., $\mathcal{H} = L^2(\mathbb{R}^d)$. In this context, it is worth noting that Lemma 2.7 as well as the RAGE theorem can be extended to operators A which are relatively compact with respect to H. We refer the interested reader to [**380**].

2.5. A scattering perspective on the ac spectrum

Let us end this chapter with an illustration of the significance of absolutely continuous (*ac*) spectrum for conduction. One can see that through the Kubo formula for conductivity which is discussed further in Chapter 13. Alternatively, the point may be conveyed through the considerations of coherent transport and scattering. This approach was emphasized by R. Landauer and M. Büttiker and in the form shown below presented by J. D. Miller and B. Derrida [**300**].

Consider the situation in which an external wire is attached at a vertex x_0 and current is sent through it towards a device in which the particle's state space is modeled by a graph, as depicted in Figure 2.1. The quantum amplitude as a function of the position is described by $\phi(u)$ along the wire ($u \in \mathbb{R}_+$) and $\psi(x)$ along the graph ($x \in \mathbb{Z}^d$), that is, by the function (ϕ, ψ) in the direct sum space $L^2(\mathbb{R}_+) \oplus \ell^2(\mathbb{Z}^d)$.

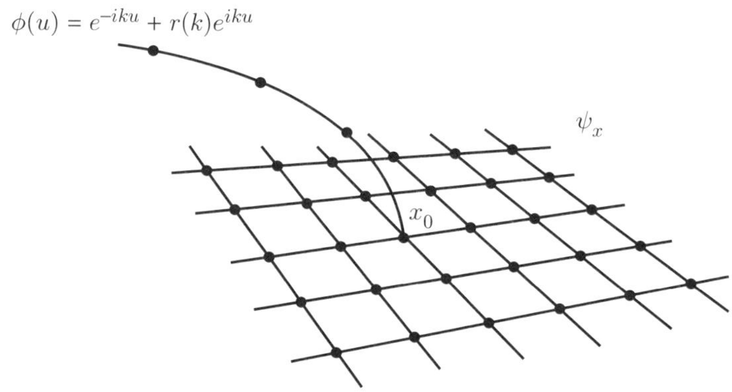

Figure 2.1. The setup for a scattering experiment in which current is sent down a wire which is attached to the graph. The reflection will be only partial ($|r| < 1$) if and only if H has an absolutely continuous spectrum at the corresponding energy range.

For the sake of concreteness we assume here that the system's time evolution is generated by the operator $\widehat{H}$ which acts as

$$\widehat{H}(\psi,\phi) = \begin{cases} -\phi''(u) & \text{along the wire,} \\ H\psi(x) - \phi'(0)\delta_{x,x_0} & \text{along the graph,} \end{cases} \tag{2.40}$$

an action which is self-adjoint on the subspace of functions which satisfy the continuity condition

$$\phi(0) \;=\; \psi(x_0)\,. \tag{2.41}$$

In a more general model of the contact the operator $\widehat{H}$ may be chosen from a multi-parameter family of self-adjoint operators, or one-parameter in the case of a point contact, with the parameter(s) characterizing details of the welding at x_0. Our main conclusion will be unaffected by that choice.

When particles are sent down the wire at energy $E = k^2$ and decay rate $\eta > 0$, the steady state wave function for observing the particle along the wire is given, in the limit $\eta \downarrow 0$, by

$$\phi(u) \;=\; e^{-iku} + r\,e^{iku}\,. \tag{2.42}$$

Here r is the complex-valued **reflection amplitude**, which depends on the operator H and the energy E. Along the graph the wave function is given by $\psi(x)$ such that jointly the two satisfy $(\widehat{H} - E)(\phi,\psi) = 0$. This translates into $(H - E)\psi(x) \;=\; \phi'(0)\delta_{x,x_0}$ or

$$\psi(x) = \phi'(0)\, \langle \delta_x \,, (H - E - i0)^{-1}\, \delta_{x_0} \rangle \tag{2.43}$$

(where we now need to invoke the sign of η, even though it was taken to zero). Combining (2.41), (2.42), and (2.43) we get

$$\langle \delta_{x_0} \,, (H - E - i0)^{-1}\, \delta_{x_0} \rangle \;=\; (-ik)^{-1}\frac{1+r}{1-r}\,, \tag{2.44}$$

and, in particular,

$$\boxed{\frac{1}{\pi}\,\mathrm{Im}\langle \delta_{x_0} \,, (H - E - i0)^{-1}\, \delta_{x_0} \rangle \;=\; \frac{1-|r|^2}{\pi k\, |1-r|^2}}\,, \tag{2.45}$$

where the term $(1 - |r|^2)$ can be recognized as the **transmission probability**.

The above expresses a relation between the spectral and the dynamical properties of H. The condition under which not all the current will be reflected back up the wire is $|r|^2 < 1$. Comparing (2.45) with (2.11) we see that this is equivalent to the positivity of the density of the absolutely continuous component of the spectral measure associated with the state δ_{x_0}.

The above discussion notwithstanding, for certain conduction issues the pure-point spectrum can also play a role. That is the case in adiabatic transport and the celebrated quantum Hall effect, which is addressed in Chapters 13 and 14. Moreover, in the absence of an ac spectrum, the divergence of the localization length, even at a single energy, plays an important role in the dynamics [**107**].

Notes

The basic discussion on the relation of spectra and dynamics can be found in [**380**] and [**283**].

The decay which is described in Theorem 2.3 has been argued for by the authors of [**223**], who were apparently unaware of the theorem of R. Strichartz [**372**] that is stated here in the bound (2.22). The latter was used by Y. Last to formulate and prove Theorem 2.3 as stated here. Based on that, I. Guaneri [**182**] presented the lower bounds on the spread of $\psi^c(t) \in \ell^2(\mathbb{G})$ in case the spectral measure is UαH. This is the subject of Exercise 2.3. Generalizations suitable for continuum models, and to α-continuous spectral measures, are discussed in [**91, 283**]. Further relations concerning the spreading of generalized eigenfunctions and dynamical properties were obtained in [**243**].

A conjectured spectrum/dynamics relation concerning specifically random Schrödinger operators is the diffusion hypothesis on which more is said in Section 13.1.

The scattering relation (2.45) was highlighted by J. D. Miller and B. Derrida [**300**]. It is reminiscent of the R. Landauer and M. Büttiker formula for conductance through a finite black box. Along with related formulas, it is discussed rigorously in [**315, 100, 101, 52, 33, 78**].

Exercises

2.1. For a self-adjoint operator H on $\ell^2(\mathbb{G})$ show that the projections onto the continuous and pure-point component in $I \subset \mathbb{R}$ admit the representations

$$\|P_I^{\mathrm{c}}(H)\,\psi\|^2 = \lim_{L\to\infty}\lim_{T\to\infty}\frac{1}{T}\int_0^T \left\|1_{\mathbb{G}\setminus\mathbb{G}_L}\, e^{-itH}P_I(H)\,\psi\right\|^2 dt\,,$$

$$\left\|P_I^{\mathrm{pp}}(H)\,\psi\right\|^2 = \lim_{L\to\infty}\lim_{T\to\infty}\frac{1}{T}\int_0^T \left\|1_{\mathbb{G}_L}\, e^{-itH}P_I(H)\,\psi\right\|^2 dt\,,$$

for any $\psi \in \ell^2(\mathbb{G})$, where $(\mathbb{G}_L)$ is an arbitrary sequence of subsets which exhaust $\mathbb{G}$ in the limit $L \to \infty$.

2.2. Let H be a self-adjoint operator in $\ell^2(\mathbb{G})$ whose matrix elements in the canonical basis $\{\delta_x\}_{x\in\mathbb{G}}$ satisfy

$$\sup_{x\in\mathbb{G}}\sum_{y\in\mathbb{G}}|H(x,y)|\,e^{\beta d(x,y)} < \infty$$

for some $\beta > 0$.

a. Show the finiteness of the speed of propagation in the sense that for some $v_0, \mu < \infty$ the probability that in the time-evolved state $\psi(t) := e^{-itH}\delta_0$ the particle would be found to have moved a distance greater than tv, with $v \geq v_0$ satisfies

$$\mathrm{Prob}_{\psi(t)}(x \notin B_{vt}) := \sum_{x\notin B_{vt}}\left|\langle\delta_x,\, e^{-itH}\delta_0\rangle\right|^2 \leq e^{-\mu t(v-v_0)}\,.$$

[This is just a one-particle version of the more general **Lieb-Robinson bound** [**288**]. It should be noted that the stated decay holds even for graphs of exponential volume growth, in which case the bound is stronger than just a statement of exponential decay of the two-point function $|\langle\delta_x,\, e^{-itH}\delta_0\rangle|$.]

b. Show that the above implies that for any $\alpha \geq 0$,

$$M_{\delta_0}(\alpha, t) \;:=\; \sum_{x \in \mathbb{G}} d(x,0)^{\alpha} \, \left|\langle \delta_x, \, e^{-itH} \delta_0 \rangle\right|^2 \leq \; C_{\alpha} \, (1 + t^{\alpha}),$$

$$\widehat{M_{\delta_0}}(\alpha, \eta) \;:= 2\eta \int_0^{\infty} \; e^{-2\eta t} \, M_{\delta_0}(\alpha, t) \, dt \leq \; \widehat{C}_{\alpha} \; (1 + \eta^{-\alpha}) \, .$$

2.3. Let H be a self-adjoint operator in $\ell^2(\mathbb{Z}^d)$ and assume the spectral measure of ψ with respect to H is UαH for some α.

a. Prove that there is some $C_d < \infty$ such that for all $b > 0$ and $\eta \in (0, b^{d/\alpha})$,

$$2\eta \int_0^{\infty} \; e^{-2\eta t} \, \mathrm{Prob}_{\psi(t)} \left(|x| < b \, \eta^{-\alpha/d} \right) \; dt \; \leq \; C_d \, b^d \, .$$

b. Show that there is some $b > 0$ such that for all $\beta > 0$ and $\eta \in (0, b^{d/\alpha})$,

$$\widehat{M_{\psi}}(\beta, \eta) \; \geq \; \frac{b^{\beta}}{2} \; \eta^{-\beta\alpha/d} \, ,$$

where $\widehat{M_{\psi}}(\beta, \eta)$ is defined analogously as in Exercise 2.2.
[This statement forms the **Guarneri bound** [**182**].]

Chapter 3

Ergodic Operators and Their Self-Averaging Properties

We now turn to random operators acting in the Hilbert space $\ell^2(\mathbb{G})$ with $\mathbb{G}$ the vertex set of a graph endowed with a vertex-transitive group of isomorphism, e.g., $\mathbb{Z}^d$ with the group of lattice shifts. In such situations one is interested in two forms of averaging: over the disorder and over the volume. The concepts introduced in this chapter include:

Ergodic operators: Ergodicity carries the important implication that averages over the volume of local quantities converge almost surely and agree with averages over the realizations.

Self-averaging of the spectra: Quantities which are affected by the disorder but almost surely assume non-fluctuating values are often referred to as self-averaging. An example is found in the spectra of ergodic operators.

Density of states measure: Ergodicity is also of relevance for the density of states (DOS) measure. This notion is discussed in two contexts: first directly for infinite-volume operators and then for suitably converging sequences of finite-volume operators, possibly with varying boundary conditions. In the amenable case, the limiting DOS turns out to be independent of the latter.

In the context of self-averaging it should be stressed that while such properties and quantities are by definition *predictable*, that does not necessarily mean easy to determine.

In addition to examples from the domain of random operators, the class of ergodic operators also includes other types, such as operators which are almost-periodic under shifts. Disorder and almost-periodicity are easily distinguished and are associated with rather different phenomena. However, the concepts of ergodicity and the results which are presented in this section apply to both.

3.1. Terminology and basic examples

Combined in the notion of ergodic operators are two structures: operators indexed by a parameter (ω) whose values range over a probability space and a symmetry group of the operator's distribution.

Definition 3.1. A (self-adjoint) **random operator** is an operator-valued weakly measurable function defined on a probability space $(\Omega, \mathcal{A}, \mathbb{P})$ which assigns to every $\omega \in \Omega$ a self-adjoint operator $H(\omega)$ acting in a common separable Hilbert space $\mathcal{H}$.

Weak measurability means that for all $f \in L^\infty(\mathbb{R})$ and all $\varphi, \psi \in \mathcal{H}$ the functions $\omega \mapsto \langle \varphi, f(H(\omega))\psi \rangle$ are measurable. It suffices to check this condition for functions of the form $f(x) = (x - z)^{-1}$ with $z \in \mathbb{C}\backslash\mathbb{R}$. We shall not dwell here on this and related questions of measurability, on which further discussion may be found in [**82, 233**].

The structure to be added to render a random operator ergodic is a group of transformations which acts ergodically on the underlying probability space.

Definition 3.2. The action of a group of measure-preserving transformations $(T_x)_{x \in I}$ on a probability space $(\Omega, \mathcal{A}, \mathbb{P})$ is said to be **ergodic** if all events $A \in \mathcal{A}$ which are invariant, i.e., $T_x^{-1} A = A$ for all $x \in I$, are of probability either zero or one.

An equivalent property is that any random variable $Y : \Omega \to \mathbb{R} \cup \{\infty\}$ which is T-invariant, i.e., $Y \circ T_x = Y$ for all $x \in I$, is almost surely constant; i.e., there exists $c \in \mathbb{R} \cup \{\infty\}$ such that

$$\mathbb{P}(Y = c) = 1; \tag{3.1}$$

cf. Exercise 3.1.

Combining the concept of random operators with ergodicity yields the following notion of ergodic operators.

Definition 3.3. Let $(\Omega, \mathcal{A}, \mathbb{P})$ be a probability space and $(T_x)_{x\in I}$ an ergodic family of measure-preserving transformation on Ω. A random operator defined over this probability space is called an **ergodic operator** if for every $x \in I$ and $\omega \in \Omega$ the operator $H(T_x\omega)$ is unitarily equivalent to $H(\omega)$.

The definition readily implies that for ergodic operators functions of $H(\omega)$ which are invariant under unitary transformations take (almost surely) non-random values. These include the operators' norms ($\|H(\omega)\|$) and, as discussed in the next section, also their spectra—when suitably defined. More can be said under additional assumptions on the structure of the ergodic operator, as discussed next.

The above definition of ergodic operators differs from the one found in some of the literature [**228, 82**], which focusses on operators acting in the Hilbert space $\mathcal{H} = \ell^2(\mathbb{G})$ associated with a vertex-transitive graph. That is a graph endowed with a group of automorphisms $\mathcal{S} = (S_x)_{x\in I}$ which act transitively on $\mathbb{G}$. A prime example would be $\mathbb{G} = \mathbb{Z}^d$ with the symmetry group of translations indexed by $x \in I = \mathbb{Z}^d$:

$$S_x\xi := \xi - x \quad \text{for all} \quad \xi \in \mathbb{Z}^d\,. \tag{3.2}$$

The standard situation further demands that the graph automorphisms $\mathcal{S}$ are represented on the probability space through an ergodic family of measure-preserving transformations and in $\ell^2(\mathbb{G})$ through unitary transformations under which gauge-invariant quantities vary covariantly.

Definition 3.4. A **standard ergodic operator** is a random operator $H(\omega)$ such that:

1. The operators act in $\mathcal{H} = \ell^2(\mathbb{G})$ with $\mathbb{G}$ a vertex set of a vertex-transitive graph endowed with a vertex-transitive group of graph automorphisms $\mathcal{S} = (S_x)_{x\in I}$.
2. The group $\mathcal{S}$ is represented by a group of measure-preserving transformations (T_x) whose action on $(\Omega, \mathcal{A}, \mathbb{P})$ is ergodic. For every $x \in I$ and $\omega \in \Omega$,

$$\boxed{H(T_x\omega) \;=\; U_{x,\omega}\, H(\omega)\, U_{x,\omega}^\dagger} \tag{3.3}$$

with a unitary operator $U_{x,\omega}$ for which for all $\psi \in \ell^2(\mathbb{G})$

$$(U_{x,\omega}\,\psi)\,(\xi) \;=\; \psi\,(S_x\xi)\; e^{i\phi_{x,\omega}(\xi)} \tag{3.4}$$

with some real-valued phase $\phi_{x,\omega}(\xi)$.

It is of interest to note that the unitary mappings which relate $H(T_x\omega)$ to $H(\omega)$ need not form a group, even in the case when they have no ω dependence (as in the second example below of magnetic Schrödinger operators).

For standard ergodic operators more can be said through Birkhoff's seminal theorem, which in the amenable context of $\mathbb{Z}^d$ states:

Proposition 3.5. *Let $(\Omega, \mathcal{A}, \mathbb{P})$ be a probability space, on which the lattice shifts of $\mathbb{Z}^d$ are represented by a group of measure-preserving transformations $(T_x)_{x\in\mathbb{Z}^d}$, which acts ergodically. Then for each $Y \in L^1(\Omega, \mathbb{P})$ the following limit exists for $\mathbb{P}$-almost all $\omega \in \Omega$ and satisfies*

$$\boxed{\lim_{L\to\infty} \frac{1}{|\Lambda_L|} \sum_{x\in\Lambda_L} Y(T_x\omega) \;=\; \mathbb{E}\left[Y\right]}\,, \tag{3.5}$$

where $\Lambda_L := (-L, L)^d \cap \mathbb{Z}^d$.

A modern proof can be found in [**268**], and extensions of the theorem to non-amenable groups including that of graph automorphisms of a regular tree graph are discussed in [**79**]. For standard ergodic operators in $\ell^2(\mathbb{Z}^d)$ the above can be applied to functions of the form $Y(\omega) := \langle \delta_0, f(H(\omega))\, \delta_0 \rangle$ with bounded measurable $f \in L^\infty(\mathbb{R})$. Useful implications will be seen in the discussion of the density of states.

The following three examples demonstrate some of the different ways in which the ergodicity criteria can be met.

Example 3.6 (Random Schrödinger operator on $\mathbb{Z}^d$)**.** The random Schrödinger operator

$$H(\omega) = -\Delta + V(\omega) \tag{3.6}$$

acting in $\ell^2(\mathbb{Z}^d)$ consists of

- the **lattice Laplacian**: $(\Delta\psi)(x) := \sum_{y:|y-x|=1} (\psi(y) - \psi(x))$,
- a **random potential**, i.e., multiplication operator: $(V(\omega)\psi)(x) = \omega(x)\psi(x)$.

A natural and convenient choice for the probability space is $\Omega = \mathbb{R}^{\mathbb{Z}^d}$, with the product topology and the product σ-algebra of Borel sets $\mathcal{B}$. Under this explicit construction, the values of the random potential at sites $\xi \in \mathbb{Z}^d$ are given by the components $\omega(\xi)$.

The symmetry group $\mathcal{S}$ consists of lattice shifts defined in (3.2), which are implemented on Ω by the mappings

$$\boxed{(T_x\,\omega)\,(\xi) \;:=\; \omega(\xi - x)\,, \quad \xi, x \in \mathbb{Z}^d}\,. \tag{3.7}$$

Among the probability measures on $(\Omega, \mathcal{B}, \mathbb{P})$ which are shift-invariant and ergodic are all the product measures, for which the $(\omega(\xi))_{\xi\in\mathbb{Z}^d}$ form independent and identically distributed (iid) random variables; cf. Exercise 3.2.

The induced action of the shifts on $H(\omega)$ corresponds to

$$H(T_x\omega) \;=\; U_x\, H(\omega)\, U_x^\dagger \tag{3.8}$$

with $(U_x)_{x\in\mathbb{Z}^d}$ the unitary transformations on $\ell^2(\mathbb{Z}^d)$, i.e.,

$$(U_x\psi)\,(\xi) \;:=\; \psi(S_x\xi) = \psi(\xi - x)\,. \tag{3.9}$$

Consequently, for any probability measure on the product space which is ergodic under the lattice shifts, $H(\omega)$ forms a standard ergodic operator.

Example 3.7 (Random Schrödinger operator on $\mathbb{Z}^2$ with constant magnetic field)**.** A quantum particle on $\mathbb{Z}^2$ with charge $q \in \mathbb{R}$ in a perpendicular magnetic field, and subject to a random potential, is described by an operator on $\ell^2(\mathbb{Z}^2)$ of the form $H^{(A)}(\omega) = -\Delta^{(A)} + V(\omega)$ with the **magnetic Laplacian**

$$\left(\Delta^{(A)}\,\psi\right)(x) \;=\; \sum_{y:\,|y-x|=1} \left(e^{-iqA(x,y)}\psi(y) - \psi(x)\right)\,. \tag{3.10}$$

Here $A(x,y) \;=\; -A(y,x)\,\delta_{|x-y|,1}$ is an antisymmetic function over the oriented nearest-neighbor edges (which represents the line integral of a **vector potential**). Its relation to the magnetic field is the discrete version of $\mathbf{B} \;=\; \mathrm{Curl}\;\mathbf{A}$:

$$B_P \;=\; \sum_{b\in\partial P} A(b) \tag{3.11}$$

where b is summed over the oriented edges which form the boundary of the plaquette P.

For such an operator to be ergodic the magnetic field should be of translation-invariant distribution or constant. For the constant field $B \in \mathbb{R}$ the magnetic Laplacian, which for the given field is unique up to gauge transformations, is in the symmetric gauge:

(3.12)

$$\begin{aligned}\left(\Delta^{(A)}\,\psi\right)(x) \;&= e^{iBqx_2/2}\psi(x_1+1,x_2) + e^{-iBqx_2/2}\psi(x_1-1,x_2)\\ &\quad + e^{-iBqx_1/2}\psi(x_1,x_2+1) + e^{iBqx_1/2}\psi(x_1,x_2-1) - 4\,\psi(x)\,.\end{aligned}$$

In this case the simple covariance relation (3.8) is replaced by

$$H^{(A)}(T_x\omega) \;=\; U_x^{(A)}\, H^{(A)}(\omega)\, \left(U_x^{(A)}\right)^\dagger, \tag{3.13}$$

where the magnetic translations $U_x^{(A)}$ combine U_x with a gauge transformation:

$$\left(U_x^{(A)}\psi\right)(\xi) \;:=\; e^{-i\varphi_A(\xi,x)}\,\psi(\xi - x)\,, \tag{3.14}$$

with $\varphi_A(\xi,x) := qB(\xi_1 x_2 - \xi_2 x_1)/2$. Due to the latter, these magnetic shifts do not form a group. However, that in itself is not detrimental for our purpose: gauge invariant functionals are shift covariant, and $H^{(A)}(\omega)$ is still a standard ergodic operator if $(T_x)_{x\in\mathbb{Z}^d}$ defined through (3.7) acts ergodically on $(\mathbb{R}^{\mathbb{Z}^2}, \mathcal{B}, \mathbb{P})$.

The next example is an **almost-periodic** operator. Though without disorder, it also fits within the framework of ergodic operators. It has an important lesson concerning the occasional duality under Fourier transform of the notions of localization and extended states, on which more is said in the Notes.

Figure 3.1. The spectrum of the Harper Hamiltonian at $\lambda = 1$. The horizontal axis depicts the energy and the vertical axis depicts the Hamiltonian parameter $\alpha \in [0,1]$. For rational α the operator is periodic and the spectrum is absolutely continuous with Bloch bands. For irrational α the spectrum is purely singular continuous. (We thank D. Osadchy and Y. Avron for the figure.)

Example 3.8 (Harper Hamiltonian). For a fixed $\alpha \in (0,1)$ and $\theta \in [0,2\pi)$ the Harper Hamiltonian, also dubbed the almost-Mathieu operator, is the discrete Schrödinger operator in $\ell^2(\mathbb{Z})$:

$$H_\alpha(\theta) = -\Delta + \lambda\, 2\cos(2\pi\alpha\,\xi + \theta) \tag{3.15}$$

with ξ ranging over $\mathbb{Z}$. Under the unitary implementation of the lattice shifts ($U_x\psi(\xi) := \psi(\xi - x)$) the Harper Hamiltonian transforms as $U_x H_\alpha(\theta) U_x^\dagger = H_\alpha(T_x\theta)$, for any $x \in \mathbb{Z}$, with

$$T_x\theta = \theta - 2\pi\alpha x_{\text{mod } 2\pi}. \tag{3.16}$$

It follows that for α irrational $H_\alpha(\theta)$ is an ergodic operator-valued function, defined over $[0, 2\pi)$, and an example of a standard ergodic operator.

It is of interest to note that for α irrational the spectrum of the Harper Hamiltonian forms a Cantor set [**36**] in the sense that its closure is totally disconnected. The spectrum's doubly fractal structure, in $E \in \mathbb{R}$ and $\alpha \in [0, 1]$, at the critical value $\lambda = 1$ forms the celebrated Hofstadter butterfly which is depicted in Figure 3.1.

To say a bit more, through considerations of the almost self-dual nature of the operator under the Fourier transform and analysis of the Lyapunov exponent associated to the solution of the eigenfunction equation, S. Aubrey and G. André [**34**] conjectured a duality principle for the almost-Mathieu operator at irrational values of α, proposing that for $\lambda > 1$ the eigenfuctions are localized, whereas for $\lambda \in [0, 1)$ they are extended—in a suitable sense. The rich collection of works which followed showed that the full story has many subtleties, but in a broad sense the Aubrey-André duality holds at any given irrational α:

- **$\lambda > 1$:** For almost every θ, $H_\alpha(\theta)$ has only pure-point spectrum [**209**] (by ergodicity its closure does not depend on θ).
- **$0 < \lambda < 1$:** For each θ, $H_\alpha(\theta)$ has purely absolutely continuous spectrum [**35**].
- **$\lambda = 1$:** For almost every θ the spectrum of $H_\alpha(\theta)$ is purely singular continuous [**177**].

The last example is the most elementary one, but the lesson it provides on the implications of ergodicity on pure-point spectra and its limitation is useful.

Example 3.9 (Ergodic potentials)**.** The potentials in the above examples are random operators themselves, acting on $\ell^2(\mathbb{G})$ through multiplication by $V(x)$. Their spectra are clearly pure point with the complete set of eigenfunctions consisting of the localized states, i.e., the canonical basis $(\delta_x)_{x\in\mathbb{G}}$ in $\ell^2(\mathbb{G})$.

In this case the set of eigenvalues consists of the countable collection of values of $V(x)$ at the given realization (indexed by ω or θ). The spectrum varies from one realization to another: two independently sampled realizations almost surely having no common eigenvalue. However, the set's closure is deterministic: it is almost surely given by the support of the distribution in the iid random case and by the full range of values of $V_\alpha(\theta) : \mathbb{Z} \to \mathbb{R}$ for $\theta \in [0, 2\pi]$.

3.2. Deterministic spectra

The following application of the above ideas to the spectra of ergodic operators dates back to L. Pastur [**322**] with extensions by H. Kunz and B. Souillard [**274**] and W. Kirsch and F. Martinelli [**233**].

Theorem 3.10 (Pastur). *The spectrum of a family of ergodic operators $H(\omega)$ is $\mathbb{P}$-almost surely non-random; i.e., there is $\Sigma \subset \mathbb{R}$ such that*

$$\boxed{\mathbb{P}\left(\sigma(H) = \Sigma\right) \;=\; 1}\,. \tag{3.17}$$

The same applies to any subset in the Lebesgue decomposition of the spectrum; i.e., there are $\Sigma^{\#} \subset \mathbb{R}$, $\# = \mathrm{ac}, \mathrm{sc}, \mathrm{pp}$, such that

$$\mathbb{P}\left(\sigma^{\#}(H) = \Sigma^{\#}\right) \;=\; 1\,, \quad \# = \mathrm{ac}, \mathrm{sc}, \mathrm{pp}\,.$$

The above non-random set Σ is called the **almost-sure spectrum** associated with $H(\omega)$ and likewise for the spectral components.

Proof of Theorem 3.10 for the spectrum. For any $E_1, E_2 \in \mathbb{R}$, the functions

$$\omega \;\mapsto\; X_{E_1,E_2}(\omega) \;:= \dim \operatorname{Range} P_{(E_1,E_2)}\left(H(\omega)\right)\,,$$

defined on the underlying probability space $(\Omega, \mathcal{A}, \mathbb{P})$, with values in $[0, \infty]$, are measurable, e.g., given by the (possibly divergent) non-negative series $X_{E_1,E_2}(\omega) = \sum_{k=1}^{\infty} \langle \psi_k\,,\, P_{(E_1,E_2)}\left(H(\omega)\right) \psi_k \rangle$ with (ψ_k) an arbitrary orthonormal basis in $\mathcal{H}$. These functions are invariant under the action of the ergodic transformations $(T_x)_{x \in I}$ since the assumed unitary equivalence (3.3) implies for any $x \in I$ and $\omega \in \Omega$

$$\dim \operatorname{Range} P_{(E_1,E_2)}\left(H(T_x\omega)\right) \;=\; \dim \operatorname{Range} P_{(E_1,E_2)}\left(H(\omega)\right)\,. \tag{3.18}$$

Ergodicity then ensures that there are constants $c_{E_1,E_2} \in [0, \infty]$ such that

$$\mathbb{P}\left(X_{E_1,E_2} = c_{E_1,E_2}\right) = 1\,.$$

The characterization (A.18) of the spectrum as the support of the spectral projections identifies

$$\Sigma \;=\; \{E \in \mathbb{R} \,|\, \text{for all } E_1, E_2 \in \mathbb{Q} \text{ with } E_1 < E < E_2,\;\; c_{E_1,E_2} > 0\}$$

as a possible candidate for the almost-sure spectrum. The reason for choosing intervals with rational endpoints becomes apparent from the fact that the event

$$\bigcap_{\substack{E_1,E_2\in\mathbb{Q}\\ c_{E_1,E_2}>0}} \{X_{E_1,E_2} = c_{E_1,E_2}\} \subset \{\sigma(H) = \Sigma\}$$

still has probability one since it is a countable intersection of such events. □

The ideas of the above proof also apply to the other components of the spectrum. The only subtle point is the measurability of the spectral projections associated to the restriction of $H(\omega)$ to the subspaces $\mathcal{H}^{\#}$, $\# = \mathrm{ac}, \mathrm{sc}, \mathrm{pp}$. We refer the interested reader to [**233, 274**].

Determining the different spectral components of an ergodic operator is in general a hard question despite the deterministic nature of the answer. In contrast, their union, i.e., the full spectrum, is much easier to identify. A useful strategy is to consider **Weyl sequences** of approximate (normalizable) eigenfunctions.

Proposition 3.11 (Weyl criterion)**.** *Let $A : \mathcal{D}(A) \to \mathcal{H}$ be a self-adjoint operator on a Hilbert space $\mathcal{H}$. Then*

$$\boxed{\sigma(A) \;=\; \Big\{E \in \mathbb{R} \,|\, \exists\, (\psi_n) \subset \mathcal{D}(A),\, \|\psi_n\| = 1 :\, \lim_{n\to\infty} \|(A-E)\psi_n\| = 0\Big\}}\,. \tag{3.19}$$

A more general statement and a proof can be found in [**380**]. A point to be appreciated here is that approximating sequences form a sturdy tool for determining the range of energies in the spectrum, but the approximation is too blunt to provide information on the nature of the actual eigenfunctions.

In applying the Weyl criterion to ergodic Schrödinger operators one finds two relevant notions of the range of the potential:

1. For the spectrum of the operator $V(\omega)$ alone, the relevant notion is the support of the distribution of the single random variable $V(0,\cdot)$:

$$\mathrm{supp}_1(\mathbb{P}) \;:=\; \{u \in \mathbb{R} \,|\, \forall \varepsilon > 0 : \mathbb{P}\left(|V(0) - u| < \varepsilon\right) > 0\}\,.$$

For any ergodic process $(V(x))_{x\in\mathbb{Z}^d}$ the closure of the set of realized values is almost surely given by this set; i.e., for $\mathbb{P}$-almost all ω,

$$\overline{\{V(x,\omega) \,|\, x \in \mathbb{Z}^d\}} \;=\; \mathrm{supp}_1(\mathbb{P})\,. \tag{3.20}$$

The proof of this assertion is left as an exercise for the reader.

2. For determining the almost-sure spectrum of $-\Delta + V(\omega)$ on $\ell^2(\mathbb{Z}^d)$ it is relevant to seek the range of values u for which $V(x,\omega)$ can be found to

uniformly approximate over arbitrarily large rectangular patches $\Lambda_L := (-L, L)^d \cap \mathbb{Z}^d$. This defines

$$\mathrm{supp}_2(\mathbb{P}) := \left\{ u \in \mathbb{R} \mid \forall \varepsilon > 0, L < \infty : \mathbb{P}\left(\sup_{x \in \Lambda_L} \{|V(x) - u|\} < \varepsilon \right) > 0 \right\}.$$

For ergodic processes the general relation is $\mathrm{supp}_2(\mathbb{P}) \subset \mathrm{supp}_1(\mathbb{P})$, with equality in the case of iid random variables and strict inequality for certain processes with strong anticorrelations.

Considerations of Weyl sequences of approximate eigenfunctions lead to the following result of H. Kunz and B. Souillard [**274**], generalization of which can be found in [**234**].

Theorem 3.12 (Kunz/Souillard). *For the family $H(\omega) = -\Delta + V(\omega)$ of ergodic operators on $\ell^2(\mathbb{Z}^d)$*

$$[0, 4d] + \mathrm{supp}_2(\mathbb{P}) \subset \sigma(H(\omega)) \subset [0, 4d] + \mathrm{supp}_1(\mathbb{P}) \tag{3.21}$$

for $\mathbb{P}$-almost all ω.

Proof. For the second inclusion note that

$$\left|\langle\psi, (-\Delta)\psi\rangle + \langle\psi, V(\omega)\psi\rangle - E\|\psi\|^2\right| \leq \|(H(\omega) - E)\psi\| \tag{3.22}$$

for any $\psi \in \mathcal{H}$. Letting ψ range over a Weyl approximating sequence for a given $E \in \sigma(H(\omega))$, the right side tends to 0 while the first expression on the left ranges only over $\sigma(-\Delta) = [0, 4d]$ and the second is restricted to

$$\sigma(V(\omega)) = \overline{\{V(x, \omega) \mid x \in \mathbb{Z}^d\}} = \mathrm{supp}_1(\mathbb{P}). \tag{3.23}$$

It readily follows that the spectrum of H is contained in the set sum

$$\sigma(H(\omega)) \subset \sigma(-\Delta) + \sigma(V(\omega)), \tag{3.24}$$

from which the upper bound follows.

For a proof of the first inclusion, we pick $\lambda \in \sigma(-\Delta) = [0, 4d]$ and a corresponding Weyl sequence $(\varphi_n) \subset \ell^2(\mathbb{Z}^d)$. Adding a suitable smooth truncation it may be further assumed that φ_n has compact support. For each $n \in \mathbb{N}$ and $\mu \in \mathbb{R}$ we consider the event

$$\Omega_n(\mu) := \left\{ \text{for some } j \in \mathbb{Z}^d\text{:} \sup_{x \in \mathrm{supp}\,\varphi_n} |V(x + j) - \mu| < \tfrac{1}{n} \right\}.$$

If $\mu \in \mathrm{supp}_2(\mathbb{P})$, these events have a non-zero probability and are invariant under the shifts $(S_x)_{x \in \mathbb{Z}^d}$. By ergodicity we conclude that for each n, $\mathbb{P}(\Omega_n(\mu)) = 1$. Therefore also

$$\Omega_0(\mu) := \bigcap_{n \in \mathbb{N}} \Omega_n(\mu)$$

is of full probability. By construction, for any $\omega \in \Omega_0(\mu)$ there is a sequence $(j_n) \subset \mathbb{Z}^d$ such that

$$\psi_n \; := \; \varphi_n \, (\cdot - j_n)$$

is a Weyl sequence for $H(\omega)$ and energy $E = \lambda + \mu$. This follows from

$$\begin{aligned} \|(H(\omega) - \lambda - \mu) \, \psi_n\| \; &\leq \; \|(-\Delta - \lambda) \, \psi_n\| + \|(V(\omega) - \mu) \, \psi_n\| \\ &\leq \; \|(-\Delta - \lambda) \, \varphi_n\| + \sup_{x \in \operatorname{supp} \varphi_n} |V(x + j_n, \omega) - \mu| \; \|\varphi_n\| \, , \end{aligned}$$

which goes to zero as $n \to \infty$. □

In the case of iid random variables $\operatorname{supp}_1(\mathbb{P}) = \operatorname{supp}_2(\mathbb{P})$ (rather obviously). Therefore Theorem 3.12 yields a complete answer:

Corollary 3.13. *In case* $(V(x))_{x \in \mathbb{Z}^d}$ *are iid with common distribution* $\rho(dv)$*,*

$$\boxed{\sigma \, (H(\omega)) \; = \; [0, 4d] + \operatorname{supp} \rho} \tag{3.25}$$

for $\mathbb{P}$*-almost all* ω*.*

3.3. Self-averaging of the empirical density of states

For a realization of a random operator in $\ell^2(\mathbb{G})$ the **local spectral measure (LSM)** at $x \in \mathbb{G}$ is the spectral measure of $H(\omega)$ associated with the vector $\delta_x \in \ell^2(\mathbb{G})$. Its average over the realizations (i.e., the mean local spectral measure) will be denoted

$$\boxed{\nu_x \; := \; \mathbb{E} \left(\langle \delta_x, P_{(\cdot)}(H) \, \delta_x \rangle \right)} \, . \tag{3.26}$$

When the probability distribution of the LSM is shift-invariant, as is the case for standard ergodic operators, ν_x does not vary with $x \in \mathbb{G}$ and the subscript may be omitted.

Finite-volume averages of the local spectral measures, over finite subsets $\Lambda \subset \mathbb{G}$, allow us to compute the trace per unit volume of functions f of H, i.e., quantities such as

$$\frac{1}{|\Lambda|} \operatorname{tr} \left[1_\Lambda \, f(H(\omega)) \right] \; = \; \frac{1}{|\Lambda|} \sum_{x \in \Lambda} \langle \delta_x, f(H(\omega)) \, \delta_x \rangle \, . \tag{3.27}$$

When the infinite-volume limit of such averages exists, it is natural to regard the corresponding limiting measure as the empirical density of states. The term "empirical" is used to convey that it corresponds to a given realization, and "density of states" is used in the sense of trace per unit volume. A somewhat more elementary notion of the density of states is discussed below.

For standard ergodic operators on amenable graphs such as $\mathbb{Z}^d$, the almost-sure existence of the infinite-volume limit is implied by Birkhoff's theorem.

Theorem 3.14 (Self-averaging of LSM). *Let $H(\omega)$ be a standard ergodic operator in $\ell^2(\mathbb{Z}^d)$. Then there exists a full measure set $\Omega_0 \subset \Omega$, $\mathbb{P}(\Omega_0) = 1$, such that for all $\omega \in \Omega_0$ and all $f \in C_0(\mathbb{R})$ the limit exists and satisfies*

$$\lim_{L\to\infty} \frac{1}{|\Lambda_L|} \operatorname{tr}\left[1_{\Lambda_L} f(H(\omega))\right] = \int_{\mathbb{R}} f(u)\, \nu(du)\,, \tag{3.28}$$

where ν is the average of the LSM.

Proof. We start from (3.27) and note that by (3.4)

$$\frac{1}{|\Lambda_L|} \sum_{x\in\Lambda_L} \langle \delta_x, f(H(\omega))\, \delta_x\rangle = \frac{1}{|\Lambda_L|} \sum_{x\in\Lambda_L} \langle \delta_0, f(H(T_x\omega))\, \delta_0\rangle\,. \tag{3.29}$$

Birkhoff's ergodic theorem (Proposition 3.5) is applicable to this expression and implies that for every $f \in C_0(\mathbb{R})$ there exists $\Omega_f \in \mathcal{A}$ with $\mathbb{P}(\Omega_f) = 1$ such that the above average converges to the non-random value

$$\mathbb{E}\left(\langle \delta_0, f(H(\omega))\, \delta_0\rangle\right) = \int f(u)\, \nu(du)\,.$$

One may conclude that for any countable collection $\mathcal{D} \subset C_0(\mathbb{R})$ this convergence holds simultaneously for a full measure set, $\Omega_0 = \bigcap_{f\in\mathcal{D}} \Omega_f$. By the Stone-Weierstrass theorem, this applies, for example, to the collection of functions $f(u) = (u-z)^{-1}$, with $\operatorname{Re} z$ and $\operatorname{Im} z \neq 0$ rational. Since this is a uniformly dense subset whose linear span is dense in the uniform topology, we conclude that with probability one, the convergence holds simultaneously for all $f \in C_0(\mathbb{R})$. This establishes (3.28). □

3.4. The limiting density of states for sequences of operators

The most intuitive notion of the density of states (DOS) is associated with self-adjoint operators over finite graphs Λ and is given by the point measure

$$\frac{1}{|\Lambda|} \sum_{E_n \in \sigma(H)} \delta_{E_n}\,.$$

Averages of functions f over this measure coincide with $\frac{1}{|\Lambda|} \operatorname{tr} f(H)$.

It may be noted that for operators in finite volumes the above DOS measure coincides (at each realization) with the average over sites of the corresponding local spectral measures. Having seen the self-averaging property of the LSM, it is only natural to expect that the DOS of the finite-volume approximants is also self-averaging, with the same limit. The following theorem asserts this and a bit more.

Theorem 3.15 (DOS for sequences of approximating operators). *Let $H(\omega)$ be a standard ergodic operator in $\ell^2(\mathbb{Z}^d)$, and let $(H_L(\omega))$ be a sequence of operators such that almost surely*

1. $1_{\Lambda_L} H_L(\omega) = H_L(\omega) 1_{\Lambda_L}$, *with* $\Lambda_L = (-L, L)^d \cap \mathbb{Z}^d$ *and* 1_{Λ_L} *orthogonal projections to* $\ell^2(\Lambda_L)$,
2. *the difference* $1_{\Lambda_L} H_L(\omega) - 1_{\Lambda_L} H(\omega)$ *is trace class, satisfying*

$$\operatorname{tr} | 1_{\Lambda_L} H_L(\omega) - 1_{\Lambda_L} H(\omega)| \leq \epsilon(L) |\Lambda_L| , \tag{3.30}$$

with some $\epsilon(L)$ *which vanishes as* $L \to \infty$.

Then there exists a set $\Omega_0 \subset \Omega$ *of full measure,* $\mathbb{P}(\Omega_0) = 1$, *such that for all* $\omega \in \Omega_0$ *and all* $f \in C_0(\mathbb{R})$ *the limit exists and satisfies*

$$\boxed{\lim_{L\to\infty} \frac{1}{|\Lambda_L|} \operatorname{tr} 1_{\Lambda_L} f(H_L(\omega)) = \int_{\mathbb{R}} f(u)\, d\nu(u)} . \tag{3.31}$$

Equation (3.31) in effect states that the **finite-volume density of states measure** given by $|\Lambda_L|^{-1} \operatorname{tr} P_{(\cdot)}(H_L(\omega) \,|\, _{\ell^2(\Lambda_L)})$ almost surely converges weakly to ν. Since $H_L(\omega)$ commutes with 1_{Λ_L}, the trace on the left side only depends on the restriction of $H_L(\omega)$ to the subspace $\ell^2(\Lambda_L)$.

Proof. It suffices to establish the convergence for resolvents, i.e., a function of the form $f(u) = (u - z)^{-1}$, with $z \in \mathbb{C}\backslash\mathbb{R}$ (Proposition A.7). For such a function we have, using the resolvent identity,

$$\begin{aligned}
&\left| \frac{1}{|\Lambda_L|} \operatorname{tr} 1_{\Lambda_L} f(H_L(\omega)) - \frac{1}{|\Lambda_L|} \operatorname{tr} 1_{\Lambda_L} f(H(\omega)) \right| \\
&\quad = \frac{1}{|\Lambda_L|} \left| \operatorname{tr} 1_{\Lambda_L} \left[\frac{1}{H_L(\omega) - z} - \frac{1}{H(\omega) - z} \right] \right| \\
&\quad = \frac{1}{|\Lambda_L|} \left| \operatorname{tr} 1_{\Lambda_L} \frac{1}{H_L(\omega) - z} [H_L(\omega) - 1_{\Lambda_L} H(\omega)] \frac{1}{H(\omega) - z} \right| \\
&\quad \leq \frac{1}{|\operatorname{Im} z|^2} \frac{1}{|\Lambda_L|} \operatorname{tr} | 1_{\Lambda_L} H_L(\omega) - 1_{\Lambda_L} H(\omega)| \\
&\quad \leq \frac{\epsilon(L)}{|\operatorname{Im} z|^2} \longrightarrow 0 \qquad (\text{for } L \to \infty).
\end{aligned} \tag{3.32}$$

We already know by Theorem 3.14 that for $\mathbb{P}$-almost all ω, $\frac{1}{|\Lambda_L|} \operatorname{tr} 1_{\Lambda_L} f(H(\omega))$ tends to $\int (u - z)^{-1} \nu(du)$. The uniform bound on the difference (3.32) permits us to conclude that $\frac{1}{|\Lambda_L|} \operatorname{tr} 1_{\Lambda_L} f(H_L(\omega))$ does also. ☐

Theorem 3.15 can be rephrased in terms of the **integrated density of states**

$$n(E) \;:=\; \nu((-\infty, E)) \tag{3.33}$$

and its finite-volume counterparts $N(H_L; E) := \mathrm{tr}\, P_{(-\infty,E)}(H_L|_{\ell^2(\Lambda_L)})$. The latter counts (including multiplicity) the number of eigenvalues of $H_L(\omega)$ when restricted to $\ell^2(\Lambda_L)$ below energy E.

Corollary 3.16. *For any ergodic operator with the structure described in Theorem* 3.15 *there exists a full measure set $\Omega_0 \subset \Omega$ such that*

$$\lim_{L\to\infty} \frac{1}{|\Lambda_L|} N(H_L(\omega); E) \;=\; n(E) \tag{3.34}$$

for all $\omega \in \Omega_0$ and all $E \in \mathbb{R}$ except the (at most countably many) points of discontinuity of n.

Proof. Since weak convergence implies the convergence of the distribution function at all points of continuity of the limit (Exercise 3.3 or [**46**]), this is an immediate consequence of Theorem 3.15. □

It may be added here that for any standard ergodic operator on $\ell^2(\mathbb{Z}^d)$ with finite-range hopping, such as the random Schrödinger operator with or without magnetic field in Examples 3.6 and 3.7, the integrated density of states n is continuous [**114**]. A proof of this statement is outlined in Exercise 3.8.

Theorem 3.15 can be applied in several ways:

1. $H_L(\omega)$ can be just the minimal modification of $H(\omega)$ obtained by decoupling Λ_L from its exterior; i.e.,

$$H_L(\omega) \;=\; 1_{\Lambda_L}\, H(\omega)\, 1_{\Lambda_L} + 1_{\Lambda_L^c}\, H(\omega)\, 1_{\Lambda_L^c}\;. \tag{3.35}$$

2. $H_L(\omega)$ restricted to $\ell^2(\Lambda_L)$ can also be taken to be any of the many finite-volume versions of the operator $H(\omega)$ with rather arbitrary choices of **boundary conditions**: free, periodic, or wired in some way. In order to relate such approximations to the infinite operator, it is convenient to set $H_L(\omega)$ to zero in the complement of Λ_L (anyway, the trace on the left side of (3.31) is independent of this choice). Prominent examples of Dirichlet and Neumann boundary conditions are discussed in the next chapter.

It should be noted that for the latter situation the theorem implies that the boundary conditions in $H_L(\omega)$ do not affect the limiting DOS, as long as the trace condition (3.30) is met.

3.5.* Statistic mechanical significance of the DOS

Before concluding this section let us note that the DOS measure ν is a quantity of great importance in condensed matter physics. For homogeneous quantum systems it represents the number of single-particle states per volume. It enters into the basic expression of thermal-equilibrium properties of the corresponding system of non-interacting particles. Among those, for a homogenous gas of non-interacting (spinless) fermions the specific free energy at inverse temperature $\beta = (kT)^{-1}$ (with k the Boltzmann constant) and particle concentration $\bar{n} > 0$ is given by

$$f(\beta, \bar{n}) := \sup_{\mu \in \mathbb{R}} \left[\mu \, \bar{n} - T \int_{\mathbb{R}} \log \left(1 + e^{(\mu - E)} \right) \, \nu(dE) \right] . \tag{3.36}$$

Presupposing that, as is shown in the next chapter, the density of states measure ν is absolutely continuous, with a density function $\nu(E) = \nu(dE)/dE$, one gets Sommerfeld's asymptotic formula for the specific free energy f at low-temperatures ($\beta \to \infty$):

$$f(\beta, \bar{n}) \sim \int_{-\infty}^{E_F} \nu(E) E \, dE - \beta^{-2} \nu(E_F) \, \pi^2/6 \tag{3.37}$$

with the Fermi energy E_F given by the solution of the equation $n(E_F) = \bar{n}$; cf. [**364**] and [**272**, Chap. 4]. As will be seen in Chapter 13, $\nu(E)$ also enters formulas for conductance.

Notes

Aubrey-André duality: To bring up-front the self-dual nature of the operator $H_\alpha(\theta)$ defined in (3.15), one may consider its action on the space $L^2(\mathbb{R})$, viewing that as a direct integral:

$$L^2(\mathbb{R}) \;=\; \int_{[0,1)}^{\oplus} \ell^2(\mathbb{Z} + t) \; dt \, . \tag{3.38}$$

The discrete second difference operator Δ of (3.15) is to be interpreted in this context as $\left[e^{ip\xi} + e^{-ip\xi} - 2 \right]$ with $p = -id/d\xi$. Thus, in its action on $L^2(\mathbb{R})$ the almost-Mathieu operator (3.15) can be viewed as

$$\widehat{H}_\alpha \;=\; 2 \left[\cos(p) + \lambda \cos(\xi) \right] - 2 \, . \tag{3.39}$$

Under the Fourier transform this operator is mapped unitarily onto an explicit linear function of its version with $\lambda \mapsto 1/\lambda$. It is then relevant to note that localization in the Fourier space, e.g., as exhibited in the extreme case of the plane wave functions $e^{ik\xi}$, corresponds to extended behavior in the real space.

The above observations play a role in the Aubrey-André duality. Yet the discussion may leave one perplexed: Since the Fourier transform acts unitarily, the spectrum of $\widehat{H}_\alpha$ at λ is of the same nature as at $1/\lambda$. However, as described in the text and made intuitive by the above uncertainty relation, the spectra of H_α at λ and λ^{-1} are of clearly different characteristics. We leave it to the reader to reflect upon the apparent paradox and make peace with it.

Density of states: The existence of the DOS measure as a limit of finite-volume quantities can be established in various ways. Aside from the argument seen in the proof of Theorem 3.15, a proof can be based on the sub- and superadditivity of the Dirchlet and Neumann, respectively, finite-volume integrated DOS as expressed in (4.35). The convergence (3.34) for these boundary conditions also follows from the Akcoglu-Krengel subadditive ergodic theorem [**268**]. For details, see [**232, 228**]. Another approach is based on the representation of $\operatorname{tr} e^{-tH}$, which coincides with the Laplace transform of the DOS, in terms of suitable integrals over random walks (or the Brownian motion in the continuous case) [**321, 82**].

For random Schrödinger operators in the continuum, i.e., on $L^2(\mathbb{R}^d)$, the DOS measure ν may be defined in a similar way [**228, 82**]. The main difference is that ν is no longer a bounded measure, and therefore weak convergence should be replaced by vague convergence. For a recent survey about the density of states, see [**237**].

Exercises

3.1. Let $X : \Omega \to \mathbb{R} \cup \{\infty\}$ be a random variable which is invariant under a family $(T_x)_{x\in I}$ of ergodic transformations.
   a. Show that the distribution function $(-\infty, \infty] \ni t \mapsto \mathbb{P}(X \le t)$ takes values in the set $\{0, 1\}$ only.
   b. Show that $\mathbb{P}(X = c) = 1$ for $c := \inf\{t \in (-\infty, \infty] \mid \mathbb{P}(X \le t) = 1\}$.

3.2. A family of measure-preserving transformations $(S_x)_{x\in\mathbb{Z}^d}$ on a probability space $(\Omega, \mathcal{A}, \mathbb{P})$ is called **mixing** if for any pair of events $A, B \in \mathcal{A}$

$$\mathbb{P}\left(A \cap S_x^{-1} B\right) \longrightarrow \mathbb{P}(A)\,\mathbb{P}(B) \quad \text{as } |x| \to \infty.$$

   a. Show that mixing implies the ergodicity of $(S_x)_{x\in\mathbb{Z}^d}$.

b. Show that the shifts (3.7) defined on the canonical realization $(\mathbb{R}^{\mathbb{Z}^d}, \mathcal{B}, \mathbb{P})$ of the probability space corresponding to iid random variables, $(\omega(\xi))_{\xi\in\mathbb{Z}^d}$, are mixing.

3.3. Let μ_n be a sequence of probability measures which converges weakly to a probability measures μ. Show that the distribution functions $F_n(t) = \mu_n(-\infty, t]$ converge at all points of continuity of the limit.

3.4. Prove that if $H(\omega)$ is a standard ergodic operator acting in $\ell^2(\mathbb{Z}^d)$ then:
a. For any Borel set $I \subset \mathbb{R}$ the range of the spectral projection $P_I(H(\omega))$ is almost surely of dimension zero or infinity.
b. The operator's discrete spectrum is almost surely empty.
[The topics are discussed in [**228, 82**].]

3.5. For the Laplacian on $\ell^2(\mathbb{Z}^2)$ with a constant magnetic field $B \neq 0$ defined in (3.12) show the following:
a. $(-\Delta^{(A)})$ is a standard ergodic operator (in the sense of Definition 3.4),
b. the magnetic translations $U_x^{(A)}$ which implement the shift covariance relation do not form a group (unless taken modulo gauge transformation).

3.6. Derive the relation of the magnetic operator defined in (3.12) to the Harper Hamiltonian at the critical value $\lambda = 1$ (3.15).
[Hint: It is convenient to first transform the operator (3.12) unitarily into one in which the vector potential depends only on the second coordinate.]

3.7. The purpose of the following question is to exemplify in a very simple context the difference between two relevant notions of pure-point spectra of ergodic operators. Let $U_{\alpha,\theta}$ be the unitary operator defined on $\ell^2(\mathbb{Z})$ by

$$(U_{\alpha,\theta}\psi)(n) \;=\; e^{i(2\pi\alpha n+\theta)}\psi(n).$$

Prove the following for any α irrational:
a. For each value of $\theta \in [0, 2\pi)$, $U_{\alpha,\theta}$ has a complete set of proper eigenfunctions, and the operator's pure-point spectrum, as defined in Section 2.1, does not vary with θ.
b. For almost every pair of realizations of this random operator, $\theta_1, \theta_2 \in [0, 2\pi)$, the collections of eigenvalues of U_{α,θ_1} are disjoint from those of U_{α,θ_1}.

3.8. Let H be a self-adjoint operator on $\ell^2(\mathbb{Z}^d)$ with a kernel $H(x,y) = \langle \delta_x, H\delta_y\rangle$ of finite range $r \in [1,\infty)$, i.e., satisfying

$$H(x,y) = 0 \quad \text{for all } x,y \in \mathbb{Z}^d \text{ with } |x-y| \geq r.$$

a. Show that for any $E \in \mathbb{R}$ and all $L > 0$

$$\operatorname{tr}\left[1_{\Lambda_L} P_{\{E\}}(H)\right] \;\leq\; |\partial\Lambda_L^{(r)}|\,,$$

where $\partial\Lambda_L^{(r)} := \{x \in \Lambda_L \mid \operatorname{dist}(x, \mathbb{Z}^d \backslash \Lambda_L) \leq r\}$.
[Hint: For any two orthogonal projections P, Q on a Hilbert space $\operatorname{tr} PQ \leq \dim Q\,\big|_P$.]

b. Explain how this implies that for any standard ergodic operator H on $\ell^2(\mathbb{Z}^d)$ with the above property the **integrated density of states** is **continuous**; i.e., for any $E \in \mathbb{R}$,

$$\mathbb{E}\left[\langle \delta_0, P_{\{E\}}(H)\delta_0\rangle\right] \;=\; 0\,.$$

[This statement and its proof go back to F. Delyon and B. Souillard [**114**].]

Chapter 4

Density of States Bounds: Wegner Estimate and Lifshitz Tails

Having introduced the density of states (DOS) in the previous section, our goal here is to present a pair of general results:

Wegner's estimate: In the presence of a random potential $(V(x))_{x \in \mathbb{G}}$ the DOS measure ν is at least as regular as the conditional distribution of the single-site values of the potential. In particular, if the conditional distribution of the potential at each site $x \in \mathbb{G}$ is absolutely continuous, then so is the DOS.

Lifshitz tails: The integrated density of states exhibits a drastic thinning at the edges of the spectrum.

The reader may also find of independent interest the techniques employed in the proofs of these statements, which include the following:

Dirichlet-Neumann bracketing: A useful tool for the proof of the convergence of volume averages for the DOS of operators with homogeneous disorder.

Temple's inequality: A useful tool for lower bounds which supplement the Rayleigh-Ritz variational principle for the ground state energy.

4.1. The Wegner estimate

Any random operator $H(\omega)$ on $\ell^2(\mathbb{G})$ comes along with a mean local spectral measure ν. In case it is absolutely continuous, i.e., of the form $\nu(dE) = \nu(E)\,dE$, its Radon-Nikodym derivative is referred to as the (local) **density of states** (DOS):

$$\boxed{\nu(E) := \frac{\nu(dE)}{dE} = \frac{\mathbb{E}\left[\langle \delta_0, P_{dE}(H)\,\delta_0\rangle\right]}{dE}} \,. \tag{4.1}$$

The DOS function $\nu(E)$ should not be confused with the DOS measure $\nu(dE)$, nor with the integrated density of states $n(E)$ which was presented in (3.33). The relation to the latter is $n(E) \;=\; \int_{-\infty}^{E} \nu(u)\; du$.

Neither the existence of the derivative in (4.1) nor its boundedness should be taken for granted. An example with singular DOS measure ν is the Landau Hamiltonian describing a particle in $\mathbb{R}^2$ in the presence of a perpendicular magnetic field [**158, 279**]. In the case of periodic systems in one dimension, the DOS is known to diverge at the band edges (cf. (4.39) below). As was pointed out by F. Wegner [**397**], the addition of a random potential with sufficiently smooth density washes away divergences of the DOS such as mentioned above.

Wegner's bound holds under the assumption that the conditional distribution of the random potential $V(x)$ at a site $x \in \mathbb{G}$, conditioned on the potential values at all other sites ($V_{\neq x} := \{V(y)\}_{y\neq x}$), is absolutely continuous with a bounded density function. This conditional probability distribution $\rho_x(dv|V_{\neq x})$ is defined so that for any $f \in C_0(\mathbb{R})$ the corresponding conditional expectation is given by

$$\mathbb{E}\left[f(V(x))\,\middle|\, V_{\neq x}\right] \;=\; \int f(v)\,\rho_x(dv\,|\,V_{\neq x})\,. \tag{4.2}$$

Our assumption is that $\rho_x(dv\,|\,V_{\neq x})$ is absolutely continuous with uniformly bounded density,

$$\rho_x(dv\,|\,V_{\neq x}) \;\le\; \|\rho\|_\infty\, dv \tag{4.3}$$

with some $\|\rho\|_\infty < \infty$, which is independent of $V_{\neq x}$. The Wegner bound, stated here initially for finite graphs, reads:

Theorem 4.1 (Wegner). *Let $H(\omega) = T + \lambda V(\omega)$ with $\lambda > 0$ be a random operator on $\ell^2(\mathbb{G})$ over a finite graph $\mathbb{G}$ with a random potential for which the conditional distribution of $V(x)$ is absolutely continuous with density $\rho(\cdot|V_{\neq x})$ uniformly bounded by $\|\rho\|_\infty < \infty$. Then the mean local spectral measure (LSM) $\nu_x = \mathbb{E}\left[\langle\delta_x, P_{(\cdot)}(H)\,\delta_x\rangle\right]$ associated with any $x \in \mathbb{G}$ is also* *absolutely continuous with*

$$\boxed{\nu_x(E) = \frac{\nu_x(dE)}{dE} \leq \lambda^{-1} \|\rho\|_\infty} \,. \tag{4.4}$$

In the presence of translation invariance, i.e., for standard ergodic operators, the measures ν_x do not depend on x, and on amenable graphs they coincide with the operator's density of state. The Wegner estimate can be obtained by applying the following pair of useful tools.

Lemma 4.2. *A non-negative measure μ on $\mathbb{R}$ is absolutely continuous with derivative satisfying*

$$\mu(u) := \frac{\mu(du)}{du} \leq B \tag{4.5}$$

if and only if for any $z \in \mathbb{C}^+$

$$\pi^{-1} \operatorname{Im} \int_{\mathbb{R}} \frac{\mu(du)}{u - z} \leq B \,. \tag{4.6}$$

Proof. To conclude (4.6) from (4.5), we use the positivity of the integrand $\operatorname{Im}(u-z)^{-1} \geq 0$ and the equality $\int \operatorname{Im}(u-z)^{-1} du = \pi$ for $\operatorname{Im} z > 0$.

In the opposite direction, using Fubini's theorem and the dominated convergence theorem one has for every pair $a, b \in \mathbb{R}$ with $\mu(\{a\}) = \nu(\{b\}) = 0$,

$$\begin{aligned} \lim_{\eta \downarrow 0} \int_a^b \int \operatorname{Im} \frac{\mu(du)}{u - E - i\eta}\, dE \\ = \int \lim_{\eta \downarrow 0} \left[\operatorname{Arg}(a - u + i\eta) - \operatorname{Arg}(b - u + i\eta)\right] \mu(du) \;=\; \pi\, \mu((a,b)) \,, \end{aligned} \tag{4.7}$$

where $\operatorname{Arg}(z) = \operatorname{Im} \operatorname{Log}(z)$ is the principal value of the complex logarithm. This allows us to conclude (4.5) from (4.6). □

Lemma 4.3. *Let $H_v = H_0 + v\, 1_{\{x\}}$ be a self-adjoint operator on $\ell^2(\mathbb{G})$ over a finite graph $\mathbb{G}$, and let $x \in \mathbb{G}$. Then, for $z \in \mathbb{C}^+$, the diagonal kernel of the resolvent has the following dependence on $v \in \mathbb{R}$:*

$$\boxed{\left\langle \delta_x, \frac{1}{H_v - z}\, \delta_x \right\rangle \;=\; \frac{1}{v - \Sigma_0(z)}} \tag{4.8}$$

with $\Sigma_0(z) = -\langle \delta_x, \frac{1}{H_0 - z}\, \delta_x\rangle^{-1}$.

Proof. Regarding $H_v - z$ as a finite-dimensional matrix, the statement can be seen as a simple consequence of Cramer's rule, by which $\langle \delta_x, \frac{1}{H_v - z}\, \delta_x\rangle = M_{x,x}(z)/[v\, M_{x,x}(z) + \det(H_0 - z)]$ with $M_{x,x}(z)$ an appropriate minor of $H_0 - z$. □

Combining the above two statements we get

Proof of Theorem 4.1. Using Lemma 4.3, for any $z \in \mathbb{C}^+$ the conditional expectation satisfies

(4.9)

$$\begin{aligned}\mathbb{E}\left[\operatorname{Im}\left\langle \delta_x, \frac{1}{H-z}\delta_x\right\rangle \,\middle|\, V_{\neq x}\right] &\leq \|\rho\|_\infty \int \operatorname{Im}\left\langle \delta_x, \frac{1}{H_x + \lambda v \, 1_{\{x\}} - z}\,\delta_x\right\rangle dv \\ &= \|\rho\|_\infty \int \operatorname{Im}\frac{1}{\lambda v - \Sigma_x(z)}\, dv \\ &= \lambda^{-1}\,\|\rho\|_\infty \, \pi\end{aligned}$$

with $\Sigma_x(z) = -\langle \delta_x, \frac{1}{H_x - z}\,\delta_x\rangle^{-1}$ where H_x is the operator obtained from H by setting the potential at x to zero. Averaging (4.9) over $V_{\neq x}$ also, we find that for any $z \in \mathbb{C}^+$,

$$\pi^{-1}\operatorname{Im}\int \frac{\nu_x(du)}{u-z} \;\leq\; \lambda^{-1}\,\|\rho\|_\infty\,. \tag{4.10}$$

The Wegner bound then follows by an application of Lemma 4.2. □

The Wegner bound (4.4) is also valid for operators on infinite graphs. To prove it in that generality one needs to explore a bit deeper the rank-one perturbation theory within the context of infinite operators. This is done in the next chapter.

4.2.* DOS bounds for potentials of singular distributions

A somewhat more general version of the DOS bound (4.4), which also applies to random potential with singular distributions, can be formulated in terms of the measure's concentration function. In order to address the case of correlated variables it is convenient to formulate it directly for the multivariate case.

Definition 4.4. For any probability measure ρ on $\mathbb{R}^{\mathbb{G}}$, we define its **concentration function**, $S_\rho : [0,\infty) \mapsto \mathbb{R}_+$, through the following uniform bound on the conditional probability distributions of the variables $V(x)$, in each case conditioned on the rest ($V_{\neq x}$):

$$S_\rho(t) \;:=\; \sup_{x\in\mathbb{G},\, u\in\mathbb{R}} \;\operatorname*{ess\,sup}_{V_{\neq x}} \rho\left([u, u+t] \,|\, V_{\neq x}\right) \tag{4.11}$$

(where the second supremum allows us to omit zero measure sets).

The boundedness of the density (4.5) is equivalent to $S_\rho(t) \leq B\,t$. It is useful to have the following extension of the notion of Hölder continuity, as defined in Definition 2.2.

Definition 4.5. The joint probability distribution of $\{V(x)\}_{x\in\mathbb{G}}$ is said to be **uniformly τ-Hölder continuous** if for some $C_\tau(\rho) < \infty$

$$S_\rho(2t) \leq C_\tau(\rho)\, t^\tau\,. \tag{4.12}$$

The Wegner bound (4.4) can be equivalently stated in the integrated form

$$\mathbb{E}\left[\langle \delta_x, P_I(H_v)\, \delta_x\rangle\right] \;\leq\; \lambda^{-1}\|\rho\|_\infty\, |I|\,, \tag{4.13}$$

which holds for all Borel sets $I \subset \mathbb{R}$, with $|I|$ the set's Lebesgue measure. The following provides a variant of this bound which also applies to potentials of singular distributions. That includes measures supported on sets of fractal dimension lower than 1, but strictly positive (excluding, in essence, mainly the case of Bernoulli variables).

Theorem 4.6. *Let $H(\omega) = T + \lambda\, V(\omega)$ with $\lambda > 0$ be a random operator on $\ell^2(\mathbb{G})$ over a finite graph $\mathbb{G}$ with a random potential for which S_ρ stands for the concentration function of its distribution. Then for all intervals $I \subset \mathbb{R}$,*

$$\boxed{\nu_x(I) \;=\; \mathbb{E}\left[\langle \delta_x, P_I(H)\, \delta_x\rangle\right] \;\leq\; (4 + \pi/\lambda)\; S_\rho(|I|)}\,. \tag{4.14}$$

It should be noted that unlike (4.13), in the last statement I is limited to intervals, to the exclusion of more general Borel subsets.

The proof employs a useful observation about single-hump functions.

Definition 4.7. A **single-hump function** on $\mathbb{R}$ is a real-valued function which for some $v_0 \in \mathbb{R}$ is monotone increasing on $(-\infty, v_0)$ and monotone decreasing on (v_0, ∞) and which vanishes at infinity.

Of specific interest to us are the following two examples:

(1) $v \mapsto |G_v(x,x;z)|$ for every $z \in \mathbb{C}\backslash\mathbb{R}$,

(2) $v \mapsto \operatorname{Im} G_v(x,x;z)$ for every $z \in \mathbb{C}^+$,

with $G_v(x,x;z) := \langle \delta_x, \frac{1}{H_v - z}\, \delta_x\rangle$ and $H_v = H_0 + v\, 1_{\{x\}}$ self-adjoint on $\ell^2(\mathbb{G})$, which from (4.8) can be immediately recognized to be single-hump functions.

Lemma 4.8. *For any bounded, integrable single-hump function $F : \mathbb{R} \mapsto [0,\infty)$, any non-negative measure ρ on $\mathbb{R}$, and any $\varepsilon > 0$,*

$$\int_{\mathbb{R}} F(u)\, \rho(du) \;\leq\; [2\|F\|_\infty \;+\; \|F\|_1/\varepsilon]\; S_\rho(\varepsilon)\,. \tag{4.15}$$

In particular,

$$\int_{\mathbb{R}} F(u)\, \rho(du) \;\leq\; 3\, \|F\|_\infty\; S_\rho\Big(\frac{\|F\|_1}{\|F\|_\infty}\Big)\,. \tag{4.16}$$

Proof. We split the integration over $\mathbb{R}$ into two parts corresponding to $(-\infty, v_0)$ and $[v_0, \infty)$. In turn, each half-line is partitioned into intervals $[x_n, x_{n+1})$ of length $\varepsilon > 0$ such that

$$(4.17)\quad \int_{[v_0,\infty)} F(u)\,\rho(du) \;=\; \sum_{n=0}^{\infty} \int_{[x_n,x_{n+1})} F(u)\,\rho(du) \\ \le\; \sum_{n=0}^{\infty} F(x_n)\,\rho\left([x_n, x_{n+1})\right) \;\le\; S_\rho(\varepsilon) \sum_{n=0}^{\infty} F(x_n)\,,$$

and similarly for the other half-line. By monotonicity, up to a factor of ε and when dropping the first term, the sum on the right side is a lower sum of the integration over $[v_0, \infty)$:

$$(4.18)\qquad \sum_{n=0}^{\infty} F(x_n) \le F(v_0) + \frac{1}{\varepsilon} \int_{v_0}^{\infty} F(u)\,du\,.$$

A similar estimate for the other half-line concludes the proof of (4.15).

The second assertion, which up to constants constitutes the optimal bound, follows by choosing $\varepsilon = \|F\|_1 \,/ \|F\|_\infty$. □

We note in passing that (4.15) immediately generalizes to functions F which can be written as a sum $F(u) = \sum_{j=1}^{N} F_j(u)$ of N single-hump functions $F_j : \mathbb{R} \mapsto [0, \infty)$ with maxima at $\{x_j\}_{j=1}^{N}$ for which for any $\varepsilon > 0$,

$$(4.19)\qquad \int_{\mathbb{R}} F(u)\,\rho(du) \;\le\; 2\, S_\rho(\varepsilon) \sum_{J=1}^{N} F_j(x_j) \;+\; \frac{S_\rho(\varepsilon)}{\varepsilon}\, \|F\|_1\,.$$

Lemma 4.8 permits us to extract the more general bound (4.14) from (4.8):

Proof of Theorem 4.6. Denoting by E the midpoint of I, one has

$$(4.20)\quad \langle \delta_x, P_I(H)\,\delta_x\rangle \;\le\; |I|\; \mathrm{Im}\left\langle \delta_x, \frac{1}{H - E - i|I|/2}\,\delta_x \right\rangle \;=:\; F(V(x))\,.$$

Since $F(v)$ has the single-hump property, with $\|F\|_\infty \le 2$ and $\|F\|_1 \le |I|\,\pi/\lambda$, the result follows by conditioning on $V_{\neq x}$ from (4.15) with $\varepsilon = |I|$. □

Thus, the LSM measure ν_x is at least as regular, in the sense of the concentration function, as the conditional distribution of the random potential $V(x)$. The above discussion yields the following bounds on the mean number of levels in an interval:

Corollary 4.9. *In the situation of Theorem* 4.6,

$$\boxed{\mathbb{E}\,[\operatorname{tr} P_I(H)] \;\le\; (4+\pi/\lambda)\,|\mathbb{G}|\,S_\rho(|I|)}\,. \tag{4.21}$$

Proof. Expanding the trace and applying (4.14), one gets

$$\mathbb{E}\,[\operatorname{tr} P_I(H)] = \sum_{x\in\mathbb{G}} \mathbb{E}\,[\langle \delta_x, P_I(H)\,\delta_x\rangle] \le (4+\pi/\lambda) \sum_{x\in\mathbb{G}} S_\rho(|I|)\,. \tag{4.22}$$

□

It is customary to apply the term **Wegner estimate** to (4.4) as well as to its various extensions. We shall therefore also apply this term to (4.14) and (4.21). The bound (4.21) guarantees that on average the mean eigenvalue spacing of H is at most $|\mathbb{G}|^{-1}$. Combined with the Chebyshev inequality,

$$\mathbb{P}\,(\operatorname{tr} P_I(H) \ge 1) \;\le\; \mathbb{E}\,[\operatorname{tr} P_I(H)]\;, \tag{4.23}$$

the bound (4.21) can be used to control the probability of the occurrence of an eigenvalue of $H(\omega)$ in any fixed interval I. Such estimates play an important role in the localization proof via the multi-scale analysis ([**367, 230**]). Some further comments on the subject can be found in this chapter's notes.

4.3. Dirichlet-Neumann bracketing

In discussing additive properties of the spectra of Schrödinger operators in large domains, such as the total number of states in a given energy range, it is often of help to subdivide the region into a disjoint union of fundamental domains (e.g., boxes) and consider an approximation of the operator by means of a direct sum of operators associated with the individual domains. The **Dirichlet-Neumann bracketing** principle refers to a pair of complementary bounds on the original operator's spectrum obtained through two such additive schemes. This tool will be of help in the analysis of the Lifshitz tails DOS asymptotics, discussed next.

A natural starting point for this purpose is to think in terms of the quadratic form associated with the operator, which for $H = -\Delta + V$ on $\ell^2(\mathbb{G})$ is $\langle\psi, H\psi\rangle = Q_{\mathbb{G}}(\psi) + \sum_{x\in\mathbb{G}} V(x)\,|\psi(x)|^2$ with $Q_{\mathbb{G}}(\psi)$ the Laplacian form

$$Q_{\mathbb{G}}(\psi) \;:=\; -\langle\psi, \Delta_{\mathbb{G}}\psi\rangle \;=\; \sum_{[x,y]\in E_{\mathbb{G}}} |\psi(y)-\psi(x)|^2\;. \tag{4.24}$$

Let $\mathcal{P} = (\mathbb{G}_n)$ be a partition of the vertex set of an arbitrary graph with vertex set $\mathbb{G}$ and edge set $E_{\mathbb{G}}$. The corresponding Neumann and Dirichlet Laplacians can be presented through their quadratic forms $Q_{\mathcal{P}}^N(\psi)$ and $Q_{\mathcal{P}}^D(\psi)$ which, following [**350**], are defined as

(4.25)

$$\begin{aligned} Q_{\mathcal{P}}^N(\psi) &:= \sum_n \sum_{[x,y]\in E_{\mathbb{G}_n}} |\psi(y)-\psi(x)|^2 , \\ Q_{\mathcal{P}}^D(\psi) &:= \sum_n \sum_{[x,y]\in E_{\mathbb{G}_n}} |\psi(y)-\psi(x)|^2 + \sum_{n\neq m} \sum_{\substack{x\in\mathbb{G}_n, y\in\mathbb{G}_m \\ [x,y]\in E_{\mathbb{G}}}} \left[|\psi(y)|^2 + |\psi(x)|^2\right] . \end{aligned}$$

The definition immediately yields the quadratic form relation:

$$\boxed{Q_{\mathcal{P}}^N(\psi) \;\le\; Q_{\mathbb{G}}(\psi) \;\le\; Q_{\mathcal{P}}^D(\psi)} \tag{4.26}$$

which holds for all $\psi \in \ell^2(\mathbb{G})$. The lower bound holds since the change from $Q_{\mathbb{G}}$ to $Q_{\mathcal{P}}^N$ is just in the omission of the contribution of edges linking different partition elements. In the upper bound the contribution of the previously omitted edges is overestimated since

$$|\psi(y)-\psi(x)|^2 \;\le\; 2\left[|\psi(y)|^2 + |\psi(x)|^2\right] . \tag{4.27}$$

The factor 2 is hidden in the (n,m) double counting in (4.25).

To translate these relations into operator language, let

$$\deg_{\mathbb{G}}(x) := |\{y\in\mathbb{G} \mid [x,y]\in E_{\mathbb{G}}\}| \tag{4.28}$$

denote the degree of vertices in the graph $\mathbb{G}$, and let $(\mathcal{P}, E_{\mathcal{P}})$ be the subgraph obtained by deleting all edges which link sites in different elements of the partition $\mathcal{P}$. Corresponding to the above quadratic forms, for a given graph $\mathbb{G}$ and partition $\mathcal{P} = (\mathbb{G}_n)$ are the following:

1. The **graph Laplacian**:

(4.29)

$$(\Delta_{\mathbb{G}}\,\psi)(x) := \sum_{y:[x,y]\in E_{\mathbb{G}}} (\psi(y)-\psi(x)) \;=\; \sum_{y:[x,y]\in E_{\mathbb{G}}} \psi(y) \;-\; (D_{\mathbb{G}}\,\psi)(x) .$$

2. The **Neumann Laplacian**:

(4.30)

$$(\Delta_{\mathcal{P}}^N\,\psi)(x) := \sum_{y:[x,y]\in E_{\mathcal{P}}} (\psi(y)-\psi(x)) \;=\; \sum_{y:[x,y]\in E_{\mathcal{P}}} \psi(y) \;-\; (D_{\mathcal{P}}\,\psi)(x) .$$

3. The **Dirichlet Laplacian**:

$$\Delta_{\mathcal{P}}^D := \Delta_{\mathcal{P}}^N + 2\,(D_{\mathcal{P}} - D_{\mathbb{G}}) , \tag{4.31}$$

where $(D_{\mathbb{G}}\psi)(x) := \deg_{\mathbb{G}}(x)\,\psi(x)$ stands for the degree operator and the correspondence of the quadratic form with operator is based on

$$Q^X_{\mathcal{P}}(\psi) \;=\; -\langle \psi\,, \Delta^X_{\mathcal{P}}\,\psi\rangle\,, \quad X = N, D, \tag{4.32}$$

subject to the usual restrictions on operator and form domains [**331**] of which there are none in the case of graphs of uniformly bounded degree.

The above Neumann ($X = N$) and Dirichlet ($X = D$) operators can also be regarded as the sums

$$\Delta^X_{\mathcal{P}} \;=\; \bigoplus_n \Delta^X_{\mathbb{G}_n} \tag{4.33}$$

whose terms $\Delta^X_{\mathbb{G}_n} = 1_{\mathbb{G}_n}\,\Delta^X_{\mathbb{G}_n}\,1_{\mathbb{G}_n}$ act in the subspaces $\ell^2(\mathbb{G}_n)$:

- $\Delta^N_{\mathbb{G}_n}$ as the regular graph Laplacian in the subgraph $\mathbb{G}_n$,
- $\Delta^D_{\mathbb{G}_n}$ as a Dirichlet-type operator with, in effect, zero boundary conditions on $\mathbb{G}\backslash\mathbb{G}_n$—modified, however, to have a doubly strong coupling to those sites.

The above terminology is borrowed from [**350**] and is not consistently used throughout the literature. The virtue of this choice should, however, be clear from (4.26), which can be restated as

Theorem 4.10 (Dirichlet-Neumann bracketing). *For any graph $\mathbb{G}$ and partition $\mathcal{P}$,*

$$-\Delta^N_{\mathcal{P}} \;\leq -\Delta_{\mathbb{G}} \;\leq\; -\Delta^D_{\mathcal{P}}$$

in the sense of positive operators on $\ell^2(\mathbb{G})$.

Readers familiar with the Dirichlet and Neumann Laplacian on $L^2(\Lambda)$ over a domain $\Lambda \subset \mathbb{R}^d$ with piecewise smooth boundary will recognize Theorem 4.10 as an analogue of the Dirichlet-Neumann bracketing available there [**329**]. Useful tools for further implications of operator monotonicity and the related Löwner theorem are mentioned in Section B.4.

Let us now turn to applications and consider Schrödinger operators restricted to $\Lambda \subset \mathbb{Z}^d$ arising from the Neumann or Dirichlet Laplacian; i.e., $H^X_\Lambda := -\Delta^X + V$ on $\ell^2(\Lambda)$. The following is an immediate consequence of Theorem 4.10.

Corollary 4.11. *Let $\Lambda \subset \mathbb{Z}^d$ be the disjoint union of Λ_j, with $j = 1, \dots, N$. Then*

$$\boxed{\bigoplus_{j=1}^N H^N_{\Lambda_j} \leq H^N_\Lambda \leq H^D_\Lambda \leq \bigoplus_{j=1}^N H^D_{\Lambda_j}}\,.$$

In the standard ergodic setup, it is an easy exercise to prove that

$$H_L^X(\omega) = -\Delta_{\Lambda_L}^X + V(\omega)\,,$$

when naturally embedded in the full Hilbert space, satisfies the requirements of Theorem 3.15 and Corollary 3.16 for both $X = N, D$. Thus the normalized eigenvalue counting function converges almost surely at all $E \in \mathbb{R}$ to the integrated density of states $n(E) = \lim_{L\to\infty} |\Lambda_L|^{-1} N(H_L^X; E)$. In view of Corollary 4.11, this convergence happens to be monotone at least on average.

Lemma 4.12. *For a standard ergodic operator $H(\omega) = -\Delta + V(\omega)$ in $\ell^2(\mathbb{Z}^d)$ one has for all $\ell > 0$ and all $E \in \mathbb{R}$*

$$\frac{1}{|\Lambda_\ell|}\,\mathbb{E}\left[N(H_\ell^D; E)\right] \le n(E) \le \frac{1}{|\Lambda_\ell|}\,\mathbb{E}\left[N(H_\ell^N; E)\right]. \tag{4.34}$$

Proof. Consider a large region Λ_L, which appears in the representation (3.34) of the integrated DOS, and assume that Λ_L is a disjoint union of smaller cubes $\Lambda_{\ell,j}$ of fixed side length $\ell > 0$. Applying Corollary 4.11 together with the min-max principle (Theorem 4.13 and (4.37)) we obtain

$$\begin{aligned}\sum_j N(H_{\ell,j}^D(\omega); E) &\le N(H_L^D(\omega); E)\\ &\le N(H_L^N(\omega); E) \le \sum_j N(H_{\ell,j}^N(\omega); E)\,.\end{aligned} \tag{4.35}$$

We now divide by $|\Lambda_L|$ and take the expectation value. Since there are $|\Lambda_L|/|\Lambda_\ell|$ terms in each of the two sums which are equally distributed, the left and right sides of the above inequality yield the respective left and right sides of (4.34). In the limit $L \to \infty$ the two terms in the middle converge to $n(E)$ by Corollary 3.16 together with Exercise 3.8 for all $E \in \mathbb{R}$. □

The previous proof was based on the **min-max principle** by E. Fischer [**154**] and R. Courant [**102**] for self-adjoint operators $A : \mathcal{D}(A) \to \mathcal{H}$ which are bounded from below; i.e., $\langle\psi, A\psi\rangle \ge a\,\|\psi\|^2$ for some $a > -\infty$ and all $\psi \in \mathcal{D}(A)$. For its formulation we denote by $E_0(A) \le E_1(A) \le \cdots \le E_n(A) \le \cdots$ the eigenvalues of A taking into account multiplicity by repeating the eigenvalues below the essential spectrum $\sigma_{\mathrm{ess}}(A)$. We adopt the convention that $E_N(A) = \inf \sigma_{\mathrm{ess}}(A)$ in case the number of these eigenvalues is $N \in \mathbb{N}_0$.

Theorem 4.13 (Min-max principle). *Let $A : \mathcal{D}(A) \to \mathcal{H}$ be a self-adjoint operator which is bounded from below. Then*

$$\boxed{E_0(A) \;=\; \inf_{\varphi\in\mathcal{D}(A),\,\|\varphi\|=1} \langle\varphi, A\varphi\rangle}\,,$$

$$\boxed{E_n(A) \;=\; \sup_{\psi_1,\dots,\psi_n\in\mathcal{H}} \;\inf_{\substack{\varphi\in\mathcal{D}(A),\,\|\varphi\|=1\\ \varphi\perp\mathrm{span}\{\psi_1,\dots,\psi_n\}}} \langle\varphi, A\varphi\rangle} \tag{4.36}$$

for any $n \le N$.

Proof. We consider the N eigenvalues $E_0, \dots, E_{N-1}$ below E_N and we denote by $\varphi_0, \dots, \varphi_{N-1}$ the corresponding orthonormal eigenfunctions. For any $n \le N$ and $\varphi \perp \mathrm{span}\{\varphi_0, \dots, \varphi_{n-1}\}$ with $\varphi \in \mathcal{D}(A)$, one has $\varphi = P_{[E_n,\infty)}(A)\varphi$ such that by the spectral representation

$$\langle\varphi\,, A\varphi\rangle = \langle\varphi\,,\, A\,P_{[E_n,\infty)}(A)\varphi\rangle = \int_{E_n}^{\infty} \lambda\,\mu_\varphi(d\lambda) \ge E_n\,\|\varphi\|^2\,.$$

Picking $\varphi_0, \dots, \varphi_{n-1}$ for $\psi_1, \dots, \psi_n$ in the supremum in (4.36) shows that the left side is bounded from above by the right side. To establish the converse inequality, we pick for $\varphi_n \perp \mathrm{span}\{\varphi_0, \dots, \varphi_{n-1}\}$ either a normalized eigenfunction corresponding to E_n or, in case the latter is the bottom of the esssential spectrum, an approximate normalized eigenfunction, i.e., $\|(A - E_n)\varphi_n\| < \varepsilon$; cf. Proposition 3.11. Then any normalized linear combination $\varphi = \sum_{j=0}^{n} \alpha_j\varphi_j$ obeys

$$\langle\varphi\,, A\varphi\rangle \;\le\; \sum_{j=0}^{n} |\alpha_j|^2\,E_j + \varepsilon \;\le\; E_n + \varepsilon\,.$$

By dimensional considerations there is at least one φ of this form which belongs to $\mathrm{span}\{\psi_1, \dots, \psi_n\}^\perp$. This completes the proof since we can choose $\varepsilon > 0$ arbitrarily small. □

The proof shows that the infima and suprema are attained as long as $E_n(A)$ is an eigenvalue. The special case $n = 0$ is called the **Rayleigh-Ritz principle**. Theorem 4.13 implies that ranked eigenvalues form monotone functions of self-adjoint operators. Thus for any pair of operators $A \le B$ and $N(A;E) = \mathrm{tr}\,P_{(-\infty,E)}(A)$ (which equals ∞ for $E > \inf\sigma_{\mathrm{ess}}(A)$) one has

$$\boxed{E_n(A) \;\le\; E_n(B)} \qquad \text{for all } n \text{ and} \tag{4.37}$$

$$\boxed{N(B;E) \;\le\; N(A;E)} \qquad \text{for all } E. \tag{4.38}$$

More generally, $\mathrm{tr}\,F(A)$ for monotone functions F form monotone functions of the operator.

4.4. Lifshitz tails for random operators

The predictable range of the spectra of random operators with iid potential often appears to be in variance with numerical calculations based on the exact diagonalization of spectra of the finite system. The reason is found in the notable phenomenon, which was pointed out in the 1960s by I. M. Lifshitz [**286**], concerning a drastic thinning of the density of states near the bottom of the spectrum. Briefly stated, for a Schrödinger operator with homogeneous random potential of short range correlations, low lying states will be associated with regions where the potential is found to attain extremely low values over large domains. While in infinite systems any such constraint will almost surely be met somewhere, the regions where such conditions are met are very rare, and consequently the density of states at the bottom of the spectrum is extremely low. In this section we shall briefly discuss the resulting asymptotics of the density of states and in particular the law which is known as the Lifshitz tail estimate.

A consequence of relevance, which is discussed in later chapters, is that in the case of the finite-dimensional lattices $\mathbb{Z}^d$ the corresponding eigenstates are localized: in essence, each being supported in some region in space where the potential exhibits a rare fluctuation of the kind discussed in this section. If one thinks of the random potential as describing a forest of trees of random heights, the low-lying states would be supported mainly within rare "forest clearings", and due to the separation of these the states will be localized.

To appreciate the next statement, one may first note that for a Schrödinger operator with a periodic potential on $\mathbb{Z}^d$, which is our main example of an ordered system, the integrated density of states n near the bottom E_0 of the spectrum vanishes in dimension d at the rate

$$n(E) \sim C\,(E - E_0)^{\frac{d}{2}} \qquad \text{as} \quad E \searrow E_0\,. \tag{4.39}$$

The power law (4.39) is rather ubiquitous: It it found not only at the bottom of the spectrum but also at the edges of spectral bands of generic periodic potentials. By an extension of this formula, the power law is often used to define for homogenous graphs a so-called spectral dimension—taking for H the graph Laplacian.

In contrast to (4.39), Lifshitz noted that in situations described below

$$\boxed{n(E) \sim C_1\, e^{-C\,[E-E_0]^{-\frac{d}{2}}} \qquad \text{as} \quad E \searrow E_0}\,. \tag{4.40}$$

The extreme behavior of the density of states, or rather the energy regimes in which it is found, is now referred to as **Lifshitz tails**. The main point in the succinct expression (4.40) is the rate at which the term in the exponent diverges. Beyond this and in constrast to (4.39) (Exercise 4.2), the expression is not to be taken too literally; i.e., C should in general be replaced by a function $C(E)$ with $\log C(E) = o(1)$, and C_1 can be replaced by a function $C_1(E)$ which behaves as a power law. Following is a more precise formulation.

4.4.1. The statement and essential bounds.

In this section we will restrict our discussion to random operators in $\ell^2(\mathbb{Z}^d)$ of the form

$$H(\omega) = -\Delta + \lambda V(\omega)\,. \tag{4.41}$$

Unless stated otherwise, we will assume here that $\lambda > 0$ and the following:

i) The random variables $(V(x))_{x\in\mathbb{Z}^d}$ are iid with common, non-trivial distribution $\rho(dv) := \mathbb{P}(V(x) \in dv)$, whose support is bounded below; i.e., $a_0 := \inf \operatorname{supp} \rho > -\infty$.

ii) For some $C, \kappa > 0$ and all $\varepsilon > 0$ sufficiently small,

$$\rho([a_0, a_0 + \varepsilon)) \geq C\varepsilon^\kappa \tag{4.42}$$

(e.g., for the Anderson potential, with V uniform distributed in $[-1, 1]$, one has $\kappa = 1$).

Under the first of these assumptions, the $\mathbb{P}$-almost sure bottom of the spectrum is at

$$\boxed{E_0 := \inf \sigma(H(\omega)) \;=\; -\lambda a_0}\,. \tag{4.43}$$

This rather intuitive fact follows from Corollary 3.13, proven above, or alternatively from the bounds derived below.

Theorem 4.14 (Lifshitz tails). *For the random operator* (4.41) *with a random potential satisfying the above assumptions*

$$\boxed{\lim_{E\searrow E_0} \frac{\ln|\ln n(E)|}{\ln[E - E_0]} = -\frac{d}{2}}\,. \tag{4.44}$$

The statement (4.44) is a weak form of (4.40). A somewhat stronger version would be

$$\lim_{E\searrow E_0} [E - E_0]^{\frac{d}{2}} \ln n(E) \;=\; -C\,. \tag{4.45}$$

For the Poisson random potentials Lifshitz tails can be proven in this strong form (4.45); see [**125**] and [**313**]. In general, however, there can be a logarithmic correction to (4.45) so that we can only expect the weak form (4.44).

This form of the asymptotics is called the double logarithmic asymptotics.

The proof presented below is based on the pair of bounds on the integrated DOS which are based on Lemma 4.12:

$$\boxed{\frac{1}{|\Lambda_\ell|}\,\mathbb{P}\left(E_0(H_\ell^D) < E\right) \;\leq\; n(E) \;\leq\; \mathbb{P}\left(E_0(H_\ell^N) < E\right)}\,. \tag{4.46}$$

Here $E_0(H_\ell^X)$ is the lowest eigenvalue of the Dirichlet ($X = D$) or Neumann ($X = N$) operator associated to H on a cube Λ_ℓ of edge length $2\ell \in (0, \infty)$, which for the bound can be arbitrary. In deducing (4.46) from (4.34), for the lower bound we use the observation that if the spectral condition is satisfied, then $N(H_\ell^D; E)$ is at least 1, and for the upper bound we note that

$$\begin{aligned} N(H_\ell^N; E) \;&\leq\; 1\left[E_0(H_\ell^N) < E\right]\; N(-\Delta_\ell^N; E_0 + E) \\ &\leq\; 1\left[E_0(H_\ell^N) < E\right]\; |\Lambda_\ell|\,. \end{aligned} \tag{4.47}$$

A lower bound on the left side in (4.46) amounts to an upper bound on the lowest Dirichlet eigenvalue and will be based on the Rayleigh-Ritz variational principle. In turn, an upper bound on the right side requires a lower bound on the lowest Neumann eigenvalue, a tool for which is presented next.

The Rayleigh-Ritz variational principle (Theorem 4.13) identifies $\langle\psi, H\psi\rangle$ as an upper bound on the ground-state energy of any Hamiltonian H. For a given choice of the (normalized) trial wave function ψ, it is usually hard to estimate the error in this variational bound. A good basic tool for such a purpose is provided by **Temple's inequality** [**378**].

Proposition 4.15 (Temple). *Let H be a self-adjoint operator with an isolated non-degenerate eigenvalue $E_0 = \inf\sigma(H)$, and let $E_1 = \inf\left(\sigma(H)\backslash\{E_0\}\right)$. Then for any $\psi \in \mathcal{D}(H)$ which satisfies*

$$\langle\psi, H\psi\rangle < E_1 \quad \textit{and} \quad \|\psi\| = 1, \tag{4.48}$$

the following bound holds:

$$\boxed{E_0 \;\geq\; \langle\psi, H\psi\rangle - \frac{\langle H\psi, H\psi\rangle - \langle\psi, H\psi\rangle^2}{E_1 - \langle\psi, H\psi\rangle}}\,. \tag{4.49}$$

Proof. By assumption we have $(H - E_1)(H - E_0) \geq 0$. Hence, for any $\psi \in \mathcal{D}(H)$ with norm one,

$$\langle H\psi, H\psi\rangle - E_1\langle\psi, H\psi\rangle - E_0\langle\psi, H\psi\rangle + E_1E_0 \geq 0\,.$$

This implies

$$E_1E_0 - E_0\langle\psi, H\psi\rangle \geq E_1\langle\psi, H\psi\rangle - \langle\psi, H\psi\rangle^2 - \left(\langle H\psi, H\psi\rangle - \langle\psi, H\psi\rangle^2\right)\,.$$

Since $E_1 - \langle \psi, H\psi \rangle > 0$, we obtain

$$E_0 \geq \langle \psi, H\psi \rangle - \frac{\langle H\psi, H\psi \rangle - \langle \psi, H\psi \rangle^2}{E_1 - \langle \psi, H\psi \rangle} \,. \qquad \square$$

4.4.2. Proof of Lifshitz tails. The following proof of Theorem 4.14 is due to W. Kirsch and B. Simon [**238**]; see also [**230**]. For convenience, and without loss of generality, it is assumed that $a_0 = 0$ and consequently also $E_0 = 0$. This can always be arranged by adding a constant to the potential. The argument makes use of (4.46) and will proceed through separate estimates for the upper and lower bounds.

The lower bound. Applying the Rayleigh-Ritz variational principle (Theorem 4.13),

$$\begin{aligned} E_0\big(H_\ell^D(\omega)\big) \; &= \inf_{\psi \in \ell^2(\Lambda_\ell)} \frac{\langle \psi, H_\ell^D(\omega)\, \psi \rangle}{\|\psi\|^2} \\ &\leq \inf_{\psi} \frac{\langle \psi, -\Delta_\ell^D \psi \rangle}{\|\psi\|^2} \; + \; \lambda \max_{x \in \Lambda_\ell} V(x; \omega). \end{aligned} \tag{4.50}$$

For an upper bound on the kinetic energy of a function which is supported in Λ_ℓ and vanishes at the boundary, it suffices to consider the "tent function"

$$\psi(x) = \ell - \|x\|_\infty \,, \qquad x \in \Lambda_\ell \,,$$

for which it is easy to see that

$$\langle \psi, -\Delta_\ell^D \psi \rangle \; = \frac{1}{2} \sum_{\substack{x,y \in \Lambda_\ell \\ |x-y|=1}} |\psi(x) - \psi(y)|^2 \; \leq \; c_0\, \ell^{-2}\, \|\psi\|^2$$

with a finite constant c_0. Thus, for any $\ell \in \mathbb{N}$,

$$E_0\big(H_\ell^D(\omega)\big) \; \leq \; \frac{c_0}{\ell^2} \; + \; \lambda \max_{x \in \Lambda_\ell} V(x; \omega). \tag{4.51}$$

For a lower bound on $n(E)$ at any given E, we take $\ell = \sqrt{\frac{2\, c_0}{E}}$, so that the kinetic energy term is at most $E/2$. It follows from (4.51) and the lower bound in (4.46) that

$$\begin{aligned} n(E) &\geq \frac{1}{|\Lambda_\ell|} \; \mathbb{P}\left(\text{for all } x \in \Lambda_\ell \,:\, \lambda V(x) < E/2\right) \\ &= \frac{1}{|\Lambda_\ell|} \; \left\{\rho\left([0, E/(2\lambda)]\right)\right\}^{|\Lambda_\ell|} \\ &\geq \; \exp\left(|\Lambda_\ell| \ln\left(C\, E^\kappa/(2\lambda)^\kappa\right) - \ln|\Lambda_\ell|\right) \,, \end{aligned} \tag{4.52}$$

using assumption (4.42), where we remind the reader that κ is the exponent occurring there. Inserting the relation of ℓ to E, we get the lower bound:

$$\liminf_{E \searrow 0} \frac{\ln|\ln n(E)|}{\ln E} \geq -\frac{d}{2} .$$

Here E should be viewed as $E - E_0$. The supplementary upper bound which is established next proves that indeed the ground state is not smaller than $0 = a_0 = E_0$.

The upper bound. The strategy for an upper bound on $n(E)$ is to use the upper bound in (4.46), i.e., $n(E) \leq \mathbb{P}\left(E_0(H_\ell) < E\right)$ with suitably chosen ℓ, which will again be of the order $E^{-1/2}$. To estimate the right side from above we need an estimate of $E_0(H_\ell(\omega))$ from below which is provided by Temple's inequality. For the test function ψ it is very convenient to use the normalized ground state of $-\Delta_\ell$, namely

$$\psi(x) = \frac{1}{\sqrt{|\Lambda_\ell|}} \qquad \text{for all } x \in \Lambda_\ell . \tag{4.53}$$

For this function $-\Delta_\ell \psi = 0$ and

$$\langle \psi, H_\ell(\omega)\, \psi \rangle = \lambda \, \langle \psi, V(\omega) \psi \rangle = \frac{\lambda}{|\Lambda_\ell|} \sum_{x \in \Lambda_\ell} V(x;\omega) , \tag{4.54}$$

$$\langle \psi, (H_\ell(\omega))^2 \, \psi \rangle = \lambda^2 \, \langle \psi, V(\omega)^2 \, \psi \rangle = \frac{\lambda^2}{|\Lambda_\ell|} \sum_{x \in \Lambda_\ell} V(x;\omega)^2 . \tag{4.55}$$

By the law of large numbers, for large ℓ the quantity in (4.54) is typically close to $\mathbb{E}(V(0)) > 0$. To apply Temple's inequality we would need

$$\frac{\lambda}{|\Lambda_\ell|} \sum_{x \in \Lambda_\ell} V(x) < E_1(H_\ell), \tag{4.56}$$

which fails for large ℓ since $E_1(H_\ell) \to 0$. The rate at which $E_1(H_\ell)$ vanishes is given by the estimate (related to the standard uncertainty principle)

$$E_1(H_\ell) \geq E_1(-\Delta_\ell) \geq \alpha \, \ell^{-2} , \tag{4.57}$$

which can be obtained by a direct calculation (Exercise 4.1).

The difficulty which is posed by the above observations can be bypassed by considering the modified, truncated, potential

$$\widetilde{V}_\ell(x;\omega) = \min\left\{ V(x;\omega), \ \frac{\alpha}{3\ell^2 \lambda} \right\} . \tag{4.58}$$

The random variables $\left(\widetilde{V}_\ell(x)\right)_{x \in \mathbb{Z}^d}$ are still independent and identically distributed. Moreover, with

$$\widetilde{H}_\ell(\omega) := -\Delta_\ell + \lambda \, \widetilde{V}_\ell(\omega)$$

on $\ell^2(\Lambda_\ell)$ one has $E_0(H_\ell(\omega)) \geq E_0(\widetilde{H}_\ell(\omega))$, by the min-max principle. With this truncated potential,

$$\langle\psi, \widetilde{H}_\ell(\omega)\,\psi\rangle = \frac{\lambda}{|\Lambda_\ell|}\sum_{x\in\Lambda_\ell}\widetilde{V}_\ell(x;\omega) \leq \frac{\alpha}{3\,\ell^2} \tag{4.59}$$

and (4.57) still applies. Hence the sine qua non condition (4.48) is now satisfied and $E_0(\widetilde{H}_\ell)$ can be estimated with the help of the Temple inequality. The energy shift which appears in (4.49) is

$$\begin{aligned}\frac{\langle\psi, (\widetilde{H}_\ell)^2\psi\rangle}{\alpha\ell^{-2} - \langle\psi, \widetilde{H}_\ell\,\psi\rangle} &\leq \frac{\frac{\lambda^2}{|\Lambda_\ell|}\sum_{x\in\Lambda_\ell}\widetilde{V}_\ell(x)^2}{(\alpha-\frac{\alpha}{3})\ell^{-2}} \\ &\leq \frac{\lambda}{2\,|\Lambda_\ell|}\sum_{x\in\Lambda_\ell}\widetilde{V}_\ell(x).\end{aligned} \tag{4.60}$$

Thus, using the Temple inequality (4.49) with ψ and $\widetilde{H}_\ell$,

$$\begin{aligned}E_0(H_\ell) \;&\geq\; E_0(\widetilde{H}_\ell) \\ &\geq\; \langle\psi, \widetilde{H}_\ell\,\psi\rangle - \frac{\langle\psi, (\widetilde{H}_\ell)^2\psi\rangle}{\alpha\ell^{-2} - \langle\psi, \widetilde{H}_\ell\psi\rangle} \\ &\geq\; \frac{\lambda}{2\,|\Lambda_\ell|}\sum_{x\in\Lambda_\ell}\widetilde{V}_\ell(x)\;.\end{aligned} \tag{4.61}$$

Collecting the above estimates we arrive at

$$n(E) \leq \mathbb{P}\left(E_0(H_\ell) < E\right) \leq \mathbb{P}\left(\frac{\lambda}{|\Lambda_\ell|}\sum_{x\in\Lambda_\ell}\widetilde{V}_\ell(x)\,\ell^2 < 2E\,\ell^2\right)\;. \tag{4.62}$$

This upper bound is somewhat similar to the lower bound (4.52). Since the distribution ρ is non-trivial, one has $\lim_{\ell\to\infty}\mathbb{E}\left[\lambda\,\widetilde{V}_\ell(x)\,\ell^2\right] = \frac{\alpha}{3}$ and hence the probability on the right is expected to be very small when we choose ℓ as the largest integer for which

$$2E\,\ell^2 \leq \frac{\alpha}{6}\,. \tag{4.63}$$

The necessary probability estimate is provided by Cramér's large-deviation bound (4.67) below. In applying Proposition 4.17, we set $Y_n = -\lambda\widetilde{V}_\ell(x)\,\ell^2$ and $N = |\Lambda_\ell|$. For E small enough and ℓ large enough chosen according to (4.63), we then obtain

$$\mathbb{P}\left(\frac{\lambda}{|\Lambda_\ell|}\sum_{x\in\Lambda_\ell}\widetilde{V}_\ell(x)\,\ell^2 < 2E\,\ell^2\right) \leq e^{-c\,|\Lambda_\ell|} = e^{-c'\,E^{-d/2}} \tag{4.64}$$

with some constants which are independent of E. We have thus shown

$$\limsup_{E\searrow 0} \frac{\ln|\ln n(E)|}{\ln E} \leq -\frac{d}{2}, \tag{4.65}$$

which completes the proof of the Lifshitz tail (4.44). □

The argument used above for the upper bound also yields the following statement whose proof we leave as Exercise 4.3.

Corollary 4.16. *In the situation of Theorem* 4.14 *there exist constants* $L_0 > 0$ *and* $b, c > 0$ *such that for any* $\lambda \in (0, 1]$ *and* $L = L_0/\sqrt{\lambda}$ *and any* $\Delta E \in [0, bL_0^{-2}]$,

$$\boxed{\mathbb{P}\left(E_0(H_L) \leq E_0 + \lambda\,\Delta E\right) \;\leq\; \exp\left(-c\,L^d\right)}, \tag{4.66}$$

where $E_0 := \inf \sigma(H(\omega))$.

This estimate shows that, as has been noted in numerical computations, it may be difficult to ascertain numerically the correct spectral edge, despite the simplicity of the mathematical proof that $E_0 = \inf \sigma(-\Delta) + \lambda a_0$.

4.5. Large deviation estimate

The following result is known as Cramér's theorem [**216**] and provides a basic version of a large-deviation principle which was invoked in the proof of the upper bound on the integrated density of states.

Proposition 4.17 (Cramér). *Let* (Y_n) *be a collection of iid random variables of mean value* μ*, whose common distribution satisfies*

$$e^{f(\tau)} \;:=\; \mathbb{E}[\exp(\tau(Y-\mu))] \;<\; \infty$$

for some $\tau > 0$*. Then, for all* $\epsilon \geq 0$ *and all* $N \in \mathbb{N}$,

$$\boxed{\mathbb{P}\left(\frac{1}{N}\sum_{k=1}^{N} Y_k \;\geq\; \mu + \epsilon\right) \leq \exp\left(-N \sup_{\tau\geq 0}\left[\,\epsilon\,\tau - f(\tau)\right]\right)}. \tag{4.67}$$

In case Y *takes values in* $[0, b]$*, one has* $\sup_{\tau\geq 0}[\,\epsilon\,\tau - f(\tau)] \geq \frac{\epsilon^2}{2b^2}$.

Proof. The exponential Chebyshev inequality yields

$$\mathbb{P}\left(\frac{1}{N}\sum_{k=1}^{N} Y_k \;\geq\; \mu + \epsilon\right) \leq \mathbb{E}\left\{\exp\left[\tau\left(\sum_{k=1}^{N} Y_k - (\mu+\epsilon)N\right)\right]\right\}$$
$$= \exp\left\{N\left[f(\tau) - \epsilon\,\tau\right]\right\}\,.$$

The function $f : [0,\infty) \to \mathbb{R}$ is monotone increasing and convex with $f(0) = 0$ and $f'(0) = 0$. Hence, its Legendre transform

$$\delta(\epsilon) := \sup_{\tau\geq 0}\left[\epsilon\,\tau - f(\tau)\right]$$

is well-defined for any $\epsilon \geq 0$.

In particular, if the range of Y is $[0, b]$, it follows that $f''(\tau) \leq b^2$ and hence $f(\tau) \leq b^2\tau^2/2$, which in turn implies

$$\delta(\epsilon) \geq \sup_{\tau\geq 0}\left[\epsilon\,\tau - \frac{b^2\tau^2}{2}\right] = \frac{\epsilon^2}{2b^2}\,. \qquad \square$$

A simple consequence of the above theorem is the two-sided bound

$$\mathbb{P}\left(\left|\frac{1}{N}\sum_{k=1}^{N} Y_k \;-\mu\right| \geq \epsilon\right)$$
$$\leq \exp\left(-N \sup_{\tau\geq 0}\left[\epsilon\,\tau - f(\tau)\right]\right) + \exp\left(-N \sup_{\tau\geq 0}\left[\epsilon\,\tau - f(-\tau)\right]\right)\,, \tag{4.68}$$

which results from (4.67) by distinguishing the cases $\frac{1}{N}\sum_{k=1}^{N} Y_k \;\geq\; \mu + \epsilon$ and $\frac{1}{N}\sum_{k=1}^{N} Y_k \;\leq\; \mu - \epsilon\,.$

4.6.* DOS bounds which imply localization

Among the finite-volume localization criteria which will be derived in Chapter 11, we shall see that a sufficient condition for localization in the energy range $E \leq E(\lambda)$ is that for some specified C, $\# > 0$ the following condition is met for some $L < \infty$:

$$L^{\#}\;\mathbb{P}\left(\{E_0(H_L) \leq E(\lambda)\}\right) \;\leq\; C\,. \tag{4.69}$$

(The localization implication is drawn there under some additional assumptions on the distribution of the random potential which need not be listed here.)

Corollary 4.16 implies that for any $\# > 0$ the condition (4.69) will be met at small enough λ for $E \;<\; E(\lambda) := \inf\sigma(-\Delta) + c\lambda$ with some $c > 0$ small, a regime which is indicated by line (a) in Figure 4.1. One may, however,

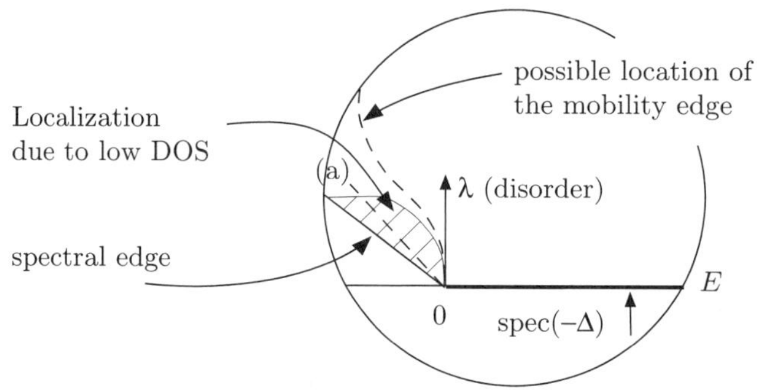

Figure 4.1. Lifshitz tail asymptotics: The line (a) indicates the regime up to which our main DOS estimate applies. Somewhat weaker DOS bounds imply that the localization condition (4.69) holds throughout a regime which looks as the hatched set here. The sketch refers to bounded iid potentials with $\mathbb{E}(V) = 0$.

expect that in any finite dimension the localization regime extends at weak disorder at least up to $E(\lambda)$ with

$$\lim_{\lambda \searrow 0} \frac{E(\lambda) - \inf \sigma(-\Delta)}{\lambda} = \mathbb{E}(V(0)) . \tag{4.70}$$

This too can be established by the method used above, albeit through DOS bounds with a less severe decay than that of the Lifshitz tails.

Theorem 4.18. *Let $H(\omega) = -\Delta + \lambda V(\omega)$ be a random operator in $\ell^2(\mathbb{Z}^d)$ with an iid random potential whose common probability distribution has sub-Gaussian decay, in the sense that*

$$\mathbb{E}\left[e^{\tau_0 V(0)^2}\right] < \infty \tag{4.71}$$

for some $\tau_0 > 0$. Then there are some $c, C, \alpha \in (0, \infty)$ such that for any $q \in (0, \frac{d}{d+4})$ and all $\lambda > 0$ small enough,

$$\mathbb{P}\left(E_0(H_L^N) \leq \lambda\, \mathbb{E}[V(0)] - c\lambda^{1+q}\right) \leq C \exp\left(-c\, L^{\varepsilon}\right) \tag{4.72}$$

where $L = L(\lambda) = \sqrt{\alpha}/\lambda^{(1-q)/2}$ and $\varepsilon := \frac{d-q(4+d)}{1-q}$.

Proof. The sub-Gaussian bound (4.71) in particular implies the following large-deviation bounds:

1. The moment-generating function of the common probability distribution of the random potential,

$$e^{f(\tau)} := \mathbb{E}[\exp(\tau(V(0) - \mathbb{E}[V(0)]))] , \quad \tau \in \mathbb{R} ,$$

satisfies $f''(\tau) \leq c$ for all $|\tau| \leq \tau_0$ with some $c < \infty$, and hence $f(\tau) \leq \tau^2 c/2$ for all $|\tau| \leq \tau_0$. The large-deviation bound (4.68) then yields

$$\begin{aligned} \mathbb{P}\left(\left| \frac{\lambda}{|\Lambda_L|} \sum_{x \in \Lambda_L} (V(x) - \mathbb{E}[V(x)]) \right| \geq \lambda^{1+q} \right) & \\ \leq 2\exp\left(-|\Lambda_L| \sup_{|\tau| \leq \tau_0} \left[\lambda^q \tau - c\tau^2/2 \right] \right) & \\ \leq 2\exp\left(-\frac{\lambda^{2q}|\Lambda_L|}{2c} \right) & \qquad (4.73) \end{aligned}$$

for any $q > 0$ and all $\lambda > 0$ small enough.

2. An exponential Chebyshev inequality shows

$$\mathbb{P}\left\{ \frac{1}{|\Lambda_L|} \sum_{x \in \Lambda_L} V(x)^2 \geq 2c_0 \right\} \leq \exp\left(-c_0 \tau_0 \, |\Lambda_L| \right) \qquad (4.74)$$

with $c_0 := \tau_0^{-1} \ln \mathbb{E}\left[e^{\tau_0 V(0)^2} \right] < \infty$.

Excluding the two events whose probability is estimated above, we may apply Temple's inequality (Proposition 4.15) with the constant trial function, (4.53). In case

$$\frac{\alpha}{L^2} > \lambda \, \mathbb{E}[V(0)] + \lambda^{1+q}\,, \qquad (4.75)$$

where $\alpha > 0$ is the constant from (4.57), the necessary condition (4.48) is met since

$$E_1(H_L) - \langle \psi, H_L \psi \rangle \;\geq\; \frac{\alpha}{L^2} - \frac{\lambda}{|\Lambda_L|} \sum_{x \in \Lambda_L} V(x) \;>\; 0\,. \qquad (4.76)$$

Through the explicit expressions (4.54) and the opposite bound on the fluctuating term, one gets the lower bound on $E_0(H_L)$:

$$E_0(H_L) \geq \lambda \, \mathbb{E}[V(0)] - \lambda^{1+q} - \frac{2c_0 \, \lambda^2}{\frac{\alpha}{L^2} - \lambda \, \mathbb{E}[V(0)] - \lambda^{1+q}}\,. \qquad (4.77)$$

Taking $\alpha/L^2 = \lambda^{1-q}$, the condition (4.75) is satisfies for all $\lambda > 0$ small enough (at any fixed value of $q \in (0,1)$). With this selection of the length scale, the exponent in the probability bound (4.73) behaves as L^{ε} with the value of ε stated below (4.72). This value is strictly positive provided $q \in (0, \frac{d}{d+4})$. (It is the latter condition which limits q in the above argument.) $\square$

The above theorem clearly implies that for any # and $q \in (0, \frac{d}{d+4})$ the localization condition (4.69) is met at energies

$$E \;<\; E(\lambda) := \lambda\, \mathbb{E}[V(0)] - c\lambda^{1+q} \tag{4.78}$$

for λ small enough.

Notes

There seems to be room for technical improvements of the DOS bounds in two directions:

Other continuously distributed potentials: The Wegner estimate does not yet take into account possible regularizing effects of the simultaneous dependence of the resolvent on many potential variables. One may expect this multi-dependence to produce effects akin to those of repeated convolutions. Thus, it seems natural to conjecture that the DOS is ac under much weaker assumptions on the single-site potential, e.g., that $\rho(dv)$ is of continuous distribution and τ-Hölder is continuous for some $\tau > 0$.

This line of thought does not apply to Bernoulli potentials, and indeed for that case it is known that for $d = 1$ and iid Bernoulli random variables $(V(x))_{x\in\mathbb{Z}}$ with values in $\{0, v\}$ with v large enough, ν has a singular component. This was conjectured in [**362**] and proven in [**81**].

Improved estimate for weak disorder: The Wegner estimate expresses just the smoothing effects of the random potential, which naturally degrades for $\lambda \to 0$. Hence the prefactor λ^{-1} is present in (4.4) even for cases where the limiting operator H_0 is of bounded DOS. It would be of interest to see a bound for which the potential regularity of the latter is also taken into account.

It may be mentioned in this context that for deterministic potentials J. Bourgain and A. Klein [**70**] established the log-Hölder continuity of an outer DOS measure (which is introduced in that work) for dimensions below three. Log-Hölder continuity of the DOS for ergodic operators was established previously by W. Craig and B. Simon [**103**]. F. Delyon and B. Souillard [**114**] provided a simple proof of continuity of the integrated density of states for ergodic Schrödinger operators on $\mathbb{Z}^d$.

Setting the above questions aside, much is known concerning the regularity of the integrated DOS under corresponding regularity assumptions on the random potential for models on the lattice or on continuous space;

see [**96, 237, 285, 392**] and the references therein. However, so far, the continuity of the integrated DOS (i.e., continuity of the measure itself) for ergodic Schrödinger operators in $\mathbb{R}^d$, albeit expected, has been hard to prove in full generality. A continuity result for the lattice case was discussed in Exercise 3.8.

There are by now various derivations of the Lifshitz tails. The subject's presentation here follows that of W. Kirsch [**230**], whom we thank for permission to use his organization of the material. Among the earlier derivations is one through the Donsker-Varadhan theory of large deviations [**125, 313, 321**]; other methods appeared in [**235, 238, 228, 350, 368, 241**]. Related results and further references can be found in [**59, 373**]. For an approach using periodic approximation see [**255, 257, 258, 259**] and the references therein.

There also exist results on the so-called internal Lifshitz tails [**299, 351, 260, 261, 309**], which occur at band edges other than the bottom of the spectrum (which in the case discussed here is the same as at the top, by a symmetry argument). Magnetic fields change the Lifshitz behavior drastically [**73, 139, 394**].

The Lifshitz tail estimate presented here is not optimal. Estimates which are asymptotically optimal in their dependence on the coupling parameter λ in case the latter is small are discussed in [**196**]. There it is shown that Theorem 4.18 remains valid for $q = 1/3$ with $d = 1$ and for $q = 1$ with $d \geq 2$. This result hence extends the regime in which exponential localization can be established at weak disorder.

Exercises

4.1. Determine the spectrum of the Neumann Laplacian $\Delta^N_{\Lambda_L}$ on a box $\Lambda_L = \{x \in \mathbb{Z}^d \mid \|x\|_\infty \leq L\}$ with $L \in \mathbb{N}$.

4.2. Let $\Lambda_L = \{x \in \mathbb{Z}^d \mid \|x\|_\infty \leq L\}$ with $L \in \mathbb{N}$. The **periodic Laplacian** on $\ell^2(\Lambda_L)$ is

$$\left(\Delta_L^{\mathrm{per}} \psi\right)(x) := \sum_{\substack{y \in \mathbb{Z}^d \\ d(x,y)=1}} \psi(y) - 2d\, \psi(x)\,, \quad x \in \Lambda_L\,,$$

where, by convention, the function $\psi : \Lambda_L \to \mathbb{C}$ is periodically extended to $\mathbb{Z}^d$; i.e., $\psi(y) = \psi(y \pm (2L+1)e_j)$ for all y and all unit vectors $e_j \in \mathbb{Z}^d$, $j \in \{1, \ldots, d\}$.

a. Determine the eigenfunctions of Δ_L^{per} and show that its eigenvalues are given by $h(k) := \sum_{j=1}^d 2\,(\cos(k_j) - 1)$ with $k \in \left(\frac{2\pi}{2L+1}\,\mathbb{Z}\right)^d \cap [0, 2\pi)^d$.

b. Show that the normalized eigenvalue counting function

$$N_L(E) := |\Lambda_L|^{-1} \operatorname{tr} P_{(-\infty,E)}(-\Delta_L^{\mathrm{per}})$$

converges as $L \to \infty$ pointwise to the free integrated density of states:

$$n(E) = \frac{1}{(2\pi)^d} \int_{[0,2\pi)^d} 1_{h(k) \leq E}\, dk\,.$$

Show also that the same limit is obtained if the periodic Laplacian is replaced by either the Dirichlet or the Neumann Laplacian on Λ_L.

c. Prove the **van Hove asymptotics** (4.39) by evaluating $n(E)$ for $E \searrow 0$.

4.3. Give a proof of Corollary 4.16.

4.4. Consider the operator $H(\omega) = T + V(\omega)$ in $\ell^2(\mathbb{G})$ with an iid random potential with single-site distribution given by

$$\rho(dv) = \frac{\lambda}{\pi} \frac{dv}{v^2 + \lambda^2}\,, \qquad \lambda > 0\,.$$

Show that the density of states defined in (4.1) equals

$$\nu(E) = \frac{1}{\pi} \operatorname{Im} \langle \delta_0, (T - E - i\lambda)^{-1}\, \delta_0 \rangle$$

for any $E \in \mathbb{R}$.

[The above observation was presented in [**290**], after which this operator is referred to as the **Llyod model**. We will see in Exercise 10.2 that the second claim in Lloyd's paper concerning the absence of localization at extreme energies is erroneous. An important lesson is that analyticity of the mean DOS does not preclude singular spectrum.]

Chapter 5

The Relation of Green Functions to Eigenfunctions

We now turn to some basic concepts and techniques for the analysis of operators with random potential. The results presented include the following:

1. an important **relation between eigenfunctions and the Green function**, based on the first resolvent formula,
2. the **spectral averaging principle** for random operators,
3. **Simon-Wolff criteria** (boosted by a zero-one law) for pure-point and for continuous spectra,
4. Schur complement and related **finite-rank perturbation formulas**,
5. **simplicity of the pure point spectrum** for operators whose random potential is of conditionally continuous probability distribution.

The discussion in this chapter applies to general operators with random potential, i.e., $H = T + \sum_{x \in \mathbb{G}} V(x)\, 1_{\{x\}}$, acting in the $\ell^2(\mathbb{G})$ space over a countable set $\mathbb{G}$, with T a bounded self-adjoint operator $1_{\{x\}} = |\delta_x\rangle\langle\delta_x|$ the one-dimensional projections associated with sites $x \in \mathbb{G}$, and V a random potential. For the latter it will be assumed that the conditional distributions of $V(x)$ conditioned on $V_{\neq x} = \{V(y)\}_{y \neq x}$ are all continuous (i.e., their probability measures have no atoms).

5.1. The spectral flow under rank-one perturbations

The random operators in the class described above have a probability distribution which is unaffected when any of the potential values is redrawn with its correct conditional probability. Important implications of that will be arrived at by considering the evolution of the spectra of under rank-one perturbations. For that we shall now focus on one-parameter families of self-adjoint operators of the form

$$H_v = H_0 + v\,|\delta_x\rangle\langle\delta_x|\,. \tag{5.1}$$

It is of pedagogical value to understand the questions discussed here first in the context of finite graphs, and then to consider the extension of the resulting theory to infinite dimensional spaces. As may be expected, the first does serve as a guide to the second, but significant adjustments are called for and are enabled through analytical tools which are of independent interest. We start with a minor detour which is of help in clarifying the significance of quantities which will be encountered next.

For an $N \times N$ matrix H, acting in $\mathbb{C}^N$, the cyclic subspace associated with a vector $\phi \in \mathbb{C}^N$ is $\mathrm{span}\{\phi, H\phi, H^2\phi, ...\}$ (where by a simple argument it suffices to limit the above collection to the first N terms). For unbounded operators one may use the following version.

Definition 5.1. For a self-adjoint operator H acting in a Hilbert space $\mathcal{H}$ and a vector $\phi \in \mathcal{H}$, the associated cyclic space is $\overline{\mathrm{span}\,\{(H-z)^{-1}\phi \,|\, z \in \mathbb{C}\backslash\mathbb{R}\}}$. In case $\mathcal{H} = \ell^2(\mathbb{G})$ and $\phi = \delta_x$ for some $x \in \mathbb{G}$, the corresponding cyclic space will be denoted $\mathcal{H}_{H,x}$.

Of particular relevance for us are the following two properties of cyclic spaces (where δ_x could be replaced by an arbitrary vector in the Hilbert space).

Lemma 5.2. *The spectrum of any self-adjoint operator H within any of its cyclic spaces generated from a single vector is simple.*

The cyclic space associated with a vector does not change under the rank-one perturbation of H by a multiple of the corresponding projection operator. In particular, for any $x \in \mathbb{G}$ and a family of operators of the form (5.1)

$$\mathcal{H}_{H_v,x} = \mathcal{H}_{H_0,x}\,, \tag{5.2}$$

for all $v \in \mathbb{R}$.

Proof. The simplicity of the restricted spectrum can be seen through the spectral representation (Proposition A.2) associated with the given vector. Details of the argument are left as Exercise 5.1.

For the second statement, by the **first resolvent equation**

$$\frac{1}{H_v - z} = \frac{1}{H_0 - z} - \frac{1}{H_v - z}\, v\, |\delta_x\rangle\langle\delta_x|\, \frac{1}{H_0 - z}\,. \tag{5.3}$$

Acting on the vector $\delta_x \equiv |\delta_x\rangle$ and gathering related terms, one obtains

$$\boxed{\frac{1}{H_v - z}\,\delta_x = \left[1 + v\,\langle\delta_x, (H_0 - z)^{-1}\delta_x\rangle\right]^{-1} \frac{1}{H_0 - z}\,\delta_x}\,. \tag{5.4}$$

In particular,

$$\frac{1}{H + v\,|\delta_x\rangle\langle\delta_x| - z}\delta_x \in \mathcal{H}_{H,x}\,, \tag{5.5}$$

from which (5.2) follows. □

It is of interest to know a number of key properties of the evolution in $v \in \mathbb{R}$ of the spectrum of H_v within the cyclic subspace $\mathcal{H}_{H_v,x} = \mathcal{H}_{H_0,x}$. A key role for that is played by the function

$$\Sigma_0(z) := -\,\langle\delta_x, (H_0 - z)^{-1}\delta_x\rangle^{-1}\,, \qquad z \in \mathbb{C}\backslash\mathbb{R}\,. \tag{5.6}$$

In the finite dimensional context it extends to a meromorphic function with simple poles on the real axis, taking the value 0 for $z \in \sigma(H_0)$. The main result is:

Theorem 5.3. *Let $\mathbb{G}$ be a finite set, H_0 a self-adjoint operator in $\ell^2(\mathbb{G})$, and H_v the one-parameter family of operators defined by (5.1) with $x \in \mathbb{G}$. Then the spectra of H_v restricted to the cyclic subspace $\mathcal{H}_{H_v,x}$ are disjoint for different values of v, and for $v \neq 0$ satisfy:*

1. *The spectrum of H_v within $\mathcal{H}_{H_v,x}$ consists of exactly those $E \in \mathbb{R}$ at which*

$$\boxed{\Sigma_0(E) = v}\,. \tag{5.7}$$

2. *The corresponding eigenvectors are proportional to*

$$\boxed{\phi_{E,x} := \frac{1}{H_0 - E}\,\delta_x}\,. \tag{5.8}$$

3. *The spectral measure of H_v associated with the vector δ_x is the point measure supported on the solutions of (5.7) assigning to each point of $\Sigma_0^{-1}(\{v\})$ the mass*

$$\mu_{v,x}(\{E\}) = \frac{|\langle\phi_{E,x}, \delta_x\rangle|^2}{\langle\phi_{E,x}, \phi_{E,x}\rangle} = \left[\Sigma_0'(E)\right]^{-1}\,. \tag{5.9}$$

The last statement is more succinctly expressed in its distributional form:

$$\boxed{\mu_{v,x}(dE) = \delta(\Sigma_0(E) - v)\,dE}\,. \tag{5.10}$$

The equivalence with (5.9) is based on the distributional relation

$$\delta(f(x)) = \sum_{\substack{x_n: \\ f(x_n)=0}} \frac{1}{|f'(x_n)|}\,\delta(x - x_n)\,. \tag{5.11}$$

Before we turn to the proof, it may be good to pause and note some key features of the picture described by Theorem 5.3. Applying the eigenfunction expansion one can see that the diagonal Green function has the following structure, in its dependence on the complex spectral parameter z,

$$G_v(x,x;z) := \langle \delta_x, \frac{1}{H_v - z}\delta_x\rangle = \sum_n \frac{p_{v,n}}{E_{v,n} - z}, \tag{5.12}$$

where $p_{v,n} := \mu_{v,\delta_x}(\{E_{v,n}\})$ for all $v \in \mathbb{R}$. It readily follows that $\Sigma_0(z)$ is of similar form, though of different behavior for $z \to \infty$. More explicitly,

$$\Sigma_0(z) = \sum_n \frac{q_n}{u_n - z} + z + C \tag{5.13}$$

with u_n ranging over the zeros of the function $E \mapsto G_0(x,x,E)$, $q_n > 0$ and $C \in \mathbb{R}$. Most relevantly, the function is piecewise monotone, increasing from $-\infty$ to $+\infty$ along a finite number of segment (as in the example sketched in Figure 5.1).

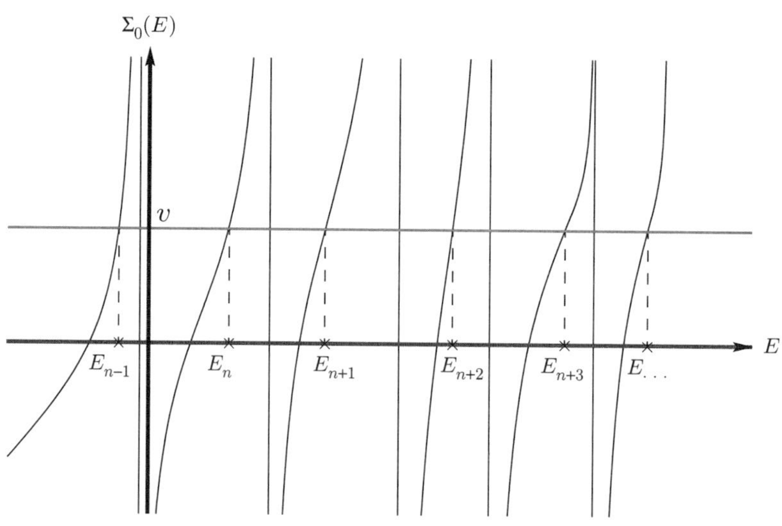

Figure 5.1. Sketch of the graph of the function $\Sigma_0(E)$ and its relation with the spectrum of H_v

One may therefore conclude the following:

i. For each $v \in \mathbb{R}$, the solution set of (5.7) is a finite collection of points, which by the first assertion of Theorem 5.3 form the spectrum of H_v

within $\mathcal{H}_{H_v,x}$. These intersection points move upward with v but remain **intertwined** with the eigenvalues of H_0 (which form the zeros of Σ_0).

ii. As v is varied from $-\infty$ to $+\infty$, exactly one spectral point will be transported across any given $E \in \mathbb{R}$; i.e., there is a single value of v for which $E \in \sigma(H_v)$.

iii. The eigenfunctions of H_v are sampled from a collection of functions which is independent of v, and whose elements are generically associated with the kernel of the resolvent operator (the kernel's second coordinate being fixed at δ_x).

iv. Since $\int_{-\infty}^{\infty} \delta(\Sigma - v) dv = 1$ for any $\Sigma \in \mathbb{R}$, integrating (5.10) over $v \in \mathbb{R}$ one gets:

$$\boxed{\int [\mu_{v,x}(I)]\, dv \;=\; \int_I dE} \,. \tag{5.14}$$

This is a useful spectral averaging principle.

It is a technically interesting task to work out the implications of the above picture for infinite systems, where the spectral gaps vanish and the above discussion does not apply. Among the simplest, and important, such generalizations is the spectral averaging principle. It is re-derived below by a more stable argument.

Proof of Theorem 5.3. 1. The eigenvalues of H_v in $\mathcal{H}_{H_v,x}$, which we already know are simple, form the singularities $(E_{v,n})$ of $G_v(x,x;z)$. By Lemma 4.3, or alternatively (5.4) which is not restricted to finite dimensions, this function bears the following relation to its unperturbed ($v = 0$) version:

$$\boxed{G_v(x,x;z) \;=\; \frac{1}{G_0(x,x;z)^{-1} + v} \;=\; [v - \Sigma_0(z)]^{-1}} \,. \tag{5.15}$$

Inspecting the singularity of the resulting function, from (5.15) one may conclude that the spectrum of H_v is given by (5.7).

2. For an energy E with $\langle \delta_x, P_{\{E\}}(H_v)\, \delta_x \rangle \neq 0$ (and hence also $\Sigma_0(E) = v \neq 0$), the function

$$\psi_{E,x} \;:=\; P_{\{E\}}(H_v)\, \delta_x \tag{5.16}$$

forms a non-zero eigenvector of H_v within the cyclic space $\mathcal{H}_{H_v,x}$. Applying now (5.4) and elementary residue calculation, for any $y \in \mathbb{G}$:

$$\begin{aligned} \langle \delta_y, \psi_{E,x} \rangle &= \langle \delta_y, P_{\{E\}}(H_v)\, \delta_x \rangle = \operatorname*{Res}_{z=E} \langle \delta_y, \frac{1}{H_v - z} \delta_x \rangle \\ &= \operatorname*{Res}_{z=E} \left[\frac{\Sigma_0(z)}{\Sigma_0(z) - v} \langle \delta_y, \frac{1}{H_0 - z} \delta_x \rangle \right] = \frac{v}{\Sigma_0'(E)} \langle \delta_y, \frac{1}{H_0 - E} \delta_x \rangle \end{aligned} \tag{5.17}$$

Thus $\psi_{E,x}$ bears the following relation to the function $\phi_{E,x}$ of (5.8)

$$\psi_{E,x} = \frac{v}{\Sigma_0'(E)} \frac{1}{H_0 - E} \delta_x = \frac{v}{\Sigma_0'(E)} \phi_{E,x} , \tag{5.18}$$

which proves the second claim.

3. The properly normalized weight that the spectral measure associated with δ_x has at energy $E \in \Sigma_0^{-1}(\{v\})$ is

$$\mu_{v,x}(\{E\}) = \frac{|\langle \phi_{E,x}, \delta_x \rangle|^2}{\langle \phi_{E,x}, \phi_{E,x} \rangle} , \tag{5.19}$$

with $\phi_{E,x}$, given by (5.8). Thus

$$\mu_{v,x}(\{E\}) = \frac{|\langle \delta_x, [H_0 - E]^{-1} \delta_x \rangle|^2}{\langle \delta_x, [H_0 - E]^{-2} \delta_x \rangle|} = \frac{1}{\Sigma_0'(E)} . \tag{5.20}$$

as claimed in (5.9) (which, as explained, is a bit more transparent in its distributional form (5.10)). □

5.2. The general spectral averaging principle

The extension of the rank-one perturbation theory of Theorem 5.3 to the infinite dimensional setup raises a number of interesting technical challenges, and opportunities for interesting math. To start, let us present a more general proof of the spectral averaging principle, which extends (5.14) to infinite graphs. The proof given above does not apply since the spectrum ceases to be discrete (in the generic situation).

For infinite graphs the functions $z \mapsto G_v(x, x; z)$ are generically no longer meromorphic. Their singularities, though still concentrated along $\sigma(H_v)$, are not confined to isolated poles, and their limiting values along the real line are not necessarily real. The more general result is obtained by focusing on the values of the Green function not along the real line but rather throughout the upper half plane, $\mathbb{C}^+$.

Off the real axis, the diagonal elements of any self-adjoint operator's Green function belong to the Herglotz-Pick class – functions which are analytic throughout the upper half plane mapping $\mathbb{C}^+$ into itself. Such functions admit the Herglotz spectral representation (Proposition B.1), which in this case coincides with the one based on the operator's spectral measure:

$$G_v(x,x;z) \;=\; \int_{\mathbb{R}} \frac{1}{u-z}\mu_{v,\delta_x}(du) \tag{5.21}$$

The related theory (cf. Appendix B) allows the following more robust derivation of the spectral averaging principle.

Theorem 5.4 (Spectral averaging)**.** *Let $\mathbb{G}$ be a countable set, H_0 a self-adjoint operator in $\ell^2(\mathbb{G})$, and H_v the one-parameter family of operators defined by* (5.1) *with $x \in \mathbb{G}$. Then for any bounded Borel set $I \subset \mathbb{R}$,*

$$\boxed{\int_{\mathbb{R}} \langle \delta_x, P_I(H_v)\,\delta_x\rangle\, dv \;=\; |I|}\,. \tag{5.22}$$

Proof. An equivalent formulation of the statement is the equality of $\int_{\mathbb{R}} \mu_{v,x}(\cdot)\,dv$ with Lebesgue measure. By the Herglotz representation theorem (Proposition B.1) this is equivalent to the equality for arbitrary $z \in \mathbb{C}^+$ of the integrals with respect to these measures of the functions

$$F_z(E) = (E-z)^{-1} - \mathrm{Re}\,(E-i)^{-1}\,. \tag{5.23}$$

Thus it suffices to show that for all $z \in \mathbb{C}^+$:

$$\int \langle \delta_x, F_z(H_v)\,\delta_x\rangle\, dv \;=\; \int F_z(E)\,dE \tag{5.24}$$

Applying the rank-one perturbation formula (5.15) (whose given derivation is valid also in the infinite dimensional case) one gets

$$\begin{aligned} \langle \delta_x, F_z(H_v)\,\delta_x\rangle \;&\equiv\; \langle \delta_x, (H_v-z)^{-1}\,\delta_x\rangle \;-\; \mathrm{Re}\,\langle \delta_x, (H_v-i)^{-1}\,\delta_x\rangle \\ &=\; \frac{1}{v-\Sigma_0(z)} \;-\; \mathrm{Re}\,\frac{1}{v-\Sigma_0(i)} \end{aligned} \tag{5.25}$$

Elementary integration shows that for any $w_1, w_2 \in \mathbb{C}^+$,

$$\int \left(\frac{1}{v-w_1} - \mathrm{Re}\,\frac{1}{v-w_2} \right) dv = i\pi\,. \tag{5.26}$$

Since both integrals in (5.24) are of this form, the claim follows. □

An immediate consequence of the spectral averaging principle, Theorem 5.4, is the bound

$$\int_{\mathbb{R}} \langle \delta_x, P_I(H_v)\,\delta_x\rangle\, \varrho(v)\, dv \;\leq\; \|\varrho\|_\infty\, |I| \tag{5.27}$$

for all $\varrho \in L^\infty(\mathbb{R})$ and all bounded Borel $I \subset \mathbb{R}$. It reproduces the Wegner estimate which was discussed in Chapter 4 in the context of finite systems. No further input was required for the extension to the concentration function bound of Theorem 4.6; hence this estimate is also valid for infinite systems.

5.3. The Simon-Wolff criterion

The challenge of extending the rank-one perturbation theory to infinite dimensional separable Hilbert spaces was addressed in the work of N. Aronszajn [**30**], building on the theory of Herglotz-Pick functions whose elements we have just encountered. The implications for random Schrödinger operators were further clarified in the work of B. Simon and T. Wolff [**363**]. The theory was shaped there into useful spectral criteria to which we now turn.

In order to extend the first part of Theorem 5.3 to countable sets $\mathbb{G}$, we start from the observation that the eigenvalues of any self-adjoint operator H within the cyclic subspace $\mathcal{H}_{H,x}$ are still characterized by the atoms of its spectral measure μ_{δ_x}. These points relate to the boundary value of the Green function through the relation :

$$\boxed{\mu_{\delta_x}(\{E\}) = \lim_{\eta\downarrow 0} -i\eta\, \langle \delta_x, (H - E - i\eta)^{-1}\delta_x\rangle} \tag{5.28}$$

for any $E \in \mathbb{R}$. (The proof is left as Exercise 5.2.)

The rank-one formula (5.15) remains valid within the infinite-dimensional setup. For $H = H_v$ it allows to express the right side of (5.28) in terms of the Herglotz-Pick function $z \mapsto \Sigma_0(z)$. However in the limit $\eta \downarrow 0$ the function $\Sigma_0(E + i\eta)$ may now exhibit more varied behavior: i) the limit may exist in $\mathbb{R}$ or diverge in modulus, as is the case in the finite-dimensional situation, ii) $\lim_{\eta\downarrow 0} \Sigma_0(E + i\eta) = \Sigma_0(E + i0)$ may exist but have a positive imaginary part, or iii) $\Sigma_0(E + i\eta)$ may not have a limit even in the sense of the one-point compactification of $\mathbb{C}$. The next proposition shows that the existence and real-valuedness of this limit at $E \in \mathbb{R}$ goes hand in hand with the condition $\mu_{v,\delta_x}(\{E\}) > 0$.

A quantity which also plays an important role is

$$\gamma_0(E) := \int \frac{\mu_{0,\delta_x}(du)}{(E-u)^2} \tag{5.29}$$

which may be infinite. The condition $\gamma_0(E) < \infty$ immediately warrants the following two properties at the given $E \in \mathbb{R}$:

i. The vector

$$\phi_{E,x} := (H_0 - E)^{-1}\delta_x \in \mathcal{H}_{H_0,x} \tag{5.30}$$

is well defined, its spectral representation in $L^2(\mathbb{R}, \mu_{0,\delta_x})$ being the square integrable function $u \mapsto (u-E)^{-1}$.

ii. The boundary value

$$G_0(x,x;E+i0) := \lim_{\eta \downarrow 0} G_0(x,x;E+i\eta) \tag{5.31}$$

exists and is real (Exercise 5.3.)

The natural extension of part of Theorem 5.3 to countable sets $\mathbb{G}$ then reads as follows:

Proposition 5.5. *Under the same assumptions as in Theorem 5.4, for any $v \neq 0$ and $E \in \mathbb{R}$ the following statements are equivalent:*

i. $\mu_{v,\delta_x}(\{E\}) > 0$,

ii. $\gamma_0(E) < \infty$ *and* $G_0(x,x;E+i0) = -v^{-1}$.

If the condition is satisfied then $\phi_{E,x}$ of (5.30) provides the unique (up to normalization and phase) proper eigenfunction of H_v within the cyclic space $\mathcal{H}_{H_v,x}$.

A proof of the main part of this statement can be found in [**363**, Thm. 4]. It is left here as a pair of exercises (Exercise 5.4 and 5.5).

Proposition 5.5 yields the following characterization of the supports of the continuous and pure-point parts of the spectral measure μ_{v,δ_x}.

Corollary 5.6. *For any $v \neq 0$,*

$$\boxed{\mu^{\mathrm{pp}}_{v,\delta_x}(\{\gamma_0 = \infty\}) = 0\,, \qquad \mu^{\mathrm{c}}_{v,\delta_x}(\{\gamma_0 < \infty\}) = 0}\,. \tag{5.32}$$

Proof. The first statement is an immediate consequence of Proposition 5.5. For a proof of the second, we note that $\gamma_0(E) < \infty$ implies that $\lim_{\eta\downarrow 0} G_0(x,x;E+i\eta) \in \mathbb{R}$ exists and is real (Exercise 5.3) such that

$$\lim_{\eta\downarrow 0} \operatorname{Im} G_v(x,x;E+i\eta) = \lim_{\eta\downarrow 0} \frac{\operatorname{Im} G_0(x,x;E+i\eta)}{|1+vG_0(x,x;E+i\eta)|^2} = 0 \tag{5.33}$$

for all E such that $\gamma_0(E) < \infty$ and $G_0(x,x;E+i0) \neq -v^{-1}$. By Proposition 5.5 aside from the countable set of atoms of μ_{v,δ_x}, the set $\{\gamma_0 < \infty\}$ coincides with such points. From the general characterization of the spectral types (Proposition B.4) we then conclude that the continuous component of μ_{v,δ_x} is for any $v \neq 0$ supported on the set $\{\gamma_0 = \infty\}$. □

Combining the characterizations (5.32) with the spectral averaging principle (5.22) we obtain for any Borel set $I \subset \mathbb{R}$

$$\int \mu^{\mathrm{pp}}_{v,\delta_x}(I)\, dv = \int \mu_{v,\delta_x}\left(I \cap \{\gamma_0 < \infty\}\right) dv = \left|I \cap \{\gamma_0 < \infty\}\right| , \tag{5.34}$$
$$\int \mu^{\mathrm{c}}_{v,\delta_x}(I)\, dv = \int \mu_{v,\delta_x}\left(I \cap \{\gamma_0 = \infty\}\right) dv = \left|I \cap \{\gamma_0 = \infty\}\right| .$$

This is known as the Simon-Wolff criterion [**363**]. An application to random operators reads as follows.

Theorem 5.7 (Simon-Wolff criterion). *Let $H(\omega) = T + V(\omega)$ be a random operator in $\ell^2(\mathbb{G})$ with a random potential whose conditional single-site distribution of $V(x)$ at $x \in \mathbb{G}$ is absolutely continuous. Then for any Borel set $I \subset \mathbb{R}$:*

1. *If for Lebesgue-almost every $E \in I$ and $\mathbb{P}$-almost all ω*

$$\lim_{\eta \downarrow 0} \sum_y |G(x, y; E + i\eta; \omega)|^2 \;<\; \infty\,, \tag{5.35}$$

then the spectral measure of $H(\omega)$ associated with δ_x has empty continuous component in I for $\mathbb{P}$-almost all ω.

2. *If for Lebesgue-almost every $E \in I$ and $\mathbb{P}$-almost all ω*

$$\lim_{\eta \downarrow 0} \sum_y |G(x, y; E + i\eta; \omega)|^2 \;=\; \infty\,, \tag{5.36}$$

then the spectral measure of $H(\omega)$ associated with δ_x has empty pure-point component in I for $\mathbb{P}$-almost all ω.

Proof. We apply the relations (5.34) for

$$H_0 = H \quad \text{and} \quad H_v = H + (v - V(x))\, 1_{\{x\}},$$

in which the potential at x is set to $v \in \mathbb{R}$. In this setting, $\gamma_0(E)$ coincides with the limit on the left side of (5.35) and (5.36).

For a proof of the first assertion we note that by assumption $|I \cap \{\gamma_0 < \infty\}| = 0$ almost surely. Hence (5.34) implies that the spectral measure of $H + (v - V(x))\, 1_{\{x\}}$ has almost surely no continuous component on I for Lebesgue-almost all v. But since the conditional distribution of $V(x)$ condition on the values of $V_{\neq x} = \{V(y) \,|\, y \neq x\}$ is absolutely continuous, this implies the result for H.

The second assertion is proven analogously. $\square$

5.4. Simplicity of the pure-point spectrum

Non-degeneracy of the point spectrum is of intrinsic interest, but it will also be of relevance in the discussions of eigenfunction correlators (Chapter 7) and of spectral statistics. The theory of rank-one perturbations facilitates the following result.

Theorem 5.8. *Let $H(\omega) = T + V(\omega)$ be a random operator on $\ell^2(\mathbb{G})$ with T bounded and self-adjoint and $V(\omega)$ a random potential such that for any $x \in \mathbb{G}$ the conditional distribution of $V(x)$ conditioned on $V_{\neq x} := \{V(u)\}_{u \neq x}$ is continuous. Then the pure-point spectrum of $H(\omega)$ is simple for $\mathbb{P}$-almost all ω.*

Our proof proceeds through:

Lemma 5.9. *Let $\mathbb{G}$ be a countable set, H_0 a self-adjoint operator in $\ell^2(\mathbb{G})$, and H_v the one-parameter family of operators defined by* (5.1) *with $x \in \mathbb{G}$. Then for any countable subset $S \subset \mathbb{R}$ and any probability measure ρ which is continuous*

$$\int \mu_{v,\delta_x}(S)\, \rho(dv) = 0 \,. \tag{5.37}$$

Proof. By the countable additivity of the measure $\int \mu_{v,\delta_x}(\cdot)\, \rho(dv)$ it suffices to prove (5.37) for one-point sets $S = \{E\}$, at arbitrary $E \in \mathbb{R}$. The contribution to the integral of the case $v = 0$ vanishes since $\rho(\{0\}) = 0$. For $v \neq 0$, by Proposition 5.5, the integrand does not vanish only if $v = -G_0(x, x; E + i0)^{-1}$. However, this point is also not charged, since ρ is a continuous measure, and hence the integral vanishes. $\square$

Proof of Theorem 5.8. For $x \in \mathbb{G}$ let

$$\Omega_x := \big\{\omega \in \Omega \mid \text{for some } E \in \mathbb{R}\text{: dim range } P_{\{E\}}(H(\omega)) \geq 2 \text{ and } P_{\{E\}}(H(\omega))\delta_x \neq 0\big\}. \tag{5.38}$$

Our goal is to prove that this set is of vanishing probability. To highlight the dependence of the random operator $H = H(\omega)$ on $V(x)$ we write it in the form

$$H =: H_0 + V(x)\, 1_{\{x\}} \,, \tag{5.39}$$

where H_0 is independent of $V(x)$ and defined through the above equation. The full Hilbert space may be presented as the direct sum of the cyclic subspace $\mathcal{H}_{H,x} = \mathcal{H}_{H_0,x}$ and its orthogonal complement:

$$\ell^2(\mathbb{G}) = \mathcal{H}_{H_0,x} \oplus \mathcal{H}_{H_0,x}^{\perp} \,. \tag{5.40}$$

If the vector δ_x is cyclic for H, then $\mathcal{H}_{H_0,x}^{\perp}$ consists of just the zero vector and the pure-point spectrum of H is simple (Lemma 5.2). In the following we therefore concentrate on the case that $\mathcal{H}_{H_0,x}^{\perp} \neq \{0\}$.

The operator H leaves both $\mathcal{H}_{H_0,x}$ and its orthogonal complement invariant. Its point spectrum is therefore the union of point spectra the operator has in the two subspaces. Two notable features of this decomposition are: i) the spectrum of H in $\mathcal{H}_{H_0,x}$ is non-degenerate, ii) the spectrum in $\mathcal{H}_{H_0,x}^{\perp}$ does not vary with $V(x)$.

Let S_x denote the set of eigenvalues of the restriction of H to $\mathcal{H}_{H_0,x}^{\perp}$. Its independence of $V(x)$ follows from the observation that the eigenfunctions ψ_E in $\mathcal{H}_{H_0,x}^{\perp}$ have to vanish at x, since for any $\varphi = f(H)\delta_x \in \mathcal{H}_{H_0,x}$

$$0 = \langle \varphi\,,\,\psi_E\rangle = \langle \delta_x\,,\,f(H)\psi_E\rangle = f(E)\,\psi_E(x)\,. \tag{5.41}$$

This implies that ψ_E is also an eigenfunction of $H_0 + \widehat{V}(x)\,1_{\{x\}}$ for any other $\widehat{V}(x) \in \mathbb{R}$ with the same eigenvalue E.

Since the set S_x is independent of $V(x)$, by Lemma 5.9 the conditional expectation of $\mu_{\delta_x}(S_x)$, conditioned on $V_{\neq x}$, is zero for each value of all the other parameters. (In this argument μ_{δ_x} is the spectral measure associated with H and the vector $\delta_x \in \ell^2(\mathbb{G})$, and use is made of the continuity of the conditional distribution of $V(x)$.) Hence

$$\mathbb{E}\left[\mu_{\delta_x}(S_x)\right] = \mathbb{E}\left[\mathbb{E}\left[\mu_{\delta_x}(S_x)\,\middle|\,V_{\neq x}\right]\right] = 0\,. \tag{5.42}$$

This means that the point spectrum of H in $\mathcal{H}_{H_0,x}$, which supports μ_{δ_x}, almost surely does not intersect S_x, or

$$\mathbb{P}(\Omega_x) \;=\; 0\,. \tag{5.43}$$

Since countable unions of null sets carry zero probability, also $\Omega_0 := \bigcup_{x\in\mathbb{G}} \Omega_x$ has this property.

On the complement of Ω_0 for each $E \in \mathbb{R}$ the vectors $P_{\{E\}}(H(\omega))\delta_x$ and $P_{\{E\}}(H(\omega))\delta_y$ at different $x, y \in \mathbb{G}$ are collinear, when non-zero. Since their collection spans the full range of $P_{\{E\}}(H(\omega))$ in $\ell^2(\mathbb{G})$, the point spectrum is simple for any ω in the complements of the null set Ω_0. □

5.5. Finite-rank perturbation theory

Much of this chapter's discussion, and some of the previous one, is built on the rank-one perturbation formula (4.8) and (5.4). We shall subsequently find use also for its finite-rank extensions.

One is often interested in the restriction of the resolvent of a self-adjoint operator H to a low-dimensional subspace, and in particular on its dependence on the operator's local structure there. E.g., the Green function

$G(x, y; z)$ can be read from $P_{x,y}(H - z)^{-1}P_{x,y}$ where $P_{x,y}$ is the projection on $\mathrm{span}\{\delta_{\mathrm{x}}, \delta_{\mathrm{y}}\}$. In such situations it is often convenient to consider the dependence of the resolvent's restriction on the values of the potential at the two sites, at given values of the rest of the variables. For this purpose, one may decompose the Hilbert space as

$$\mathcal{H} = P\mathcal{H} \oplus Q\mathcal{H}\,, \tag{5.44}$$

with P the orthogonal projection on the finite-dimensional subspace of interest and $Q = (1 - P)$ its complementary projection, and write H in the corresponding block form as

$$H \;=\; \begin{pmatrix} H_P & T \\ T^\dagger & H_Q \end{pmatrix} \;=:\; PHP + \widehat{H}\,. \tag{5.45}$$

It is then of interest to have a simple picture of the dependence of $P(H - z)^{-1}P$ on the parameters of H split into the two sets:

i. the few parameters which affect PHP, which form the **local part** of the operator

ii. the multitude of the other parameters on which the operator H depends, i.e. which affect $\widehat{H}$.

In such situations the following two useful identities hold, for any $z \in \mathbb{C}\backslash\mathbb{R}$:

$$\begin{aligned} P\,(H - z)^{-1}\,P \;&=\; \left[H_P + \left[P\left(\widehat{H} - z\right)^{-1} P\right]_P^{-1}\right]_P^{-1} \\ &=\; \left[H_P - z - T\,(H_Q - z)^{-1}\,T^\dagger\right]_P^{-1}, \end{aligned} \tag{5.46}$$

with the subscript on $[\ldots]_P^{-1}$ indicating that the inverse is on $P\mathcal{H}$ only.

The above expressions for $P\,(H - z)^{-1}\,P$ are often found referred to, alternatively, as the **Krein**, **Feshbach**, or **Schur formula**. It can be deduced from the following theorem (Exercise 5.7).

Theorem 5.10. *Let H be a self-adjoint operator in some Hilbert space $\mathcal{H}$ and let P be an orthogonal projection onto a closed subspace on which we define*

$$K(z) \;:=\; P\,(H - z)^{-1}\,P\,, \qquad z \in \mathbb{C}\backslash\mathbb{R}\,. \tag{5.47}$$

Then for any $z \in \mathbb{C}\backslash\mathbb{R}$ and any bounded self-adjoint operator of the form $A = PAP$,

(1) *the operator $1 + A\,K(z)$ is invertible on $P\mathcal{H}$,*

(2) *on $P\mathcal{H}$ one has the identity*

$$\boxed{(H + A - z)^{-1}\,P = (H - z)^{-1}\,P\,[1 + A\,K(z)]_P^{-1}\,P}\,. \tag{5.48}$$

Proof. Abbreviating $R(z) := (H + A - z)^{-1}$ and $r(z) := (H - z)^{-1}$, the resolvent identity yields

$$r(z)\,P = \big[R(z) + R(z)\,A\,r(z)\big]P = R(z)P\,[1 + A\,K(z)]\,P\,, \tag{5.49}$$

and, in particular, $K(z) = PR(z)P\,(1 + A\,K(z))\,P$. As a consequence, any vector $\varphi \in P\mathcal{H}$ in the kernel of $(1 + A\,K(z))$ is also in the kernel of $K(z)$ and hence $\varphi = (1 + A\,K(z))\varphi = 0$; i.e., $1 + A\,K(z)$ is injective on $P\mathcal{H}$. To establish surjectivity, let $\psi \in P\mathcal{H}$ and set $\varphi = \psi - AR(z)\,\psi \in P\mathcal{H}$. The resolvent identity then implies $(1 + A\,K(z))\varphi = \psi$; i.e., ψ is in the range of $1+A\,K(z)$ on $P\mathcal{H}$. This concludes the proof of the invertibility of $1+A\,K(z)$ on $P\mathcal{H}$.

The identity (5.48) thus follows from (5.49) by multiplication with this inverse from the right. □

In discussing (5.46), it is also useful to note that the function $\mathbb{C}\backslash\mathbb{R} \ni z \mapsto K(z) = P(H-z)^{-1}P$ is a matrix-valued version of a Herglotz-Pick function on the finite-dimensional space $P\mathcal{H}$. Its restriction to $\mathbb{C}^+$ is analytic and satisfies $\operatorname{Im} K(z) > 0$ for all $z \in \mathbb{C}^+$.

Proposition 5.11. *For any self-adjoint operator H on a Hilbert space $\mathcal{H}$, the operators $K(z) = P(H - z)^{-1}P$, defined for $z \in \mathbb{C}\backslash\mathbb{R}$ on a finite-dimensional subspace $P\mathcal{H}$ share the the following properties:*

1. *$K(z)$ is invertible on $P\mathcal{H}$ for every $z \in \mathbb{C}\backslash\mathbb{R}$ and the negative inverse $-K(z)^{-1}$ has the Herglotz-Pick property, $-\operatorname{Im} K(z)^{-1} > 0$ for all $z \in \mathbb{C}^+$.*
2. *The boundary values*

$$K(E + i0) := \lim_{\eta\downarrow 0} K(E + i\eta)\,, \qquad K(E + i0)^{-1} := \lim_{\eta\downarrow 0} K(E + i\eta)^{-1} \tag{5.50}$$

exist and are finite for Lebesgue-almost all $E \in \mathbb{R}$.

The first part of this proposition is left as an exercise (cf. Exercise 5.7). In case $P\mathcal{H}$ is one-dimensional, the function $z \mapsto K(z)$ is a Herglotz-Pick function and the second claim above is the de la Vallée-Poussin theorem (Proposition B.3). A proof of the second part of Proposition 5.11 in the general case can be based on the polarization identity and the fact that for any $\psi \in P\mathcal{H}$ and any matrix-valued Herglotz function K with $\operatorname{Im} K > 0$, the function $z \mapsto \langle \psi, K(z)\, \psi \rangle$ is a Herglotz-Pick function. (For more information, see [**175**].)

For the explicit implications of Theorem 5.10 for rank-one and rank-two perturbations, it will be convention to change the notation slightly and in each case apply the symbol H_0 for that part of H which does not involve the parameters under consideration.

I. Rank-one perturbation formula. For

$$\boxed{H := H_0 + V(x)\; 1_{\{x\}}}\,,$$

the Green function at $z \in \mathbb{G}$,

$$G(x; y; z) := \langle \delta_x, (H - z)^{-1} \delta_y \rangle \tag{5.51}$$

has the following dependence on $V(x)$:

a. From $x \in \mathbb{G}$ to any other site $y \in \mathbb{G}$,

$$\boxed{G(x; y; z) = \frac{G_0(x, y; z)}{1 + V(x)\; G_0(x, x; z)} =: \frac{1}{V(x) - \Sigma_0(z)} \frac{G_0(x, y; z)}{G_0(x, x; z)}} \tag{5.52}$$

(a special case of (5.48), with $P = 1_{\{x\}}$).

b. For more general $u, w \in \mathbb{G}$,

$$\begin{aligned} G(u, w; z) &= G_0(u, w; z) - V(x)\, G_0(u, x; z)\, G(x, w; z) \\ &= G_0(u, w; z) - V(x)\, \frac{G_0(u, x; z)\, G_0(x, w; z)}{1 + V(x)\, G_0(x, x; z)} \end{aligned} \tag{5.53}$$

(the first equality derives from the resolvent equation and the second one is a consequence of (5.52)).

II. Rank-two perturbation formula. For

$$\boxed{H := H_0 + V(x)\; 1_{\{x\}} + V(y)\; 1_{\{y\}}}\,, \tag{5.54}$$

the dependence of the 2×2 set of Green function elements on the pair of potential values $\{V(x), V(y)\}$ is of the form

(5.55)

$$\boxed{\begin{pmatrix} \langle \delta_x, (H - z)^{-1} \delta_x \rangle & \langle \delta_x, (H - z)^{-1} \delta_y \rangle \\ \langle \delta_y, (H - z)^{-1} \delta_x \rangle & \langle \delta_y, (H - z)^{-1} \delta_y \rangle \end{pmatrix} = \begin{pmatrix} V(x) - \sigma(x; z) & -\tau(x, y; z) \\ -\tau(y, x; z) & V(y) - \sigma(y; z) \end{pmatrix}^{-1}}$$

where $\sigma(\cdot, z)$ and $\tau(\cdot, z)$ do not depend on $\{V(x), V(y)\}$.

5.6.* A zero-one boost for the Simon-Wolff criterion

In the Simon-Wolff spectral criterion it is required to know that the relevant event, (5.35) or (5.36), occurs with probability one. It is therefore useful to know that in a broad class of models for almost every $E \in \mathbb{R}$ one of the two conditions occurs with probability one. In such cases, to employ the criterion it suffices to establish just non-zero probability for one of the two complementary conditions, at specified E. The boost which this zero-one law gives will be of help in our discussion of resonant delocalization in Chapter 15.

Let V be a random potential, i.e., a stochastic process over the graph $\mathbb{G}$. For each $\Lambda \subset \mathbb{G}$, we denote by $\mathcal{B}_\Lambda$ the minimal σ-algebra of subsets $A \subset \Omega$ for which $\omega \mapsto 1_A(\omega)$ is a measurable function of $\{V(y)\}_{y\in\Lambda}$.

Definition 5.12. 1. A random variable $F : \Omega \mapsto \mathbb{R}$ is **measurable at infinity** if for each finite $\Lambda \subset \mathbb{G}$, F is measurable with respect to $\mathcal{B}_{\Lambda^c}$.

2. A random potential over a graph $\mathbb{G}$ is said to have the K**-property** if any random variable which is measurable at infinity is constant almost surely.

The simplest example of potentials with the K-property are those for which the $\{V(y)\}$ are independent random variables.

The following result, which is proven in [**23**], is then of relevance.

Theorem 5.13 (Zero-one law). *Let $H(\omega) = T + V(\omega)$ be a self-adjoint random operator on $\ell^2(\mathbb{G})$ with T bounded and $V(\omega)$ a random potential with the K-property and such that for each $x \in \mathbb{G}$, the conditional single-site distribution of $V(x)$ conditioned on $V_{\neq x} = \{V(y)\}_{y\neq x}$ is continuous. Then for Lebesgue-almost every $E \in \mathbb{R}$,*

$$\boxed{\mathbb{P}\left(\lim_{\eta\downarrow 0}\sum_y |G(x,y;E+i\eta)|^2 < \infty\right) \quad \textit{equals either } 0 \textit{ or } 1} . \tag{5.56}$$

Since the proof of Theorem 5.13, which can be found in [**23**], takes more lines than is justified here, let us just outline the key steps of the argument.

As a preparatory step, one may note that for Lebesgue-almost every E the quantity in (5.56) can be rewritten as

$$\begin{aligned} &\lim_{\eta\downarrow 0}\sum_y |G(x,y;E+i\eta,\omega)|^2 \\ &\qquad = \left|\left\langle \delta_x, \frac{1}{H(\omega)-E-i0}\,\delta_x\right\rangle\right|^2 \lim_{\eta\downarrow 0} \operatorname{Im} \mathcal{F}(\omega, E+i\eta) \end{aligned} \tag{5.57}$$

where $\mathcal{F}(\omega, E+i\eta) := \frac{-1}{\eta}\langle\delta_x, \frac{1}{H(\omega)-E-i\eta}\,\delta_x\rangle^{-1}$ and

$$\operatorname{Im}\mathcal{F}(\omega, E+i\eta) = \frac{\langle\delta_x, \frac{1}{(H(\omega)-E)^2+\eta^2}\,\delta_x\rangle}{\left|\langle\delta_x, \frac{1}{H(\omega)-E-i\eta}\,\delta_x\rangle\right|^2}. \tag{5.58}$$

By a combination of the monotone convergence principle for the numerator and the de la Vallée-Poussin theorem (Proposition B.3) for the denominator, one checks that for each realization of the potential the last limit in (5.57) exists (in $\dot{\mathbb{R}} := [0, +\infty]$) for almost every E. The exceptional values of $(\omega, E) \in \Omega \times \mathbb{R}$ amount to a subset of zero (product) measure. By Fubini's theorem it follows that for a full (Lebesgue) measure of energies E the collection of potentials at which E is exceptional is itself of zero measure with respect to $\mathbb{P}$. In proving Theorem 5.13 our analysis will now be restricted to such a regular set of $E \in \mathbb{R}$.

Based on the above considerations, Theorem 5.13 can be directly deduced from the following:

Theorem 5.14. *Under the assumptions of Theorem* 5.13, *for each* $E \in \mathbb{R}$ *the event*

$$\mathcal{K}_x(E) := \left\{\omega \in \Omega \mid \lim_{\eta\downarrow 0} \frac{-1}{\eta}\operatorname{Im}\left\langle\delta_x, \frac{1}{H(\omega)-E-i\eta}\,\delta_x\right\rangle^{-1} < \infty\right\} \tag{5.59}$$

where the existence of the limit should be taken as part of the condition is of probability either 0 *or* 1.

A key ingredient for the proof is the following lemma.

Lemma 5.15 ([**23**]). *Let*

$$F_n(V) = \frac{a_n V + b_n}{c_n V + d_n} \tag{5.60}$$

be a sequence of Möbius functions with the property that for all $V \in \mathbb{R}$,

1. $\operatorname{Im} F_n(V) \geq 0$ *and*
2. $\operatorname{Im} F_n(V)$ *converges to a limit within* $[0, \infty]$ *(allowing the value* $+\infty$*).*

Then, $\lim_{n\to\infty} \operatorname{Im} F_n(V)$ *is finite or infinite simultaneously for all, except at most one, value of* $V \in \mathbb{R}$.

From the rank-two formula (5.55) we learn that, for any specified $y \in \mathbb{G}$, in its dependence on $V(y) = V$ the quantity $\mathcal{F}(\omega, E + i/n)$ satisfies the assumptions stated in Lemma 5.15. It then implies that when the potential is regenerated through the refreshment of any single value of V the validity of the event $\mathcal{K}_x$ does not change—except possibly over zero measure sets. Using induction on the size, $|\Lambda| < \infty$, one can then conclude that for any finite subset $\Lambda \subset \mathbb{G}$ with $x \in \Lambda$ the conditional expectations satisfy the almost-sure equality

$$\mathbb{E}\left(1_{\mathcal{K}_x} \,|\, \mathcal{B}_{\Lambda^c} \right)(\omega) \;=\; 1_{\mathcal{K}_x}(\omega)\,. \tag{5.61}$$

A notable implication is that $1_{\mathcal{K}_x}(\omega)$ is measurable at infinity (in the Lebesgue sense). Hence under the K-assumption (of Theorem 5.14) it is either almost surely 0 or almost surely 1.

The boosted Simon-Wolff criterion implies that under the conditions of Theorem 5.13:

1. The real line is covered up to a zero measure subset by the **disjoint union** $\mathcal{C}_x \cup \mathcal{P}_x$ of the non-random sets

$$\begin{aligned} \mathcal{C}_x &:= \{E \in \mathbb{R} \,|\, \mathbb{P}\left(\gamma_x(E) = \infty\right) = 1\}\,, \\ \mathcal{P}_x &:= \{E \in \mathbb{R} \,|\, \mathbb{P}\left(\gamma_x(E) < \infty\right) = 1\}\,. \end{aligned} \tag{5.62}$$

2. With probability one $\mathcal{C}_x$ serves as a support for the continuous spectrum of $H(\omega)$, in the sense that
   i. $\mu^{\mathrm{pp}}_{\delta_x}(\mathcal{C}_x;\omega) \;=\; 0$ for $\mathbb{P}$-almost all ω,
   ii. for any $\varepsilon > 0$ and Lebsgue-almost every $E \in \mathcal{C}_x$,

$$\mu^{\mathrm{c}}_{\delta_x}((E-\varepsilon, E+\varepsilon);\omega) \;>\; 0 \quad \text{for } \mathbb{P}\text{-almost every } \omega. \tag{5.63}$$

3. $\mathcal{P}_x$ supports the pure-point spectrum of $H(\omega)$ together with the real part of the resolvent set. In particular,

$$\mu^{\mathrm{c}}_{\delta_x}(\mathcal{P}_x;\omega) \;=\; 0 \quad \text{for } \mathbb{P}\text{-almost all } \omega. \tag{5.64}$$

Furthermore, the condition $\lim_{\eta\downarrow 0}\langle \delta_x, \frac{1}{H(\omega)-E-i\eta}\,\delta_x\rangle \,>\, 0$, in which existence of the limit is part of the statement, is also measurable at infinity ([**205**, Cor. 1.1.3]). Denoting

$$\mathcal{A}_x \;:=\; \left\{ E \in \mathbb{R} \,|\, \mathbb{P}\left(\operatorname{Im}\langle \delta_x, \frac{1}{H-E-i0}\,\delta_x\rangle > 0 \right) = 1 \right\} \tag{5.65}$$

and observing that $\mathcal{A}_x \subset \mathcal{C}_x$ (which follows from the spectral representation), one may add the following to the above:

4. The support of the continuous spectrum also admits a non-random disjoint decomposition with
   i. $\mathcal{A}_x$ providing an almost-sure support of the absolutely continuous spectrum,
   ii. $\mathcal{C}_x \backslash \mathcal{A}_x$ serving as an almost-sure support of the singular continuous spectrum,

For point 4 we recall that the spectral measure's absolutely continuous component is $\mu^{\mathrm{ac}}_{\delta_x}(dE) \;=\; \pi^{-1}\operatorname{Im}\langle \delta_x, \frac{1}{H-E-i0}\,\delta_x\rangle\, dE$.

Thus, the boosted Simon-Wolff criterion implies a deterministic measure-theoretic decomposition of the spectra into pure types. It should, however, be appreciated that $\mathcal{C}_x$ and $\mathcal{P}_x$ need not coincide with the topological definition of the continuous and pure-point spectra, which were discussed in Section 3.2, since these sets may not be closed.

Notes

The spectral averaging principle appeared, in slightly different forms, in the work of S. Kotani [**264**] and more explicitly in [**352**].

An interesting implication of Theorem 5.3 on the spectral effects of rank-one perturbations whose extension to the infinite dimensional setup was not discussed here is the resulting spectral transport bound. A useful extension of this notion to the infinite dimensional operators is found in the Krein spectral shift function, which is discussed in [**267, 57, 356**] and references therein.

Background on the expansions of self-adjoint operators in terms of generalized eigenfunctions, a topic which is not discusses here, can be found in [**53, 326**]. For further applications of the rank-one perturbation theory we refer to the review [**354**].

The simplicity of point spectrum was proven for the standard Anderson model on $\mathbb{Z}^d$, with potentials with absolutely continuous conditional distribution, in the work of B. Simon [**352**]. Further results in this vane were derived by V. Jaksic and Y. Last in [**206**], where simplicity is established also for more general singular spectra, in the context of general graphs. In both works a key role is played by the analysis of the relation between cyclic spaces. For finite systems simplicity of the spectrum can be established also through the Minami estimate [**301**] (cf. Chapter 17). Additional results on this topic were presented in [**249, 308**]. The proof given here (Theorem 5.8) presents a different approach, and replaces the requirement of absolute continuity by just a continuity condition.

Among its many uses, the rank-two perturbation formula (5.55) is of help also in the proof of the celebrated *non-crossing rule* which states that the occurrence of spectral degeneracy for hermitian matrices of variable entries requires the adjustment of more than one parameter. The rule was noted by F. Hund and formulated by J. von Neumann and E. Wigner [**393**].

The zero-one law boost for the Simon-Wolff criterion (whose input conditions are required to hold almost surely) makes it applicable for proofs of resonant delocalization, where the main argument initially yields only statements which hold at positive probability. In this part of the presentation we followed [**23**], where the omitted proof of Lemma 5.15 can be found.

The existence of a non-random set which almost surely serves as an essential support of the absolutely continuous spectrum and which is almost surely void of a singular spectrum was noted as a side result by V. Jaksic and Y. Last [**205**]. Presented there as the main result are conditions allowing one to conclude equivalence of the spectral measures associated with different sites $x \in \mathbb{G}$.

Exercises

5.1. Prove that for any energy E in the pure-point spectrum of a self-adjoint operator H on $\ell^2(\mathbb{G})$ there is at most one eigenfunction within each of the cyclic spaces $\mathcal{H}_{H,x}$.

5.2. Establish for any self-adjoint operator H on $\ell^2(\mathbb{G})$ the following relations:

$$\mu_{\delta_x}(\{E\}) = \lim_{\eta \downarrow 0} \eta \operatorname{Im} G(x, x; E + i\eta)\,, \quad 0 = \lim_{\eta \downarrow 0} \eta \operatorname{Re} G(x, x; E + i\eta)$$

for any $E \in \mathbb{R}$.

5.3. Show that if $\gamma_0(E) = \int (u - E)^{-2} \mu_{0,\delta_x}(du) < \infty$ at $E \in \mathbb{R}$, then

$$G_0(x, x; E + i0) = \lim_{\eta \downarrow 0} \int \frac{\mu_{0,\delta_x}(du)}{u - E - i\eta}$$

exists and is real valued.

5.4. Prove the first part of Proposition 5.5.
Hint: A possible path towards the proof, along the line of reasoning in [**354**], is to use the rank-one relation (5.52) in order to show that for any $E \in \mathbb{R}$ and $v \neq 0$:
a. $\mu_{v,\delta_x}(\{E\}) > 0$ implies $\lim_{\eta \downarrow 0} G_0(x, x; E + i\eta) = -v^{-1}$ and $\gamma_0(E) < \infty$.
b. $\lim_{\eta \downarrow 0} G_0(x, x; E + i\eta) = v^{-1}$ implies

$$\lim_{\eta \downarrow 0} \eta \, \frac{G_v(x, x; E + i\eta)}{G_0(x, x; E + i\eta)} = iv \, \mu_{v,\delta_x}(\{E\})$$

and

$$\lim_{\eta \to 0} \operatorname{Im} \frac{G_0(x, x; E + i\eta)}{\eta\, G_v(x, x; E + i\eta)} = v\, \gamma_0(E)\,.$$

c. If $\gamma_0(E) > 0$ and $\Sigma_0(E + i0) = v$, then

$$\mu_{v,\delta_x}(\{E\}) = \frac{1}{v^2\, \gamma_0(E)} > 0\,.$$

5.5. Prove the second part of Proposition 5.5.

5.6. Give a direct proof of (5.52) (without invoking Theorem 5.10).

5.7. Let P be a finite-dimensional orthogonal projection on some Hilbert space $\mathcal{H}$ on which H is an operator of block form (5.45) with self-adjoint operators H_P, H_Q and a bounded operator T.

a. Show that $P(\widehat{H} - z)^{-1}P$ with $\widehat{H} := H - PHP$ is invertible on $P\mathcal{H}$ for all $z \in \mathbb{C}\backslash\mathbb{R}$.

b. Show how (5.46) can be deduced using Theorem 5.10.

Chapter 6

Anderson Localization Through Path Expansions

In this chapter we present a proof of the **Anderson localization** for the high-disorder regime using **path expansion** methods and a **fractional moment** (FM) estimate. The path expansions presented below are also of independent interest, yielding other insights into the structure of Green functions of Schrödinger operators.

6.1. A random walk expansion

Ever since Feynman's expansion for the quantum mechanical unitary evolution operator e^{-itH}, random walk expansions have served as a source of intuition about the quantum evolution [**151**] and on occasions as an aid in proofs [**357**]. For rigorous analysis it is often more fruitful to consider expansions for the resolvent operator, which of course is related to the former, e.g., through

$$(H-z)^{-1} = i\int_0^\infty e^{-it(H-z)}\,dt\,, \quad z\in\mathbb{C}^+\,. \tag{6.1}$$

Our discussion takes place in the context of the prototypical Schrödinger operator on $\ell^2(\mathbb{G})$:

$$H = T + U\,, \tag{6.2}$$

with T the hopping term and U a real-valued multiplication operator, i.e., a potential defined on the vertex set $\mathbb{G}$ of a graph with the natural metric

$d : \mathbb{G} \times \mathbb{G} \to [0, \infty)$. In case T is associated with the graph Laplacian Δ it is convenient to include the operator's diagonal terms in the potential, e.g., writing the graph Laplacian of $\mathbb{Z}^d$ as $\Delta = -T - 2d$. In general, the norm of T is estimated by (Exercise 6.1)

$$\|T\| \leq \sup_{x\in\mathbb{G}} \sum_{y\in\mathbb{G}} |T(x,y)| =: \|T\|_{\infty,\infty}\,, \tag{6.3}$$

where $T(x,y) := \langle \delta_x, T\delta_y \rangle$ is the kernel of T. The expression on the right side is the norm of T as a mapping within the Banach space of bounded functions, $\ell^\infty(\mathbb{G})$.

We start with an expansion which is of limited convergence. However, its partial resummation, based on the loop-erasure algorithm, yields an expansion of $G(x,y;z)$ in terms associated with self-avoiding walks which in finite systems for each $x, y \in \mathbb{G}$ is convergent for all z in the operator's resolvent set.

Theorem 6.1 (RW expansion). *For $z \in \mathbb{C}^+$ with $\operatorname{Im} z > \|T\|$ the Green function of the self-adjoint operator $H = T + U$ on $\ell^2(\mathbb{G})$ admits the absolutely convergent expansion*

$$G(x,y;z) = \sum_{\gamma:x\mapsto y} (-1)^{|\gamma|}\, T(\gamma(0),\gamma(1))\; T(\gamma(1),\gamma(2)) \cdots T(\gamma(|\gamma|-1),\gamma(|\gamma|)) \prod_{k=0}^{|\gamma|} \frac{1}{U(\gamma(k)) - z} \tag{6.4}$$

where γ ranges over finite paths which start at $x \in \mathbb{G}$ and end at $y \in \mathbb{G}$ and $|\gamma|$ denotes the number of paths' steps.

Proof. Starting from the resolvent equation and iterating it for N steps, one gets

$$\begin{aligned}\frac{1}{T+U-z} &= \frac{1}{U-z} - \frac{1}{U-z}\, T\, \frac{1}{T+U-z} \\ &= \sum_{k=0}^{N-1} \frac{1}{U-z}\left[(-T)\frac{1}{U-z}\right]^k \\ &\quad - \frac{1}{U-z}\left[(-T)\frac{1}{V-z}\right]^{N-1} T\,\frac{1}{T+U-z}.\end{aligned} \tag{6.5}$$

The series is absolutely convergent in the norm sense since, under the assumption,

$$\left\| T\frac{1}{U-z} \right\| \leq \frac{\|T\|}{\operatorname{Im} z} < 1 \tag{6.6}$$

uniformly for all potentials U. □

6.2. Feenberg's loop-erased expansion

The Green function also admits the following expansion in terms of **self-avoiding walks** (SAW).

Theorem 6.2 (Feenberg expansion). *The Green function of the self-adjoint operator $H = T + U$ on $\ell^2(\mathbb{G})$ admits the expansion*

$$G(x, y; z) = \sum_{\hat{\gamma}:x \mapsto y}^{(SAW)} (-1)^{|\hat{\gamma}|}\, T(\hat{\gamma}(0), \hat{\gamma}(1))\; T(\hat{\gamma}(1), \hat{\gamma}(2)) \\ \cdots T(\hat{\gamma}(|\hat{\gamma}| - 1), \hat{\gamma}(|\hat{\gamma}|)) \prod_{k=0}^{|\hat{\gamma}|} \langle \delta_{\hat{\gamma}(k)}, \frac{1}{H_{\hat{\gamma},k} - z}, \delta_{\hat{\gamma}(k)} \rangle , \tag{6.7}$$

where the summation ranges over self-avoiding paths with a finite number $|\hat{\gamma}|$ of steps which start at $x \in \mathbb{G}$ and end at $y \in \mathbb{G}$ and $H_{\hat{\gamma},k}$ is the restriction of H to $\ell^2(\mathbb{G} \setminus \bigcup_{j<k}\{\hat{\gamma}(j)\})$ (which in case $k = 0$ is to be interpreted as the full operator H).

On any graph the expansion is convergent for $\operatorname{Im} z > \|T\|$. *On finite graphs it is convergent for all $z \in \mathbb{C}^+$.*

Proof. For each finite walk γ in $\mathbb{G}$, the associated loop-erased walk $\hat{\gamma}$ is defined inductively:

1. $\hat{\gamma}(0) \;=\; \gamma(0)$.
2. For each $j \in [0, |\gamma|)$,

$$\hat{\gamma}(j + 1) \;=\; \gamma\left(1 + \max\{\, k \in [0, |\gamma|] : \gamma(k) = \hat{\gamma}(j)\}\right) . \tag{6.8}$$

A partial resummation of (6.4) over terms γ with a common loop-erased backbone $\hat{\gamma}$ yields the Feenberg expansion (6.7) in case $\|T\| < \operatorname{Im} z$. To arrive at (6.7) note that the contributions of loops starting and ending in $\hat{\gamma}(k)$, which have the property that they avoid all vertices $\hat{\gamma}(j)$ with $j < k$, are summed up to $\langle \delta_{\hat{\gamma}(k)}, (H_{\hat{\gamma},k} - z)^{-1}, \delta_{\hat{\gamma}(k)} \rangle$ using (6.4).

In the case of finite graphs $\mathbb{G}$, the number of self-avoiding walks is finite and hence the summation in (6.7) consists only of a finite number of terms. The analyticity of both sides of (6.7) in $z \in \mathbb{C}^+$ then extends the identity to the full upper half-plane. □

It is worth emphasizing the following key features of the expansion, which are further explained and exploited in the next two sections:

1. In the product seen in (6.7) for each $j \in [0, |\hat{\gamma}|]$ the potential at $\hat{\gamma}(j)$ affects only the terms with $k \leq j$. This facilitates an elementary proof of exponential localization in a high-disorder regime, which is shown next.

2. On a graph with no loops, i.e., a tree graph or a union of trees, any given pair of sites is linked by at most one self-avoiding path. This leads to a useful factorization formula and through it to exponential decay which is characterized by a Lyapunov exponent.

The expansion (6.7), which seems to go back to E. Feenberg [**148, 149**], was quoted and applied in Anderson's seminal paper [**27**], where it is also called the **locator expansion**. For a complete proof of localization one needs to deal with the not entirely trivial issue of small denominators. These are to be expected since the resolvent operator is considered here at energies within the spectrum. This issue is handled here through the kernel's **fractional moments** [**8**].

6.3. A high-disorder localization bound

We now specify the discussion to operators of the form

$$H(\omega) = T + \lambda V(\omega) \tag{6.9}$$

acting in $\ell^2(\mathbb{G})$, with the following true:

1. $\mathbb{G}$ is a finite regular graph whose vertices are of constant degree $\deg_{\mathbb{G}}$.
2. T is a **nearest-neighbor operator** satisfying

$$|T(x,y)| = \begin{cases} 1 & \text{if } d(x,y) = 1, \\ 0 & \text{otherwise.} \end{cases} \tag{6.10}$$

3. The random potential $\{V(x,\omega)\}$ is iid with single-site distribution which is absolutely continuous with bounded density $\varrho \in L^\infty(\mathbb{R})$.

The Feenberg expansion facilitates the following exponential localization bound for the Green function at large values of the disorder parameter $\lambda \in \mathbb{R}$.

Theorem 6.3 (Localization at high disorder). *For the random operator* (6.9), *of the above structure, let λ be large enough so that for some $s \in (0,1)$,*

$$\frac{|\lambda|^s}{\deg_{\mathbb{G}}} > \sup_{\zeta \in \mathbb{C}} \int \frac{\varrho(v)}{|v-\zeta|^s}\, dv \;=: C_s\,. \tag{6.11}$$

Then for any $E \in \mathbb{R}$, $\operatorname{Im} z > 0$, and $x, y \in \mathbb{G}$,

$$\boxed{\mathbb{E}\left[|G(x,y;z)|^s\right] \;\le\; A\, e^{-\mu d(x,y)} + \left(\frac{C_s}{|\lambda|^s} - A\right) \delta_{x,y}} \tag{6.12}$$

with

$$A := \frac{C_s}{\deg_{\mathbb{G}}} \frac{1}{|\lambda|^s - C_s \deg_{\mathbb{G}}} \quad \textit{and} \quad e^{-\mu} := \frac{C_s \deg_{\mathbb{G}}}{|\lambda|^s} \;(<1)\,. \tag{6.13}$$

Let us first offer some intuitive remarks on the path expansion and on the above result.

From the RW expansion (6.4) it is rather easy to see that the sum $G(x, y; E + i\eta)$ attenuates over regions where

$$|\lambda V(u) - (E + i\eta)| > \deg_{\mathbb{G}} . \tag{6.14}$$

A **small-denominator problem** pops up at sites where this condition is not satisfied. Through backtracking over such sites the expansion may fail to be absolutely summable even for $\eta > 0$, where $|G(x, y; E + i\eta)| < 1/\eta < \infty$.

For finite graphs the divergence is not present in the Feenberg expansion (6.7) which involves a finite number of terms (not more than the number of self-avoiding paths from x to y). In addition, in the expansion (6.7) the accumulating single-site factors of (6.4) are replaced by path-dependent quantities which are useful to rewrite (using the rank-one formula (5.52)) as

$$\left\langle \delta_{\hat\gamma(k)}, \frac{1}{H_{\hat\gamma,k} - z}, \delta_{\hat\gamma(k)} \right\rangle = \frac{1}{\lambda V(\hat\gamma(k)) - \Gamma_{0,k}(z)} . \tag{6.15}$$

The intuitive condition for attenuation (6.14) is replaced now by a more complicated one, but one can still see that at high disorder it will be satisfied at most of the visited sites. Yet a small-denominator problem still remains.

To cope with the latter, a useful tool is found in the fractional moments of the terms contributing to the sum. Noting that $\Gamma_{0,k}(z)$ is independent of the value of the potential at the site $\hat\gamma(k)$, we find that the conditional expectation of fractional moments, with $s \in (0, 1)$, of the term (6.15) is finite (though not for $s \geq 1$; cf. Exercise 6.2):

(6.16)

$$\mathbb{E}\left[\left| \left\langle \delta_{\hat\gamma(k)}, \frac{1}{H_{\hat\gamma,k} - z}, \delta_{\hat\gamma(k)} \right\rangle \right|^s \,\Big|\, V_{\neq x} \right] \leq \sup_{\Gamma \in \mathbb{C}} \int \frac{\varrho(v)}{|\lambda v - \Gamma|^s} \, dv = \frac{C_s}{|\lambda|^s} < \infty .$$

This observation together with the dependence structure of the values of the random potential on a path $\hat\gamma$ directly yields

Lemma 6.4. *For the random operator* (6.9) *under the above assumption for each self-avoiding path* $\hat\gamma$*,*

$$\mathbb{E}\left(\prod_{k=0}^{|\hat\gamma|} \left| \langle \delta_{\hat\gamma(k)}, (H_{k,\hat\gamma} - z)^{-1} \delta_{\hat\gamma(k)} \rangle \right|^s \right) \leq \left[\frac{C_s}{|\lambda|^s} \right]^{|\hat\gamma|+1} \tag{6.17}$$

for any $s \in (0, 1)$ *and* $|\lambda| > 0$.

Proof. We successively integrate out the values of the potential along the path $\hat\gamma$ starting from $V(\hat\gamma(0))$. The conditional expectations corresponding to the kth step are bounded using (6.16) since the terms with $j > k$ in the product are independent of $V(\hat{\gamma}(k))$. $\square$

Combined with natural bounds on the number of simple random walks, this yields one of the more elementary derivations of exponential decay at high disorder:

Proof of Theorem 6.3. Using the fact that for any collection of $a_j \in \mathbb{C}$ and $s \in [0,1]$

$$\boxed{\Big(\sum_j |a_j|\Big)^s \leq \sum_j |a_j|^s} \tag{6.18}$$

in the Feenberg expansion (6.7) and applying (6.17), one gets for $x \neq y$ and $s \in (0,\tau)$

$$\mathbb{E}\left[|\mathbb{G}(x,y;z)|^s\right] \leq \sum_{k=\operatorname{dist}(x,y)}^{\infty} \deg_{\mathbb{G}}^{k-1} \left[\frac{C_s}{|\lambda|^s}\right]^{k+1} \tag{6.19}$$

where the first factor is a bound on the number of self-avoiding walks of $k \geq 1$ steps with given endpoints, and the second factor is from (6.17). Upon summation one gets the claimed bound (6.12) in case $x \neq y$. In case $x = y$ we use the a priori bound (6.16) to complete the proof. $\square$

Extensions of the high-disorder bounds to more general distributions are discussed in Chapters 8 and 10. Theorem 6.3 does not yet directly address the question of (exponential) bounds on the kernel $\langle \delta_x, \exp(-itH)\,\delta_y\rangle$ of the time-evolution operator. In the next chapter we should see that these readily follow.

Before moving on to that subject, let us present some other useful implications of the Feenberg expansion.

6.4. Factorization of Green functions

In various situations of interest the hopping operator is $T = -A$ with the **adjacency operator** of the graph, which acts as

$$\boxed{(A\psi)(x) := \sum_{y:\, d(x,y)=1} \psi(y)}\,. \tag{6.20}$$

For graphs $\mathbb{G}$ with a uniformly bounded degree the adjacency operator A is bounded, and on graphs with constant degree it coincides up to an additive constant with the graph Laplacian; cf. (4.29).

In the next application of the random walk expansions, we derive factorization properties of the Green function of the operator

$$H = -A + U \quad \text{in } \ell^2(\mathbb{G}) \tag{6.21}$$

associated to two fixed vertices $x \neq y$ in two situations:

Bottleneck: There is some triplet (u_-, u, u_+) of vertices with $u \neq x, y$ such that every simple path $\gamma : x \to y$ passes through the triplet with the $u_\pm$ being the unique forward $(+)$ and backward $(-)$ neighbor of u on the sequence of vertices visited by γ.

Decorated line segment: There is only one simple path connecting x and y. Its vertices will be denoted by $x = u_0, u_1, \ldots, u_n = y$.

If there is a bottleneck, then every self-avoiding path $\hat{\gamma} : x \to y$ has to pass through the triplet (u_-, u, u_+) in that order and hence uniquely decomposes into two parts $\hat{\gamma}_1 : x \to u$ and $\hat{\gamma}_2 : u_+ \to y$. Resumming the terms in (6.7) corresponding to the second part, we get

$$G(x, y; z) \;=\; G(x, u; z)\, G_+(u_+, y; z)\,. \tag{6.22}$$

The second factor stands for the Green function of the operator's restriction $H^+_{u_+}$ to that part of the graph which contains u_+ and which results from deleting u in $\mathbb{G}$. The identity is derived first for all $\operatorname{Im} z > \|A\|$ and then by the analyticity of both sides in $z \in \mathbb{C}^+$ for all such values. A similar argument can now be applied to the first term on the right side of (6.22), which equals

$$G(x, u; z) \;=\; G_-(x, u_-; z)\, G(u, u; z) \tag{6.23}$$

where the first factor now stands for the Green function of the operator's restriction $H^-_{u_-}$ to that other part of the graph which contains u_- and which results from deleting u in $\mathbb{G}$. Summarizing, we have thus derived the following result:

Corollary 6.5 (Bottleneck factorization)**.** *For the operator* (6.21) *and any fixed pair $x \neq y$ which is connected through a bottleneck (u_-, u, u_+), the Green function factorizes*

$$G(x, y; z) \;=\; G_-(x, u_-; z)\, G(u, u; z)\, G_+(u_+, y; z) \tag{6.24}$$

for all $z \in \mathbb{C}$.

In the special case that there is only one self-avoiding path $\hat{\gamma} : x \to y$ which visits the vertices $x = u_0, u_1, \ldots, u_n = y$, the resummation strategy which lead to (6.22) can be iterated n times. When applying it to the Green function $G_+(x, y; z)$ associated to the operator's restriction H_x^+ to all vertices of the graph which are ahead of x, we find a product of diagonal Green functions of the form

$$\Gamma_+(u; z) := \langle \delta_u, (H_u^+ - z)^{-1} \delta_u \rangle \tag{6.25}$$

appearing. The plus in the above formulas indicates the direction (ahead of u on the path) in the operator's restriction. Pursuing the above strategy we thus arrive at the following result.

Corollary 6.6 (Decorated line segment factorization). *For the operator* (6.21) *and any fixed pair $x \neq y$ for which there is a unique simple path with vertices $x = u_0, u_1, \ldots, u_n = y$, the Green function is the product:*

$$G_+(x, y; z) = G_+(x, x; z) \prod_{k=1}^{n} \Gamma_+(u_k; z) \tag{6.26}$$

for all $z \in \mathbb{C}^+$.

The proof is left as Exercise 6.7. Let us note that proofs of both Corollary 6.5 and Corollary 6.6 can also be based on the resolvent equation, as an alternative to the Feenberg expansion.

Notes

The random-walk aspect of the simple localization proof for the Green function at high disorder has also been emphasized in [**197, 376**].

Since there is a path-based expansion for $\langle \delta_x, \exp(-itH)\, \delta_y \rangle$ (Exercise 6.4), one could ask why there has not been a direct proof of exponential localization based on such methods—not even for the high-disorder regime. The reason is that it is rather difficult to obtain bounds which hold uniformly for large times t, which is the regime necessary in order to read the spectrum from the evolution operators. Progress was nevertheless made in [**339**], where a uniform lower bound was obtained for the averaged return probability $\mathbb{E}\left[|\langle \delta_x, \exp(-itH)\, \delta_x \rangle|^2\right]$ for certain one-dimensional systems, using path-counting methods.

Other relevant partial results can be found in [**140, 141, 142**], which present a partially renormalized expansion of the evolution operator in the weak disorder limit yielding **quantum diffusion** (i.e., opposite bounds to localization) up to times of order $O(\lambda^{-2-\kappa})$ with a small $\kappa > 0$.

Exercises

6.1. Prove the expression (6.3) for the operator norm of T.

6.2. Show that for all $s \in (0,1)$ and all bounded probability densities $\varrho \in L^\infty(\mathbb{R})$,

$$C_s = \sup_{\zeta \in \mathbb{C}} \int \frac{\varrho(v)}{|v-\zeta|^s}\, dv \le \frac{(2\|\varrho\|_\infty)^s}{1-s}\,.$$

Show also that C_s diverges for all $s \ge 1$.

6.3. Give an operator-theoretic proof of Theorem 6.2.
[Hint: Use the resolvent equation to successively eliminate those terms in H which connect to vertices previously visited starting from x.]

6.4. Let $\mathbb{G}$ be a graph of common degree $\deg_{\mathbb{G}}$.

a. Derive the path expansion for the unitary evolution generated by the graph Laplacian Δ:

$$\langle \delta_x, e^{it\Delta} \delta_y \rangle = e^{-it \deg_{\mathbb{G}}} \sum_{\gamma : x \mapsto y} \frac{(it)^{|\gamma|}}{|\gamma|!}$$

where γ ranges over paths of a finite number $|\gamma|$ of steps which start at $x \in \mathbb{G}$ and end at $y \in \mathbb{G}$. What is the radius of convergence of this series in t?

b. Show how the kernel $e^{-it[-\Delta+U]}(x,y)$ can be presented as an average over random walks on $\mathbb{G}$ of path-dependent complex amplitudes.

c. Reformulate the above expansions in terms of integrals of path-dependent amplitudes over a **continuous-time** random walk process, of walks on $\mathbb{G}$ whose waiting times are independent and with exponential distribution. (Alternatively stated, the jump instances form a Poisson process on the time axis.)
[The solution can be found in [**82**, Ch. II.3].]

6.5. Derive an inverse relation to (6.1) based on complex contour integration.

6.6. For a graph $\mathbb{G}$ and a finite subset $\Lambda \subset \mathbb{G}$, using the Feenberg expansion (6.7), find an expression for $G(x, y; z)$ for sites $x, y \in \Lambda$ in terms of
  i. the kernel of $(H_\Lambda - z)^{-1}$ over $\Lambda \times \Lambda$,
  ii. the kernel of $(H - z)^{-1}$ restricted to $\partial_+\Lambda \times \mathbb{G}$, with $\partial_+\Lambda$ the outer boundary of Λ.

[The relation is derived in Chapter 11 by other means and is applied there for finite-volume localization criteria; cf. (11.10).]

6.7. Prove the factorization relation (6.26).

Chapter 7

Dynamical Localization and Fractional Moment Criteria

A quantity which is very convenient for expressing dynamical and spectral localization is provided by the eigenfunction correlator. Its use in the context of random operator theory dates back to [**274**] in the one-dimensional situation and to [**5**] in the more general setup. The **eigenfunction correlator** is defined for any self-adjoint operator H in $\ell^2(\mathbb{G})$ as the total variation of the spectral measure associated with $x, y \in \mathbb{G}$; i.e., for any Borel set $I \subset \mathbb{R}$,

$$\boxed{Q(x,y;I) \;=\; \sup_{\substack{F\in C(\mathbb{R})\\ \|F\|_\infty\leq 1}} |\langle \delta_x, P_I(H)\, F(H)\, \delta_y\rangle|} \,. \tag{7.1}$$

The name reflects this quantity's relation to the normalized eigenfunctions of H, which in case the operator has only pure-point spectrum in I is

$$Q(x,y;I) \;=\; \sum_{E\in\sigma(H)\cap I} \left|\left\langle \delta_x\,,\, P_{\{E\}}(H)\,\delta_y\right\rangle\right| \;=\; \boxed{\sum_{E\in\sigma(H)\cap I} |\varphi_{E,x}(x)|\,|\varphi_{E,x}(y)|} \,. \tag{7.2}$$

with $\varphi_{E,x}$ the normalized eigenfunctions of H in $\mathcal{H}_{H,x}$, the cyclic subspace associated with δ_x (Definition 5.1), or the zero vector in case there is no proper eigenfunction in this subspace:

$$\varphi_{E,x} := \begin{cases} P_{\{E\}}(H)\,\delta_x \,/\, \sqrt{\langle \delta_x\,,\, P_{\{E\}}(H)\,\delta_x\rangle}\,, & \text{if } \mu_{\delta_x}(\{E\}) > 0\,, \\ 0 & \text{if } \mu_{\delta_x}(\{E\}) = 0\,. \end{cases} \tag{7.3}$$

Clearly $Q(x, y; I) \leq 1$, and in the cases of interest the diagonal terms are not much smaller than this bound. Relevant information would be found in the decay rate of the off-diagonal terms.

In this chapter we present the following:

The eigenfunction correlator as a tool: which is particularly useful for the proof of dynamical and spectral localization.

Its relations to the fractional moments of the Green function: by use of which, dynamical localization may be established in suitable regimes.

A pair of cutoffs: which are used in analyzing operators in the infinite-volume setup. These are (i) shifting the energy parameter off the real axis and (ii) initial restriction of the analysis to finite volumes.

As the main results let us highlight here the following two statements: (i) for any $s \in (0, 1)$ there is $C_s < \infty$, with which for any bounded open set $I \subset \mathbb{R}$,

$$\boxed{\mathbb{E}\left[Q(x, y; I)\right] \;\leq\; C_s \;\liminf_{|\eta|\downarrow 0} \;\int_I \mathbb{E}\left[|G(x, y; E + i\eta)|^s\right] dE} \tag{7.4}$$

and (ii) for any sequence of finite subgraphs $\mathbb{G}_L$ with $\mathbb{G}_L \nearrow \mathbb{G}$ (i.e., which in the limit cover the graph),

$$\boxed{\mathbb{E}\left[Q(x, y; I, s)\right] \;\leq\; C_s \;\liminf_{L\to\infty} \;\int_I \mathbb{E}\left[|G_L(x, y; E)|^s\right] dE}\,. \tag{7.5}$$

These are presented more completely in Theorem 7.7 and Theorem 7.11.

7.1. Criteria for dynamical and spectral localization

The significance of the eigenfunction correlator for dynamical localization is immediate from (7.1) which in particular entails

$$\boxed{\sup_{t\in\mathbb{R}} |\langle \delta_x,\, P_I(H)\, e^{-itH}\,\delta_y\rangle| \leq Q(x, y; I)}\,. \tag{7.6}$$

For a quantum particle moving under the dynamics generated by a random operator $H(\omega)$ on a graph $\mathbb{G}$ with metric d, **exponential dynamical**

localization in I refers to a condition of the form

$$\sup_{t\in\mathbb{R}} \sum_{y\in\mathbb{G}:\, d(x,y)\geq R} |\langle \delta_x,\, P_I(H(\omega))\, e^{-itH(\omega)}\, \delta_y\rangle|^2 \;\leq\; A_x(\omega)\, e^{-\mu R} \tag{7.7}$$

on the quantum probability of finding the particle at any time outside a ball of radius R about x, in case it was started there and then filtered by energy.

Definition 7.1. A random operator H is said to exhibit **strong exponential dynamical localization in** I if

$$\boxed{\sum_{y\in\mathbb{G}:\, d(x,y)\geq R} \mathbb{E}\left[\sup_{t\in\mathbb{R}} |\langle \delta_x,\, P_I(H)\, e^{-itH}\, \delta_y\rangle|^2\right] \;\leq\; A\, e^{-\mu R}} \tag{7.8}$$

at some $\mu > 0$ and $A < \infty$.

The above condition readily implies that (7.7) holds with $A_x(\omega)$ almost surely finite at the given inverse dynamical localization length $\mu > 0$ (independent of the realization of the potential and of the energy within the range I). For random operators with sufficiently regular on-site randomness such a condition was first established in [**5**] through control on the eigenfunction correlator.

As is evident from (7.6), a sufficient condition for strong exponential dynamical localization in I is

$$\sum_{y\in\mathbb{G}:\, d(x,y)\geq R} \mathbb{E}\left[Q(x,y;I)^2\right] \;\leq\; A\, e^{-\mu R}\,. \tag{7.9}$$

In the next section we will elaborate on how such a bound may be established through suitable bounds on the Green function.

Spectral localization in a given energy regime $I \subset \mathbb{R}$ requires H to almost surely have **only pure-point spectrum in** I. As is apparent from the RAGE theorem, spectral localization may be deduced from dynamical localization, and for that purpose it even suffices if (7.7) holds with $\mu = 0$ (Exercise 7.1). Clearly, a sufficient condition for that may also be presented in terms of the eigenfunction correlator.

Theorem 7.2 (Condition for spectral localization)**.** *Suppose the eigenfunction correlator Q corresponding to a self-adjoint operator H in $\ell^2(\mathbb{G})$ and a Borel set $I \subset \mathbb{R}$ is square summable:*

$$\sum_{y\in\mathbb{G}} Q(x,y;I)^2 < \infty \qquad \text{for all } x \in \mathbb{G}. \tag{7.10}$$

Then the spectrum of H within I is either empty or of pure-point type, with eigenprojections satisfying the uniform bound: $\left|\left\langle \delta_{x_0}, P_{\{E\}}(H)\, \delta_x\right\rangle\right| \leq Q(x_0, x; I)$.

Proof. We start from the RAGE representation of the projection onto the continuous component of the spectrum of H within I (cf. Exercise 2.1):

$$\|P_I^c(H)\,\psi\|^2 = \lim_{L\to\infty}\lim_{T\to\infty}\frac{1}{T}\int_0^T \left\|1_{\mathbb{G}\setminus\mathbb{G}_L}\, e^{-itH} P_I(H)\,\psi\right\|^2 dt \tag{7.11}$$

with $\left(\mathbb{G}_L\right)_{L\in\mathbb{N}}$ any sequence of finite subsets which exhaust $\mathbb{G}$. Since the vectors $(\delta_x)_{x\in\mathbb{G}}$ form a basis, it suffices to prove that the above quantity vanishes in case $\psi = \delta_x$ for all $x \in \mathbb{G}$:

$$\begin{aligned}\left\|1_{\mathbb{G}\setminus\mathbb{G}_L}\, e^{-itH} P_I(H)\,\psi\right\|^2 &= \sum_{y\notin\mathbb{G}_L} \left|\langle\delta_y, e^{-itH} P_I(H)\,\delta_x\rangle\right|^2 \\ &\leq \sum_{y\notin\mathbb{G}_L} |Q(x,y;I)|^2 \overset{L\to\infty}{\longrightarrow} 0\,.\end{aligned} \tag{7.12}$$

The inequality is a consequence of (7.1) and the limit vanishes under the square summability assumption.

The bound on the eigenprojections then follows from (7.2). □

Bounds on the eigenfunction correlator may be used to establish decay properties of the eigenfunctions with energies in the specified range. The discussion is simplified by the assumption that the point spectrum is simple. It is therefore relevant to recall here that in Theorem 5.8 it was shown that for a large class of random Schrödinger operators this condition is satisfied. The eigenfunctions' decay will be stated with the help of the following terminology.

Definition 7.3. A **weight function** over a graph $\mathbb{G}$ is a function $g : \mathbb{G} \to [1,\infty)$ for which $\sum_{x\in\mathbb{G}} g(x)^{-1} = 1$. Given a weight function and a normalized $\varphi \in \ell^2(\mathbb{G})$, a point $x_\varphi \in \mathbb{G}$ is said to be a **localization center** for φ if

$$|\varphi(x_\varphi)|^2 \geq g(x_\varphi)^{-1}\,. \tag{7.13}$$

An example of a weight function for $\mathbb{G} = \mathbb{Z}^d$ is $g(x) = N_g\,(1+|x|)^{d+1}$ with $N_g < \infty$ the normalization constant. Any normalized $\varphi \in \ell^2(\mathbb{G})$ has a localization center, since otherwise $\sum_{x\in\mathbb{G}} |\varphi(x)|^2 < \sum_{x\in\mathbb{G}} g(x)^{-1} = 1$, contradicting the normalization condition. For an explicit construction of a localization center of φ, one may order the vertices $\mathbb{G}$ and choose the first $x \in \mathbb{G}$ such that $|\varphi(x)|^2 \geq g(x)^{-1}$.

Theorem 7.4 (Eigenfunction localization). *Let $H(\omega)$ be a random operator on $\ell^2(\mathbb{G})$ and $I \subset \mathbb{R}$ a Borel set over which*

$$\sum_{x\in\mathbb{G}} g(x)^{-1} \sum_{y\in\mathbb{G}} e^{\nu\, d(x,y)}\, \mathbb{E}[Q(x,y;I)^2] < \infty \tag{7.14}$$

for some $\nu > 0$. Then for $\mathbb{P}$-almost every ω, for each of the simple eigenvalues $E \in I$ of $H(\omega)$ there is a localization center $x_E(\omega) \in \mathbb{G}$ such that the normalized eigenfunction $\varphi_E(\cdot;\omega)$ satisfies, for all $y \in \mathbb{G}$:

$$|\varphi_E(y;\omega)|^2 \leq A(I;\omega)\, g(x_E(\omega))^2\, e^{-\nu\, d(y, x_E(\omega))} \tag{7.15}$$

with $A(I,\cdot) \in L^1(\Omega, \mathbb{P})$.

One may note that the factor $g(x_E)^2$ superimposed on the exponential decay somewhat degrades the bound for functions with distant localization centers. Some form of such a correction (though perhaps milder) is necessary, in order to account for the natural fluctuations which occur in random systems [**112, 111**].

Proof of Theorem 7.4. Denoting

$$A(I,\omega) := \sum_{x\in\mathbb{G}} g(x)^{-1} \sum_{y\in\mathbb{G}} e^{\nu d(x,y)} Q(x,y;I;\omega)^2 \tag{7.16}$$

the assumption (7.14) implies that $A(I,\omega) < \infty$ for almost every ω. By Theorem 7.2 H has almost surely only pure-point spectrum in I. Furthermore, for any simple eigenvalue $E \in I$ the corresponding normalized eigenfunction φ_E satisfies

$$|\varphi_E(x;\omega)|^2\, |\varphi_E(y;\omega)|^2 \leq Q(x,y;I;\omega)^2 \leq\; A(I;\omega)\, g(x)\, e^{-\nu d(x,y)} \tag{7.17}$$

for all $x, y \in \mathbb{G}$. In particular this applies to $x = x_E(\omega)$, the localization center of the eigenfunction. With this choice, and the lower bound (7.13), one obtains the claimed inequality (7.15). □

7.2. Finite-volume approximations

It is convenient to present an operator H in $\ell^2(\mathbb{G})$ as the limit of its finite-volume restrictions $H_L = 1_L\, H\, 1_L$ in $\ell^2(\mathbb{G}_L)$, with 1_L projections to a sequence of finite subgraphs $\mathbb{G}_L$ which asymptotically cover $\mathbb{G}$. A notion of convergence which applies in such situations is

Definition 7.5. A sequence of operators H_L on a common Hilbert space is said to converge to H in the **strong resolvent sense** if for any $\psi \in \ell^2(\mathbb{G})$ and any $z \in \mathbb{C}\backslash\mathbb{R}$,

$$\lim_{L\to\infty} \left\|(H-z)^{-1}\psi - (H_L - z)^{-1}\psi\right\| \;=\; 0\,. \tag{7.18}$$

Strong resolvent convergence is equivalent [**331**] to weak convergence of the spectral measures associated with fixed vectors; i.e.,

$$\lim_{L\to\infty} \langle \delta_x\,,\, F(H_L)\, \delta_y\rangle \;=\; \langle \delta_x\,,\, F(H)\, \delta_y\rangle \tag{7.19}$$

for bounded and continuous $F \in C_0(\mathbb{R})$ and all $x, y \in \mathbb{G}$. That does not imply the convergence of the eigenprojections (except under additional restrictions), let alone that of the eigenfunctions. It also does not exclude the possibility that certain eigenvalues of H_L with eigenfunction near the boundary disappear in the limit $L \to \infty$. Nevertheless, one has the following lower semicontinuity property.

Proposition 7.6. *Let (Q_L) be a sequence of eigenfunction correlators corresponding to a sequence of operators H_L in $\ell^2(\mathbb{G})$ which converge to H in the strong resolvent sense. Then for any $x, y \in \mathbb{G}$ and any open set $I \subset \mathbb{R}$,*

$$\boxed{Q(x, y, I) \; \le \; \liminf_{L\to\infty} Q_L(x, y, I)} \, . \tag{7.20}$$

For random operators one may add that the above implies, by Fatou's lemma,

$$\mathbb{E}\left[Q(x, y, I)\right] \; \le \; \liminf_{L\to\infty} \mathbb{E}\left[Q_L(x, y, I)\right] . \tag{7.21}$$

Proof. If $I \subset \mathbb{R}$ is open, Lusin's theorem ("every measurable function is nearly continuous", in Littlewood's informal formulation) allows us to restrict the supremum in (7.1) to continuous functions, for which (7.19) holds. Thus

(7.22)
$$\begin{aligned} Q(x, y; I) = & \sup_{\substack{F \in C_c(I) \\ \|F\|_\infty \le 1}} |\langle \delta_x, \, F(H)\, \delta_y \rangle| \le \liminf_{L\to\infty} \sup_{\substack{F \in C_c(I) \\ \|F\|_\infty \le 1}} |\langle \delta_x, \, F(H_L)\, \delta_y \rangle| \\ & \le \liminf_{L\to\infty} Q_L(x, y; I) \, . \end{aligned}$$

□

Consequently, one may employ the following strategy for a proof of strong exponential dynamical localization, and by implication spectral localization with exponentially decaying eigenfunctions.

1. Establish that the finite-volume random operators H_L satisfy uniformly in L bounds of the form

$$\mathbb{E}\left[Q_L(x, y; I)^2\right] \; \le \; A\, e^{-\mu\, d(x,y)} \, , \tag{7.23}$$

   with some $A < \infty$, $\mu \in (0, \infty)$, which is exponentially summable in the sense that for some $0 < \mu' \le \mu$,

$$\sup_{x \in \mathbb{G}} \sum_{y \in \mathbb{G}} e^{-\mu' d(x,y)} < \infty \, . \tag{7.24}$$

2. The inequality (7.21) allows us then to conclude that the limiting random operator $H(\omega)$ satisfies (7.9) and hence exhibits strong dynamical localization in I.

3. The above also allows us to conclude that the spectrum of H in I is either almost surely empty or of pure-point type. In case the latter is simple, the corresponding normalized random eigenfunctions φ_E with $E \in I$ satisfy (7.15) with $\nu = \mu - \mu'$.

An alternative strategy is to derive suitable bounds directly for the infinite-volume operator H. For that, instead of the initial restriction to finite volumes it is convenient to use the "cutoff" provided by considering the Green function with the energy parameter shifted off the real axis. This technique is described next.

7.3. The relation to the Green function

The correlator's analysis is facilitated by the relations to the Green function to which we now turn. The relations are presented in two forms, suitable to the alternative approaches to the proof of localization that are described above. The one employing complex energy regularization of the Green function is somewhat inspired by, yet different from, the arguments in [**371, 178**]. The finite-volume regularization is similar to the one initially employed in [**5**]; however, the results are presented here in a further streamlined form.

7.3.1. Complex-energy regularization. As was mentioned in (7.22), for open sets $I \subset \mathbb{R}$,

$$Q(x,y;I) = \sup_{\substack{F \in C_c(I) \\ \|F\|_\infty \leq 1}} |\langle \delta_x, \, F(H)\, \delta_y \rangle| \, . \tag{7.25}$$

Since for any $F \in C_c(I)$ and $E \in \mathbb{R}$

$$F(E) \;=\; \lim_{\eta \downarrow 0} \frac{1}{\pi} \int F(E') \, \mathrm{Im} \left(E' - E - i\eta \right)^{-1} dE' \, ,$$

we may conclude the bound

$$Q(x,y;I) \;\leq\; \liminf_{\eta \downarrow 0} \frac{1}{\pi} \int_I \left| \langle \delta_x \, , \mathrm{Im}\, (H - E - i\eta)^{-1} \delta_y \rangle \right| \, dE \, . \tag{7.26}$$

The fact that the integrand on the right is limited to just the imaginary part of the resolvent is important. Without this restriction the limit will (generically) diverge if there is an eigenvalue in I (for reasons which will be further discussed in Section 8.1). However, one may relate the right side to the Green function's fractional moments.

Theorem 7.7 (Correlator bound I). *Let $H(\omega) = H_0 + V(\omega)$ be a random operator on $\ell^2(\mathbb{G})$ with a random potential whose conditional single-site* *distribution* $\rho_x(dv \,|\, V_{\neq x})$ *is absolutely continuous with a density* $\varrho_x(v \,|\, V_{\neq x})$ *satisfying for some* $s \in (0,1)$ *and* $C < \infty$

$$\sup_{v\in\mathbb{R}}(1+|v|^s)\, \varrho_x(v \,|\, V_{\neq x}) \leq C \tag{7.27}$$

almost surely and for all $x \in \mathbb{G}$. *Then the averaged eigenfunction correlator satisfies, for any bounded open set* $I \subset \mathbb{R}$,

$$\boxed{\mathbb{E}\left[Q(x,y;I)\right] \;\leq\; C_s \liminf_{|\eta|\downarrow 0} \int_I \mathbb{E}\left[|G(x,y;E+i\eta)|^s\right] dE} \tag{7.28}$$

with a finite constant $C_s < \infty$.

Proof. Starting from (7.26) and applying Fatou's and Fubini's theorems we have

$$Q(x,y;I) \;\leq\; \liminf_{\eta\downarrow 0} \frac{1}{\pi} \int_I \mathbb{E}\left[\left|\langle \delta_x \,, \operatorname{Im}(H-E-i\eta)^{-1}\delta_y\rangle\right|\right] \, dE\,. \tag{7.29}$$

By the Cauchy-Schwarz inequality applied to the positive definite kernel of $\operatorname{Im}(H-z)^{-1} \geq 0$, for every $z \in \mathbb{C}^+$,

$$\left|\langle \delta_x \,, \operatorname{Im}(H-z)^{-1}\delta_y\rangle\right| \leq \sqrt{\left|\langle \delta_x \,, \operatorname{Im}(H-z)^{-1}\delta_x\rangle\right| \, \left|\langle \delta_y \,, \operatorname{Im}(H-z)^{-1}\delta_y\rangle\right|}\,. \tag{7.30}$$

Splitting the integrand in (7.29) into powers $(1-s)$ and s and applying (7.30) to just the s power we get

$$\begin{aligned}
&\left(\int_I \mathbb{E}\left[\left|\langle \delta_x \,, \operatorname{Im}(H-E-i\eta)^{-1}\delta_y\rangle\right|\right] \, dE\right)^2 \\
&\leq \int_I \mathbb{E}\left[\left|\langle \delta_x \,, \operatorname{Im}(H-E-i\eta)^{-1}\delta_x\rangle\right|^{1-s} \left|\langle \delta_x \,, \operatorname{Im}(H-E-i\eta)^{-1}\delta_y\rangle\right|^s\right] dE \\
&\quad\times \int_I \mathbb{E}\left[\left|\langle \delta_y \,, \operatorname{Im}(H-E-i\eta)^{-1}\delta_y\rangle\right|^{1-s} \left|\langle \delta_x \,, \operatorname{Im}(H-E-i\eta)^{-1}\delta_y\rangle\right|^s\right] dE,
\end{aligned} \tag{7.31}$$

where we again applied the Cauchy-Schwarz inequality, but this time with respect to the double integral over the probability space and the energy. The two factors on the right side will now be treated similarly. Using (6.18) for the first factor we estimate

$$\left|\langle \delta_x \,, \operatorname{Im}(H-E-i\eta)^{-1}\delta_y\rangle\right|^s \leq \frac{1}{2^s}\left(|G(x,y;E+i\eta)|^s + |G(x,y;E-i\eta)|^s\right) \tag{7.32}$$

and average with respect to the random variable $V(x)$. The ratio $G(x,y;z)/G(x,x;z)$ is independent of $V(x)$ for any $z \in \mathbb{C}\backslash\mathbb{R}$. It therefore coincides with the ratio

$$\frac{G(x,y;z)}{G(x,x;z)} = \frac{G_\nu(x,y;z)}{G_\nu(x,x;z)}$$

of the Green function of the operator $H_\nu := H + (\nu - V(x))1_{\{x\}}$, in which the potential at x is changed to the value ν. As a consequence, in its dependence on $V(x)$ the integrand in the first factor on the right in (7.31) (after the application of (7.32)) is bounded by

$$\begin{aligned}(7.33)\quad &\frac{|\operatorname{Im}\Gamma(z)|^{1-s}}{|V(x)-\Gamma(z)|^{2-2s}}\,\frac{|\nu-\Gamma(z)|^s}{|V(x)-\Gamma(z)|^s}\,|G_\nu(x,y;z)|^s\\ &\le (|\nu|^s+|\Gamma(z)|^s)\,\frac{|\operatorname{Im}\Gamma(z)|^{1-s}}{|V(x)-\Gamma(z)|^{2-2s}}\,\frac{1}{|V(x)-\Gamma(z)|^s}\,|G_\nu(x,y;z)|^s\end{aligned}$$

with $\Gamma(z)$ the inverse negative Green function at x of the operator H in which the potential $V(x)$ is set to zero. Conditioning on all random variables aside from $V(x)$, the conditional expectation is estimated with the help of Lemma 7.8 below, which, e.g., for the first term yields

$$(7.34)\quad \mathbb{E}\left[|G(x,y;E+i\eta)|^s \;\middle|\; \{V(y)\}_{y\neq x}\right] \;\le\; C_s\,(|\nu|^s+1)\;|G_\nu(x,y;E+i\eta)|^s\,.$$

The resulting inequality may now be averaged over ν. For its probability distribution we pick $(|\nu|^s+1)^{-1}\varrho_x(v\,|\,V_{\neq x})/\int(|\nu|^s+1)^{-1}\varrho_x(v\,|\,V_{\neq x})dv$ as the density. Note that the normalization is bounded away from zero. This yields the claim for the first factor.

For an application of the above inequality to the factors involving $E-i\eta$, we also note that $\left|\langle\delta_x\,,\operatorname{Im}(H-z)^{-1}\delta_x\rangle\right| = \left|\langle\delta_x\,,\operatorname{Im}(H-\overline{z})^{-1}\delta_x\rangle\right|$ for any $z\in\mathbb{C}\backslash\mathbb{R}$. This concludes the proof of the assertion. □

The above proof relied on the following averaging bound.

Lemma 7.8. *Let $0\le\varrho$ with $\sup_{v\in\mathbb{R}}(1+|v|^s)\varrho(v)<\infty$ for some $s\in(0,1)$. Then there exists some constant $C_s(\varrho)<\infty$ such that for any $\gamma\in\mathbb{C}\backslash\mathbb{R}$,*

$$(7.35)\qquad \max\{1,|\gamma|^s\}\int\frac{|\operatorname{Im}\gamma|^{1-s}}{|v-\gamma|^{2-2s}}\,\frac{\varrho(v)}{|v-\gamma|^s}\,dv\;\le\;C_s(\varrho).$$

Proof. We distinguish the cases $|\gamma|\le 1$ and $|\gamma|>1$. In the first case, we estimate

$$\begin{aligned}(7.36)\qquad \int\frac{|\operatorname{Im}\gamma|^{1-s}}{|v-\gamma|^{2-2s}}\,\frac{\varrho(v)}{|v-\gamma|^s}\,dv \;&\le\; \|\varrho\|_\infty\int\frac{|\operatorname{Im}\gamma|^{1-s}}{|v-\gamma|^{2-s}}\,dv\\ &=\;\|\varrho\|_\infty\int\frac{dv}{(v^2+1)^{1-\frac{s}{2}}}<\infty\,.\end{aligned}$$

In the second case, using the triangle inequality to estimate $|\gamma|^s \leq |v-\gamma|^s + |v|^s$, we get

$$|\gamma|^s \int \frac{|\mathrm{Im}\,\gamma|^{1-s}}{|v-\gamma|^{2-2s}} \frac{\varrho(v)}{|v-\gamma|^s}\, dv \leq \int \frac{|\mathrm{Im}\,\gamma|^{1-s}}{|v-\gamma|^{2-2s}}\, \varrho(v)\, dv + \int \frac{|\mathrm{Im}\,\gamma|^{1-s}}{|v-\gamma|^{2-2s}} \frac{|v|^s\, \varrho(v)}{|v-\gamma|^s}\, dv\,. \tag{7.37}$$

The second term on the right side is estimated by pulling out

$$\rho_s := \sup_{v \in \varrho} |v|^s \varrho(v) < \infty.$$

The remaining integral is then estimated as in (7.36). The first term on the right side in (7.37) is upper bounded by

$$\rho_s \int \frac{|\mathrm{Im}\,\gamma|^{1-s}}{|v-\gamma|^{2-2s}} \frac{dv}{|v|^s} \leq \rho_s \sup_{\alpha \in \mathbb{R}} \int \frac{1}{[(v-\alpha)^2+1]^{1-s}} \frac{dv}{|v|^s} < \infty\,.$$

This completes the proof (whose gymnastics may perhaps be justified by the simplicity of the proven relation). □

7.3.2. Finite-volume regularization. To emphasize the naturalness of the relation discussed in this section it may be good to point out that for **finite systems** the eigenfunction correlator is a singular limit of fractional moments of the Green function.

Proposition 7.9. *Let H be self-adjoint on $\ell^2(\mathbb{G})$ with $|\mathbb{G}| < \infty$. Then for any bounded interval $I \subset \mathbb{R}$,*

$$\boxed{Q(x,y;I) \;=\; \lim_{s \nearrow 1} \frac{1-s}{2} \int_I |G(x,y;E)|^s\, dE}\,. \tag{7.38}$$

Proof. We decompose $I = \bigcup_n \overline{I}_n$ into open non-intersecting intervals (I_n) about each one of the eigenvalues (E_n). It suffices to show that the right side in (7.38), when restricting the integration to I_n, converges to $|\langle \delta_x, P_{\{E_n\}}(H)\, \delta_y \rangle|$ as $s \to 1$. Using the spectral representation for G this follows from

$$\lim_{s \nearrow 1} \frac{1-s}{2} \int_{I_n} \frac{dE}{|E_m - E|^s} \;=\; \delta_{mn}\,, \tag{7.39}$$

where the case $m = n$ is derived by explicit calculation. □

In practice, the limit $s \nearrow 1$ in (7.38) is hard to control. However, information can be extracted through interpolation from bounds involving fixed $0 < s < 1$. For that, it is fruitful to consider the following **interpolated**

eigenfunction correlators for operators H on a finite-dimensional Hilbert space. These are defined as

$$\boxed{Q(x,y;I;s) := \sum_{E\in I\cap\sigma(H)} \left|\langle \delta_x\,,\, P_{\{E\}}(H)\,\delta_x\rangle\right|^{1-s} \left|\langle \delta_x\,,\, P_{\{E\}}(H)\,\delta_y\rangle\right|^{s}} \tag{7.40}$$

with $I \subset \mathbb{R}$ ranging over Borel sets and $s \in [0,1]$ serving as the interpolation parameter. At the boundary points, the interpolated eigenfunction correlator reduces to the previously defined quantity:

$$Q(x,y;I;0) = Q(x,x;I) \quad \text{and} \quad Q(x,y;I;1) = Q(x,y;I)\,. \tag{7.41}$$

The following are further elementary properties:

1. In its dependence on s, the interpolated correlator is a log-convex function. I.e., for any $\lambda \in [0,1]$ and $s_\lambda := (1-\lambda)s_0 + \lambda s_1$,

$$Q(x,y;I;s_\lambda) \;\le\; Q(x,y;I;s_0)^{(1-\lambda)}\, Q(x,y;I;s_1)^{\lambda} \tag{7.42}$$

   (as is implied by either Jensen or Hölder inequality).

2. For any $0 \le s \le \beta \le 1$,

$$Q(x,y;I;\beta) \;\le\; Q(x,y;I;s)^{\frac{1-\beta}{1-s}} \le 1 \tag{7.43}$$

   (a simple consequence of (7.42)).

3. The eigenfunction correlator is bounded by the one corresponding to any fractional interpolation parameter $s \in [0,1]$:

$$Q(x,y;I) \;\le\; \sqrt{Q(x,y;I;s)\,Q(y,x;I;s)}\,. \tag{7.44}$$

The latter relation is proven by making use of the Cauchy-Schwarz inequality for the kernel of $P_{\{E\}}(H)$

$$\left|\langle \delta_x\,,\, P_{\{E\}}(H)\,\delta_y\rangle\right| \le \left|\langle \delta_x\,,\, P_{\{E\}}(H)\,\delta_x\rangle \, \langle \delta_y\,,\, P_{\{E\}}(H)\,\delta_y\rangle\right|^{1/2}\;, \tag{7.45}$$

followed by another application of the Cauchy-Schwarz inequality with respect to the summation on the spectral values (in which $\langle \delta_x\,,\, P_{\{E\}}(H)\,\delta_x\rangle$ is split into two factors and (7.45) is applied to only the $(1-s)$ power of this kernel).

As a step towards the second correlator bound we have the following identity for finite systems.

Lemma 7.10 (Correlator identity). *Let $H_\nu := H + (\nu - V(x))\, 1_{\{x\}}$ stand for the one-parameter family of self-adjoint operators on $\ell^2(\mathbb{G})$ of a finite graph, in which the potential at $x \in \mathbb{G}$ is changed to $v \in \mathbb{R}\backslash\{V(x)\}$. Then for any $s \in [0,1]$ and any Borel set $I \subset \mathbb{R}$,*

$$\boxed{Q(x,y;I;s) \;=\; \int_I \left|\langle \delta_x \,,\, (H_\nu - E)^{-1}\, \delta_y\rangle\right|^s |V(x) - \nu|^s\, \mu_{\delta_x}(dE)} \tag{7.46}$$

where μ_{δ_x} is the spectral measure associated with H and δ_x.

Proof. For any finite system $|\mathbb{G}| < \infty$, the integrals with respect to the spectral measure are given by the sum

$$\int f(E)\mu_{\delta_x}(dE) = \sum_{E\in\sigma(H)} f(E)\,\langle\delta_x, P_{\{E\}}(H)\delta_x\rangle.$$

In that case, according to Theorem 5.3, for any $E \in \sigma(H)$ with

$$\langle\delta_x, P_{\{E\}}(H)\delta_x\rangle > 0$$

the eigenfunctions at $x, y \in \mathbb{G}$ are related to the Green function of H_v:

$$\begin{aligned} \frac{\left|\langle\delta_x \,,\, P_{\{E\}}(H)\,\delta_y\rangle\right|}{\left|\langle\delta_x \,,\, P_{\{E\}}(H)\,\delta_x\rangle\right|} &= \frac{\left|\langle\delta_x \,,\, (H_v - E)^{-1}\,\delta_y\rangle\right|}{\left|\langle\delta_x \,,\, (H_v - E)^{-1}\,\delta_x\rangle\right|} \\ &= |V(x) - v|\, \left|\langle\delta_x \,,\, (H_v - E)^{-1}\delta_y\rangle\right| . \end{aligned} \tag{7.47}$$

Here the last equality results from the characterization of the spectrum of H as provided in Theorem 5.3. The claim then readily follows. □

For the averaged quantity, the above enables the following analogue of Theorem 7.7.

Theorem 7.11 (Correlator bound II). *For operators of the structure described in Theorem 7.7 on finite graphs, $|\mathbb{G}| < \infty$, for any $s \in (0,1)$ there exists $C_s < \infty$ such that*

$$\boxed{\mathbb{E}\left[Q(x,y;I,s)\right] \;\le\; C_s \int_I \mathbb{E}\left[|G(x,y;E)|^s\right] dE} \tag{7.48}$$

for all $x, y \in \mathbb{G}$ and $I \subset \mathbb{R}$.

It may be noted that through (7.44) (and an application of the Cauchy-Schwarz inequality) the right side of (7.48) also bounds the regular mean eigenfunction correlator.

Proof. Conditioning on the values of the potential at all sites other than x, the conditional average of the left side of (7.46) is bounded according to

$$\begin{aligned}(7.49)\quad \mathbb{E}\left[Q(x,y;I;s)\,\middle|V_{\neq x}\right] &= \mathbb{E}\left[|V(x)-\nu|^s \int_I \left|\langle \delta_x\,,\,(H_\nu - E)^{-1}\,\delta_y\rangle\right|^s \mu_{\delta_x}(dE)\,\middle|V_{\neq x}\right] \\ &\le \sup_{u\in\mathbb{R}}\left[(|u|^s + |\nu|^s)\,\rho(u\,|\,V_{\neq x})\right] \int_I \left|\langle \delta_x\,,\,(H_\nu - E)^{-1}\,\delta_y\rangle\right|^s dE \\ &\le C_s\,(1+|\nu|^s) \int_I \left|\langle \delta_x\,,\,(H_\nu - E)^{-1}\,\delta_y\rangle\right|^s dE\,.\end{aligned}$$

Here the second inequality is by Hölder inequality and the spectral averaging principle (5.22). The resulting bound may now be averaged over ν. For its probability distribution we again pick

$$(|\nu|^s+1)^{-1}\varrho_x(v\,|\,V_{\neq x})/\int (|\nu|^s+1)^{-1}\varrho_x(v\,|\,V_{\neq x})dv$$

as the density. Averaging over the remaining random variables yields the claim. □

7.4. The ℓ^1-condition for localization

In summary, the relations presented here show that for a random operator (of structure given in Theorem 7.7) a sufficient condition for exponential dynamical localization in an open set I is that for some $s \in (0,1)$ one of the following conditions holds:

1. $\displaystyle\sup_{|\eta|>0} \int_I \mathbb{E}\left[|G(x,y;E+i\eta)|^s\right] dE \;\le\; A\,e^{-\mu\, d(x,y)}$ or

2. $\displaystyle\liminf_{L\to\infty} \int_I \mathbb{E}\left(\left|\langle \delta_x\,,\,(H_L - E)^{-1}\,\delta_y\rangle\right|^s\right) dE \;\le\; A\,e^{-\mu\, d(x,y)}$,

with a distance $d(x,y)$ for which (7.24) holds. Since the parameter $s \in (0,1)$ can be taken arbitrarily close to 1, this condition is referred to as a **marginally-ℓ^1-criterion** on the decay of the Green function.

In comparison, the Simon-Wolff criterion (Theorem 5.7) for spectral localization involves an ℓ^2-criterion. Another criterion for spectral localization in which an ℓ^2-sum, rather than ℓ^1, is found in Theorem 7.2. In fact, to deduce localization it is not necessary to go via the eigenfunction correlator. Rather, for random operators satisfying a Wegner estimate in the sense that $\mathbb{E}[\mu_{\delta_x}(dE)] \le C\,dE$ for some $C < \infty$ and all $x \in \mathbb{G}$, the almost-sure absence of continuous spectrum in an interval I is concluded by establishing that for every $x \in \mathbb{G}$,

$$\boxed{\sum_{y \in \mathbb{G}} \limsup_{\eta \downarrow 0} \eta \int_I \mathbb{E}\left[\left| \langle \delta_y \,, (H - E - i\eta)^{-1} \delta_x \rangle \right|^2 \right] dE \;<\; \infty} \,. \tag{7.50}$$

This follows by replacing the Cesàro average in (7.11) by an Abel average for which (2.26) is readily available. For details, see Exercise 7.2 or [**20**]. This line of reasoning for a proof of spectral localization has been promoted in [**178**].

Thus it is natural to wonder whether the marginally-ℓ^1-criterion is an optimal test for localization. Our answer is yes—marginal-ℓ^1 is the correct decay rate for the averaged Green function's moments:

1. In the regime of localization the second moments diverge due to large deviations. Since no average is involved in the Simon-Wolff criterion, the effect is bypassed there. However, when the mean is taken at spectral parameter off the real axis, i.e., $\eta \neq 0$ as in (7.50), the extra factor of η which appears in the bound in effect reduces the summand's power there.
2. The relevance of the ℓ^1-condition, versus other powers, is seen most clearly in the analysis of the phase diagram of the tree graph, which is discussed in Chapter 16. On trees and some related graphs, the failure of the $\ell^{1+o(1)}$-summability of the averaged Green function implies absolutely continuous spectrum under weak disorder. The reason is related to the conditions for resonant delocalization, a topic which is discussed is Chapter 15.

Notes

The approach to dynamical localization through eigenfunction correlators which are estimated via fractional moments of the resolvent was introduced in [**5**] and further elaborated upon in [**11**] and [**9, 117**], the latter for random Schrödinger operators over the continuum $\mathbb{R}^d$. This may be regarded as a combination of the fractional moment method with the spectral averaging principle which was highlighted in the Simon-Wolff criterion. The effectiveness of the approach is related to the underlying Herglotz-Pick structure of the resolvent as a function of both the energy and (separately) the values of the potential, combined with some general properties of functions in this class. The latter form an interesting subject of harmonic analysis. Some of its directly relevant statements are discussed in the next chapter, in Section 8.2, and in Appendix B.

The first proofs of localization for random Schrödinger operators in $d > 1$ dimensions were enabled by the multi-scale analysis which was introduced to the subject by J. Fröhlich and T. Spencer [**165**]. Initially, the method established the almost-sure uniform boundedness of the displacement's second moment $M_\psi(2,t)$ and the absence of absolutely continuous spectrum [**293**]. Proofs of pure-point spectrum emerged shortly, with different arguments all empowered by the analysis in [**165**]: J. Fröhlich, F. Martinelli, E. Scoppola, and T. Spencer [**164**]; F. Delyon, Y. Lévy, and B. Soulliard [**115**]; and B. Simon and T. Wolff [**363**].

The multi-scale method proceeds by iteratively establishing, under suitable initial conditions, rapidly decaying bounds on the probability distribution of the resolvent studied on a sequence of scales, whose size grows faster than exponentially—a condition which plays a role in the analysis. That approach is also applicable to quasi-periodic operators and is reminiscent of the Kolmogorov-Arnold-Moser (KAM) method, which was actually developed in the context of perturbations of quasi-periodic dynamical systems.

A notable strength of the approach, which was simplified [**126, 127**] and also applied in the continuum [**94, 95, 152, 239, 240, 155**] as well as to Bernoulli potentials with a breakthrough result by J. Bourgain and C. Kenig [**69**], is its robustness. The method's initial limitations have been related to the fact that by proceeding at rates faster than exponential, it initially produced suboptimal bounds (with only stretched exponential decay) for the behavior of the Green function on the localization-length scale and for the probability estimates. Improvements which incorporated some of the lessons of the single-scale analysis were made by F. Germinet and A. Klein in [**169, 170**].

Textbook accounts, and surveys of the subject with further references, may be found in [**367, 68, 230, 247**].

Exercises

7.1. Show that strong dynamical localization in $I \subset \mathbb{R}$ in the sense of (7.8) with $\mu = 0$ implies spectral localization in I.

7.2. Let H be a random operator in $\ell^2(\mathbb{G})$ satisfying a Wegner estimate in the sense that $\mathbb{E}[\mu_{\delta_x}(dE)] \leq C\, dE$ for some $C < \infty$ and all $x \in \mathbb{G}$.

i. Show that for any $f \in L^2(\mathbb{R})$ and any $\psi \in \ell^2(\mathbb{G})$

$$\lim_{\eta \downarrow 0} \eta \sum_{y \in \mathbb{G}} \left| \int \mathbb{E}\left[\left|\left((H - E - i\eta)^{-1} f(H)\psi\right)(y)\right|^2\right] dE \right.$$
$$\left. - \int |f(E)|^2 \, \mathbb{E}\left[\left|\left((H - E - i\eta)^{-1}\psi\right)(y)\right|^2\right] dE \right| = 0 \,.$$

ii. Explain how (7.50) implies the almost-sure absence of continuous spectrum in I.

iii. Show that under the conditions of Theorem 7.7 one also has

$$\eta \, \mathbb{E}\left[|G(x, y; E + i\eta)|^2\right] \le C_s \, \mathbb{E}\left[|G(x, y; E + i\eta)|^s\right] .$$

Explain how the assumptions of Theorem 7.7 imply (7.50).

7.3. In the situation of Theorem 7.7 show that if $\rho \in L^\infty(\mathbb{R})$ does not satisfy (7.27), one nevertheless has

$$Q(x, y; I) \le C_s \liminf_{|\eta| \downarrow 0} \left(\int_I \mathbb{E}\left[\frac{|G(x, y; E + i\eta)|^s}{|G(x, x; E + i\eta)|^s}\right] dE \right)^{1/2}$$
$$\times \left(\int_I \mathbb{E}\left[\frac{|G(x, y; E + i\eta)|^s}{|G(y, y; E + i\eta)|^s}\right] dE \right)^{1/2} .$$

7.4. a. Show that the iterpolated eigenfunction correlators $Q_{L,v}$ of the one-parameter family of self-adjoint operators $H_\nu := H + \nu \, 1_{\{x\}}$ is related to the Green function G of H through

$$\int Q_\nu(x, y; I; s) \, \frac{dv}{|v|^s} = \int_I |G(x, y; E)|^s \, dE$$

for every $s \in (0, 1)$.

b. In the situation of Theorem 7.11 suppose that the conditional single-site distribution ρ satisfies

$$\rho \le \gamma_s \int \rho(\cdot + v) \, \frac{dv}{|v|^s}$$

for some $s \in (0, 1)$ and $\gamma_s \in (0, \infty)$ instead of (7.27). Under this condition derive a bound of the form

$$\mathbb{E}\left[Q(x, y; I; s)\right] \le \gamma_s \int_I \mathbb{E}\left[|G(x, y; E)|^s\right] dE \,.$$

Chapter 8

Fractional Moments from an Analytical Perspective

At two points of our main narrative we have encountered arguments for which it is of help to know that the Green function has exponentially decaying finite fractional moments (FM) of power $0 < s < 1$:

The high-disorder bound (6.12)**:** in which exponential decay was extracted from the Feenberg expansion (Section 6.3). The use of fractional moments allowed us to address by simple means the delicate small-denominator problem.

The eigenfunction correlator bound (7.48)**:** for which the quantities of interest are the moments with respect to integrals over the energy.

A prelude for each of these bounds is the observation that for potentials with absolutely continuous distribution the corresponding fractional moments are finite. In this chapter we present some general results which imply that for $s \in (0, 1)$,

$$\mathbb{E}\left[|G(x,x;E+i\eta)|^s\right] \le \frac{(2\|\varrho\|_\infty/|\lambda|)^s}{1-s} < \infty \tag{8.1}$$

and

$$\int_I |G(x,y;E+i0)|^s \, dE \le 4\,\frac{\mu(\mathbb{R})^s}{1-s}\,|I|^{1-s} < \infty, \tag{8.2}$$

the latter holding for any bounded $I \subset \mathbb{R}$ and any self-adjoint operator. These may be contrasted with the observation that, as was stressed already, for $s \geq 1$ the corresponding moments diverge throughout the regime of localization.

Also presented are the following:

non-diagonal extension: of (8.1) for operators with random potential for which the conditional distributions of the single-site potential are uniformly Hölder continuous (see (8.35)),

decoupling estimates: for certain products of Green functions (Corollary 8.4).

The last subject (Section 8.4) may be skipped at first reading.

8.1. Finiteness of fractional moments

The key observation which explains the validity of (8.1) for the Green function of a random operator is the the Krein formula (5.52), which makes explicit the dependence of $G(x,x;z)$ (for any $z \in \mathbb{C}$) on $V(x)$. More explicitly, let H be an operator of the form

$$H = T + \lambda \sum_{x \in \mathbb{G}} V(x)\ 1_{\{x\}} \tag{8.3}$$

acting in the $\ell^2(\mathbb{G})$ space over some countable set $\mathbb{G}$, with T a self-adjoint operator, $\lambda \in \mathbb{R}$, and $V(x)$, $x \in \mathbb{G}$, real variables. Then, by the rank-one perturbation formula (5.52), for any $z \in \mathbb{C} \backslash \mathbb{R}$,

$$G(x,x;z) \ = \ \frac{1}{\lambda V(x) - \Sigma_0(z)} \tag{8.4}$$

with $\Sigma_0(z) = -G(x,x;z)^{-1}\big|_{V(x)=0}$, which does not depend on $V(x)$.

Thus, the event $\{|G(x,x;z)| > t\}$ requires $V(x)$ to assume a value within distance $1/t$ from $\Sigma_0(z)$, of which it is independent in the random iid case. Hence, in terms of the concentration function S_ρ of Definition 4.4

$$\mathbb{P}\left(|\lambda|\,|G(x,x;z)| > t \,|\, V_{\neq x}\right) \leq S_\rho(2/t)\,. \tag{8.5}$$

The bound (8.1) then follows using the layer-cake representation of the expectation value

$$\boxed{\mathbb{E}\left(|G(x,x;E+i\eta)|^s\right) \ = \ \int_0^\infty \ \mathbb{P}\left(|G(x,x;z)| > t\right)\, d(t^s)} \tag{8.6}$$

combined with $S_\rho(2/t) \ \leq \ 2\|\varrho\|_\infty\, t^{-1}$.

The bound can also be extended to potentials with singular probability distributions (excluding, however, the Bernoulli case). For that purpose we recall Definition 4.5: a probability distribution ρ is said to be uniformly τ-Hölder continuous, for some $\tau \in (0,1]$, if for some $C_\tau(\rho) < \infty$ and all $\delta \geq 0$

$$S_\rho\,(2\delta) \;\leq\; C_\tau(\rho)\,\delta^\tau \tag{8.7}$$

(a condition which is of interest mainly for small δ).

Combining the bounds (8.5) and (8.7) with (8.6), one arrives at

Theorem 8.1 (Finiteness of FM I). *Let $H_v := H_0 + v\, 1_{\{x\}}$ be a family of self-adjoint operators in $\ell^2(\mathbb{G})$, and let ρ be a uniformly τ-Hölder continuous probability distribution with some $\tau \in (0,1]$. Then, for any $z \in \mathbb{C}\backslash\mathbb{R}$ and any $s \in (0,\tau)$,*

$$\boxed{\int |G_v(x,x;z)|^s\,\rho(dv) \;\leq\; \frac{\tau\,\mathrm{C}_\tau(\rho)^{s/\tau}}{\tau - s} \;=:\; C_{s,\tau} \;<\; \infty}\,. \tag{8.8}$$

8.2. The Herglotz-Pick perspective

The dependence of the Green function on the value of $V(x)$ is (up to a sign difference) similar to its dependence on the energy parameter E, in each case being presentable as the Stieltjes transform of a finite positive Borel measure μ on $\mathbb{R}$:

$$F(z) = \int \frac{\mu(du)}{u - z}\,, \qquad z \in \mathbb{C}^+\,. \tag{8.9}$$

We shall now show that fractional moment bounds hold quite generally over functions in this class. The key estimate is the following bound on the Lebesgue measure of the level sets of functions in the above class.

Proposition 8.2. *For any positive measure with $\mu(\mathbb{R}) < \infty$ and any $\eta \geq 0$ the function defined by (8.9) satisfies*

$$\boxed{\big|\{E \in \mathbb{R} \,\big|\, |F(E + i\eta)| \geq t\}\big| \;\leq\; \frac{4\,\mu(\mathbb{R})}{t}} \tag{8.10}$$

for all $t > 0$.

Furthermore, in case the measure μ is purely singular, the following **equality** *holds for the level sets of $F(E + i0) = \lim_{\eta\downarrow 0} F(E + i\eta)$:*

$$\big|\{E \in \mathbb{R} \,\big|\, F(E + i0) \geq t\}\big| \;=\; \frac{\mu(\mathbb{R})}{t}\,. \tag{8.11}$$

Before giving the short proof, let us add two comments:

1. The equality (8.11) was presented in the year 1857 by G. Boole [**63**] in the context of functions of the form

$$F(E) \;=\; \sum_{k=1}^{N} \frac{p_k}{u_k - E} \tag{8.12}$$

where $\{u_j\} \subset \mathbb{R}$ and $p_1, p_2, \ldots \geq 0$ form a collection of positive weights; that is, $\mu = \sum_j \delta_{u_j}$ and $\mu(\mathbb{R}) = \sum_j p_j$. Boole's derivation of (8.11) for this discrete case is given in the Notes.

2. Functions F which can be presented as in (8.9) form a subclass of the Herglotz-Pick functions, discussed in Appendix B, with the following additional properties:

$$\lim_{\eta\to\infty} F(i\eta) = 0 \quad \text{and} \quad \lim_{\eta\to\infty} \eta \;\operatorname{Im} F(i\eta) = \mu(\mathbb{R})\,. \tag{8.13}$$

Proof of Proposition 8.2. Let us start with a proof of (8.11) since it is simpler, yet it offers a blueprint for (8.10). For singular measures μ, the boundary values $F(E+i0)$ are real for almost every $E \in \mathbb{R}$. One has the following convenient representation for the indicator function on $\mathbb{R}$:

$$I[F(E+i0) > t] \;=\; \pi^{-1}\,\phi_t\,(F(E+i0)) \tag{8.14}$$

where $\phi_t\,(F(E+i0))$ is the boundary value of the harmonic function which is defined for $F \in \mathbb{C}^+$ by

$$\phi_t\,(F) \;:=\; \pi - \operatorname{Im}\operatorname{Log}(F-t) \;=\; \operatorname{Arctan} \frac{\operatorname{Im} F}{t - \operatorname{Re} F} \tag{8.15}$$

with Log the principal branch of that function. The value of $\phi_t(F)$ is the angle depicted in Figure 8.1, which provides a geometric explanation of (8.14). Using the monotone and dominated convergence theorems, it allows us to cast the quantity of interest in the form

$$\begin{aligned} |\{E \in \mathbb{R} \,|\, F(E+i0) \geq t\}| \;&=\; \lim_{\eta\to\infty} \int \frac{\eta^2}{E^2+\eta^2}\; 1[F(E+i0) \geq t]\, dE \\ &=\; \lim_{\eta\to\infty} \int \frac{\pi^{-1}\,\eta^2}{E^2+\eta^2}\, \phi_t\,(F(E+i0))\; dE \\ &=\; \lim_{\eta\to\infty}\, \lim_{\varepsilon\downarrow 0}\; \eta \int \frac{\pi^{-1}\,\eta}{E^2+\eta^2}\, \phi_t\,(F(E+i\varepsilon))\; dE\,. \end{aligned} \tag{8.16}$$

To evaluate the last integral, we note that for any $\varepsilon > 0$ the function $z \mapsto \phi_t\,(F(z+i\varepsilon))$ is bounded on $\mathbb{C}^+$ and harmonic there (being the imaginary part of an analytic function). Recognizing that the other factor in the

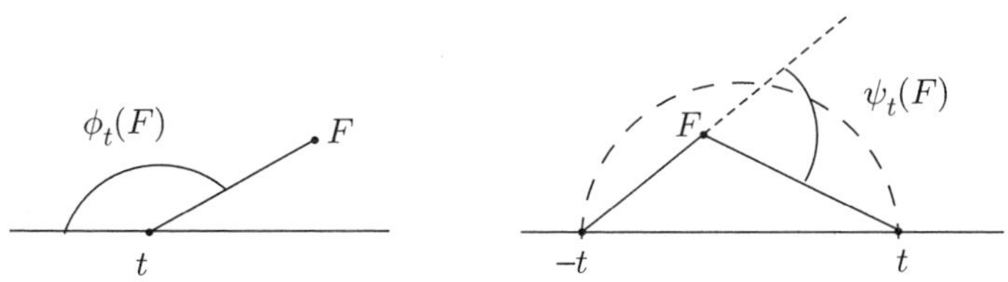

Figure 8.1. The two harmonic functions used in the proof of Proposition 8.2.

integrand is the Poisson kernel, one gets

$$\int \frac{\pi^{-1}\eta}{E^2+\eta^2}\,\phi_t\left(F(E+i\varepsilon)\right)dE \;=\; \phi_t\left(F(i(\eta+\varepsilon))\right). \tag{8.17}$$

The last function is continuous in $\eta > 0$ and thus in the limit $\varepsilon \downarrow 0$ the right side converges to $\phi_t\left(F(i\eta)\right)$. The successive limit $\eta \to \infty$, determined through the explicit expression for $\phi_t(F)$ in (8.15) and the limiting behavior of F (8.13), yields for any $t \neq 0$,

$$\lim_{\eta\to\infty} \eta\,\phi_t\left(F(i(\eta+\varepsilon))\right) = \lim_{\eta\to\infty} \frac{\eta\,\mathrm{Im}\,F(i\eta)}{t} = \frac{\mu(\mathbb{R})}{t}, \tag{8.18}$$

which proves the claim (8.11).

To prove (8.10) we proceed similarly and write

(8.19)
$$|\{E\in\mathbb{R}\,|\,|F(E+i0)|\geq t\}| \;=\; \lim_{\eta\to\infty}\lim_{\varepsilon\downarrow 0}\int \frac{\eta^2}{E^2+\eta^2}\,1[|F(E+i\varepsilon)|\geq t]\,dE\,.$$

Considering the angle view of the segment $[-t,t]$ from points $F\in\mathbb{C}^+$, we define the following bounded harmonic function over $\mathbb{C}^+$:

(8.20)
$$\psi_t(F) := \;\pi - \mathrm{Im}\,\mathrm{Log}\left(\frac{F-t}{F+t}\right) \;=\; \mathrm{Arctan}\,\frac{\mathrm{Im}\,F}{t-\mathrm{Re}\,F} \;+\; \mathrm{Arctan}\,\frac{\mathrm{Im}\,F}{t+\mathrm{Re}\,F}\,.$$

This angle equals $\pi/2$ on the upper semicircle, where $\mathrm{Im}\,F>0$ and $|F|=t$, and thus it provides an **upper bound** on the indicator function:

$$1[|F|\geq t] \;\leq\; \frac{2}{\pi}\,\psi_t(F) \qquad \text{for} \quad \mathrm{Im}\,F>0\;. \tag{8.21}$$

The arguments used in (8.17) in this case yield the inequality

$$\begin{aligned} |\{E\in\mathbb{R}\,|\,|F(E+i0)|\geq t\}| \;&\leq\; 2\lim_{\eta\to\infty}\lim_{\varepsilon\downarrow 0}\int \frac{\pi^{-1}\eta^2}{E^2+\eta^2}\,\psi_t\left(F(E+i\varepsilon)\right)dE \\ &=\; \lim_{\eta\to\infty} 2\,\psi_t\left(F(i\eta)\right) \;=\; 4\,\frac{\mu(\mathbb{R})}{t}, \end{aligned} \tag{8.22}$$

thus yielding (8.13). □

The weak L^1-estimate (8.10) implies finiteness of the factional moments:

$$\boxed{\int_I |F(E+i0)|^s \, dE \leq \frac{(4\mu(\mathbb{R}))^s}{1-s} \, |I|^{1-s}} \tag{8.23}$$

for any $s \in (0,1)$ and any bounded $I \subset \mathbb{R}$. The proof employs the layer-cake representation:

$$\int_I |F(E+i0)|^s \, dE \;=\; \int_0^\infty \int_I 1[|F(E+i0)| \geq t] dE \, d(t^s)\,. \tag{8.24}$$

Bounding the inner integral by $|I|$ for $t \leq 4\mu(\mathbb{R})/|I|$ and by (8.10) for $t > 4\mu(\mathbb{R})/|I|$, one gets (8.23).

8.3. Extension to the resolvent's off-diagonal elements

Theorem 8.1 does not yet address the FMs of the off-diagonal terms of the Green function at specified E. A bound on such terms is obtained upon the integration over the corresponding **pair of random variables**, associated with the two sites.

Theorem 8.3 (Finiteness of FM II). *Let H_0 be a self-adjoint operator in $\ell^2(\mathbb{G})$ and suppose $\rho_2(dudv)$ is a probability measure on $\mathbb{R}^2$ whose conditional distributions are uniformly τ-Hölder continuous with some $\tau \in (0,1]$; i.e., for some $C < \infty$ and all $u, v \in \mathbb{R}$ and $t \geq 0$,*

$$\rho_2\left((v-t, v+t) \;\middle|\; u\right) \leq Ct^\tau\,, \qquad \rho_2\left((v-t, v+t) \;\middle|\; v\right) \leq Ct^\tau\,. \tag{8.25}$$

Then the Green function of

$$H_{u,v} := H_0 + u\, 1_{\{y\}} + v\, 1_{\{x\}}$$

acting in $\ell^2(\mathbb{G})$ satisfies the following for any $z \in \mathbb{C}\backslash\mathbb{R}$:

1. *For all $t > 0$,* $\quad \rho_2\left(\{(u,v) \in \mathbb{R}^2 \mid |G_{u,v}(x,y;z)| > t\}\right) \;\leq\; C\, 2^{\tau+1}\, t^{-\tau}$.
2. *For all $s \in (0,\tau)$,*

$$\int |G_{u,v}(x,y;z)|^s \, \rho_2(dudv) \;\leq\; \frac{\tau}{\tau - s}\left(2^{\tau+1}C\right)^{\frac{s}{\tau}} \;=\; 2^{\frac{s}{\tau}(\tau+1)} C_{s,\tau}\,. \tag{8.26}$$

Proof. The proof is based on the rank-two version of the Krein formula (5.48):

$$P\,(H_{u,v} - z)^{-1} P \;=\; \left[\begin{pmatrix} u & 0 \\ 0 & v \end{pmatrix} + A(z)\right]^{-1}, \tag{8.27}$$

where $A(z) =: \begin{pmatrix} \alpha & \gamma \\ \delta & \beta \end{pmatrix}$ is the 2×2-matrix which coincides with the inverse of $P(H_0 - z)^{-1}P$ on $P\ell^2(\mathbb{G}) := \operatorname{span}\{\delta_x, \delta_y\}$. We may assume without loss of generality that $z \in \mathbb{C}^+$ and thus $\operatorname{Im} A(z) \le 0$, or, equivalently, $\operatorname{Im}\alpha \le 0$ and

(8.28)
$$\det(\operatorname{Im} A(z)) = \operatorname{Im}\alpha\ \operatorname{Im}\beta + \tfrac{1}{4}(\gamma - \overline{\delta})(\delta - \overline{\gamma}) = \operatorname{Im}\alpha\ \operatorname{Im}\beta - \tfrac{1}{4}|\gamma - \overline{\delta}|^2 \ge 0 .$$

Using Camer's rule and the bound $|w|^{-1} \le (\max\{|\operatorname{Re} w|, |\operatorname{Im} w|\})^{-1}$ for all $w \in \mathbb{C}\backslash\{0\}$, one gets

$$(8.29)\qquad \begin{aligned} |G_{u,v}(x,y;z)| &= \frac{|\gamma|}{|(u-\alpha)(v-\beta) - \gamma\delta|} \\ &\le \frac{|\gamma|}{\max\{|\hat{u}\,\hat{v} - \sigma|\,,\ |\hat{u}\operatorname{Im}\beta + \hat{v}\operatorname{Im}\alpha + \operatorname{Im}(\gamma\delta)|\}}\,, \end{aligned}$$

where $\hat{u} := u - \operatorname{Re}\alpha$, $\hat{v} := v - \operatorname{Re}\beta$, and

$$(8.30)\qquad \sigma := \operatorname{Im}\alpha \operatorname{Im}\beta + \operatorname{Re}(\gamma\delta) \ge \tfrac{1}{2}\left(|\gamma|^2 + |\delta|^2\right) - \tfrac{1}{4}|\gamma - \overline{\delta}|^2 \ge 0 .$$

The inequality is due to (8.28). We now distinguish two cases.

If $|\gamma| \ge |\gamma - \overline{\delta}|$, then $\sigma \ge |\gamma|^2/4$ and thus using (8.29)

$$(8.31)\quad \begin{aligned} &\rho_2\left(\{(u,v) \in \mathbb{R}^2 \mid \ |G_{u,v}(x,y;z)| > t\}\right) \\ &\quad\le\ \rho_2\left(\{(u,v) \in \mathbb{R}^2 \mid \ |\hat{u}\,\hat{v} - \sigma| \ <\ 2\sqrt{\sigma}/t\}\right) \\ &\quad\le\ \rho_2\left(\{(u,v) \in \mathbb{R}^2 \mid \ \min\{|\hat{u} - f_\sigma(\hat{v})|\,, |\hat{v} - f_\sigma(\hat{u})|\}\ <\ 2/t\}\right) \\ &\quad\le\ 2C\,2^\tau/t^\tau . \end{aligned}$$

Here $f_\sigma(x) := \frac{\sigma}{x}\min\{1, \frac{x^2}{\sigma}\}$ and the second inequality is a consequence of the estimate

$$(8.32)\qquad |\hat{u}\,\hat{v} - \sigma| \ge \sqrt{\sigma}\min\{|\hat{u} - f_\sigma(\hat{v})|\,, |\hat{v} - f_\sigma(\hat{u})|\} .$$

The last inequality in (8.31) then follows from the assumed uniform τ-Hölder continuity of the conditional distributions. For a proof of (8.32) we distinguish the cases $\hat{u}^2 \ge \sigma$ for which $\sqrt{\sigma}|\hat{v} - f_\sigma(\hat{u})| \le |\hat{u}\,\hat{v} - \sigma|$, and analogously for $\hat{v}^2 \ge \sigma$, and the case $\hat{u}^2, \hat{v}^2 \le \sigma$, for which $|\hat{u}\,\hat{v} - \sigma|^2 = \sigma|\hat{u} - \hat{v}|^2 + (\sigma - \hat{u}^2)(\sigma - \hat{v}^2) \ge \sigma|\hat{u} - \hat{v}|^2$.

If $|\gamma| < |\gamma - \overline{\delta}|$, then $|\gamma| < 2\sqrt{\operatorname{Im}\alpha\ \operatorname{Im}\beta}$, by (8.28), and consequently (8.29) yields

(8.33)
$$\begin{aligned} &\rho_2\left(\{(u,v) \in \mathbb{R}^2 \mid \ |G_{u,v}(x,y;z)| > t\}\right) \\ &\quad\le\ \rho_2\left(\left\{(u,v) \in \mathbb{R}^2 \mid |\hat{u}\operatorname{Im}\beta + \hat{v}\operatorname{Im}\alpha + \operatorname{Im}(\gamma\delta)|\ <\ 2\sqrt{\operatorname{Im}\alpha\ \operatorname{Im}\beta}/t\right\}\right) \\ &\quad\le\ C\,2^\tau/t^\tau . \end{aligned}$$

The last inequality follows from the assumed uniform τ-Hölder continuity of the conditional distribution of that coefficient, u or v, which multiplies the maximum of $\operatorname{Im}\alpha$ and $\operatorname{Im}\beta$. This completes the proof of the first assertion.

The second assertion follows from the first by the layer-cake representation similarly to (8.6). □

The proof of the existence of the off-diagonal moments which we presented goes back to [**178**]. There are several other techniques proving the same result under additional assumptions. One particularly elegant one, which only works for finite graphs though, is the subject of Exercise 8.4. Other references can be found in the Notes.

For random operators the above theorem can be recast in the following form.

Corollary 8.4. *Let V be a random potential whose conditional single-site distributions are uniformly τ-Hölder continuous for some $\tau \in (0,1]$ and consider the random operator* (8.3)*. Then for any $s \in (0,\tau)$ and $|\lambda| > 0$,*

$$\boxed{\sup_{z\in\mathbb{C}\backslash\mathbb{R}} \mathbb{E}\left[|G(x,x;z)|^s \,|\, V_{\neq x}\right] \leq \frac{C_{s,\tau}}{|\lambda|^s}}\,, \tag{8.34}$$

$$\boxed{\sup_{z\in\mathbb{C}\backslash\mathbb{R}} \mathbb{E}\left[|G(x,y;z)|^s \,|\, V_{\neq x,y}\right] \leq \frac{C_{s,\tau}}{|\lambda|^s}\, 2^{\frac{s}{\tau}(\tau+1)}}\,, \tag{8.35}$$

where the conditional expectation associated with the random variables at $x, y \in \mathbb{G}$ is abbreviated as $\mathbb{E}[\,\cdot\,] := \mathbb{E}\big[\;\cdot\;\big|V_{\neq x,y}\big]$.

The above results have been formulated for Green functions at complex energies only. They extend to real energies in the setting of Corollary 8.4 since for all $x, y \in \mathbb{G}$, all $s \in (0,\tau)$, and Lebesgue-almost all $E \in \mathbb{R}$ one has the almost-sure identity:

$$\lim_{\eta\downarrow 0} \mathbb{E}\left[|G(x,y;E+i\eta)|^s \;|\, V_{\neq x,y}\right] = \mathbb{E}\left[|G(x,y;E+i0)|^s \;|\, V_{\neq x,y}\right]. \tag{8.36}$$

Note that the limit on the right side almost surely exists for Lebesgue-almost all $E \in \mathbb{R}$.

In the case of finite graphs $\mathbb{G}$, for which the operators have only finitely many spectral values on the real axis, the fractional moments (8.8) and (8.26) exist and are bounded for almost all $z \in \mathbb{R}$. In fact, this holds for all $z \in \mathbb{R}$ in this case since the fractional moments have a continuous extension to the real axis (Exercise 8.1).

8.4.* Decoupling inequalities

A key step in various proofs of localization is the decoupling of expectation values of products of Green functions, or of a product of the Green function with a local value of the potential. In certain situations, we get away with geometric decoupling: a term used when some of the terms depend in an essential way on variables which do not appear in others. For other situations the bounds presented next will be of help.

Since the Green function is a fractional linear function of the individual potential values, the following condition on the probability distribution of $V(x)$ is of relevance for decoupling the conditional expectation of $|G_\Lambda(x,x;z)|^s$ and $|V(x)|^s$, at a given vertex x.

Definition 8.5. A probability distribution ρ on $\mathbb{R}$ is said to be **σ-moment regular** if for all $s \in (0,\sigma)$ the fractional integrals

$$\phi_s(\gamma) := \int \frac{1}{|v-\gamma|^s}\,\rho(dv)\,, \qquad \psi_s(\gamma) := \int \frac{|v|^s}{|v-\gamma|^s}\,\rho(dv) \tag{8.37}$$

with $\gamma \in \mathbb{C}$ are of uniformly bounded ratio:

$$D_{s,1}(\rho) := \sup_{\gamma\in\mathbb{C}} \frac{\psi_s(\gamma)}{\phi_s(\gamma)} < \infty\,. \tag{8.38}$$

The condition is trivially satisfied if ρ is of bounded support. For the unbounded case, we have the following pair of conditions.

Definition 8.6. A probability distribution ρ on $\mathbb{R}$ is said to be **locally uniformly τ-Hölder continuous** with $\tau \in (0,1]$ if there is some $C < \infty$ such that for all $u \in \mathbb{R}$ and all $\delta \in (0,1]$

$$\rho([u-\delta,u+\delta]) \le C\,\delta^\tau\,\rho([u-1,u+1])\,. \tag{8.39}$$

It is said to have a **regular q-decay** with $q \ge 1$ if there is some $C < \infty$ such that for all $u \in \mathbb{R}$

$$\rho([u-1,u+1]) \le \frac{C}{1+|u|^q}\,. \tag{8.40}$$

In case $q = 1$ we just call the decay regular.

Our main sufficiency criterion for σ-moment regularity is

Theorem 8.7 (Decoupling I). *Let ρ be a locally uniformly τ-Hölder continuous probability measure on $\mathbb{R}$ with $\tau \in (0,1]$ and suppose that there is $\sigma \in (0,\tau]$ such that*

i. *$\int |v|^s \rho(dv) < \infty$ for all $s \in (0,\sigma)$ and*
ii. *ρ has a regular $(1+\sigma)$-decay.*

Then ρ is σ-moment regular. Moreover, if ρ is of bounded support, then the strict inequality holds:

$$D_{s,1}(\rho) < \sup\{|v|^s \,|\, v \in \operatorname{supp} \rho\} \,. \tag{8.41}$$

Another variant on the decoupling theme allows us to decouple integrals which also involve functions of the form

$$\gamma_s(\alpha, \beta, \gamma) := \int \frac{|v-\alpha|^s}{|v-\beta|^s \, |v-\gamma|^s} \, \rho(dv) \tag{8.42}$$

with $\alpha, \beta, \gamma \in \mathbb{C}$ and

$$\xi_s(\alpha, \beta) \;:=\; \int \frac{|v-\alpha|^s}{|v-\beta|^s} \, \rho(dv) \,. \tag{8.43}$$

Theorem 8.8 (Decoupling II). *Let ρ be a locally uniformly τ-Hölder continuous probability measure on $\mathbb{R}$ for some $\tau \in (0,1]$ and fix $s \in (0, \tau/2)$ such that*

i. *$\int |v|^{2s} \rho(dv) < \infty$ and*
ii. *ρ has a regular $(1+2s)$-decay.*

Then

$$D_{s,2}(\rho) \;:=\; \sup_{\alpha,\beta,\gamma \in \mathbb{C}} \frac{\gamma_s(\alpha,\beta,\gamma)}{\xi_s(\alpha,\beta) \, \phi_s(\gamma)} \;<\; \infty \,. \tag{8.44}$$

The rest of this section is devoted to the proof of the above two statements. To rank the pair of requirements in Definition 8.6, one may note the following:

1. Any locally uniformly τ-Hölder continuous measure ρ is uniformly τ-Hölder continuous in the sense of Definition 2.2. In case ρ has compact support the converse is also true. Therefore, (8.39) essentially puts an additional requirement on the tail of the distribution. In case $\rho(dv) = \varrho(v)\, dv$ with $\varrho \in L^\infty(\mathbb{R})$, then (8.39) holds with $\tau = 1$ provided that its density $\varrho(v)$ is monotone decreasing for all sufficiently large $|v|$; cf. Exercise 8.5.

2. By the finiteness of the measure, it suffices to check (8.40) for sufficiently large $|u|$. The following sufficient conditions show that regular q-decay is a rather mild requirement:
   i. In case $\rho(dv) = \varrho(v)\, dv$, a sufficient condition for (8.40) is the bound $|\varrho(v)| \le C/(1+|v|^q)$. Any bounded density which is eventually monotone decreasing satisfies this inequality with $q = 1$.

ii. If ρ has a qth moment, i.e.,

$$m_\rho(q) := \int |v|^q \, \rho(dv) < \infty\,, \tag{8.45}$$

then (8.40) is implied through the bound

$$\rho\left([u-1,u+1]\right) \le m_\rho(q)/(|u|-1)^q \qquad \text{for } |u|>1.$$

Hence, within the class of absolutely continuous measures with bounded density, the 1-moment regular distributions include all compactly supported ones and all densities which are eventually monotone decreasing at a rate faster than 2. In particular, this covers the Cauchy as well as the Gaussian distribution.

The proof of Theorem 8.7, which will be given below, concerns the functions $\phi_s(\gamma)$ and $\psi_s(\gamma)$. The uniform boundedness in γ of their ratio will be deduced through the following two properties:

1. **continuity** as a function of $\gamma \in \mathbb{C}$ and
2. **asymptotic bounds** at infinity:

$$0 < \liminf_{|\gamma|\to\infty} |\gamma|^s \phi_s(\gamma) \le \limsup_{|\gamma|\to\infty} |\gamma|^s \phi_s(\gamma) < \infty\,, \tag{8.46}$$

and similarly for $\psi_s(\gamma)$.

Since the integral $\psi_s(\gamma)$ is of a similar nature, we focus the discussion by addressing these properties for $\phi_s(\gamma)$.

Lemma 8.9. *Let ρ be a uniformly τ-Hölder continuous, finite measure on $\mathbb{R}$ for some $\tau \in (0,1]$. Then the function $\mathbb{C} \ni \gamma \mapsto \phi_s(\gamma)$ is continuous for all $s \in (0,\tau)$ and there is some $c > 0$ such that*

$$\liminf_{|\gamma|\to\infty} |\gamma|^s \phi_s(\gamma) \;\ge\; c \int \rho(dv)\,. \tag{8.47}$$

If ρ is even locally uniformly τ-Hölder continuous and has a regular decay, then for all $s \in (0,\tau)$,

$$\limsup_{|\gamma|\to\infty} |\gamma|^s \, \phi_s(\rho) < \infty\,. \tag{8.48}$$

Proof. The continuity of ϕ_s is established be estimating the singular part of the integral. For any $\varepsilon > 0$,

$$\begin{aligned}\int \frac{1[|v-\gamma|\le\varepsilon]}{|V-\gamma|^s}\,\rho(dv) &= \int_0^\infty \frac{s}{t^{1+s}}\,\rho\left\{|v-\gamma| \le \min(t,\varepsilon)\right\}\,dt \\ &\le \frac{\tau\, C}{\tau - s}\,\varepsilon^{\tau - s}\,.\end{aligned} \tag{8.49}$$

Since the remaining integral, $\int \frac{1[|v-\gamma|>\varepsilon]}{|v-\gamma|^s}\,\rho(dv)$, is seen to be continuous in γ by the dominated convergence theorem, this concludes the proof of continuity.

For a proof of the limiting relation (8.47), we pick $R_\rho \in (0,\infty)$ large enough such that $\int 1[|v| \leq R_\rho]\,\rho(dv) \geq \frac{1}{2}\int \rho(dv)$ and estimate

$$\phi_s(\gamma) \geq \int \frac{1[|v| \leq R_\rho]}{|\gamma - v|^s}\,\rho(dv) \;\geq\; \frac{1}{|\gamma|^s + R_\rho^s}\int 1[|v| \leq R_\rho]\,\rho(dv)\,. \tag{8.50}$$

This competes the proof of (8.47).

For a proof of (8.48) we write the integral $\phi_s(\gamma)$ in layer-cake form:

$$\begin{aligned}\phi_s(\gamma) &= \int_0^\infty \frac{s}{t^{1+s}}\,\rho(|v-\gamma|<t)\,dt \\ &\leq \int_0^{T_\gamma} \frac{s}{t^{1+s}}\rho(|v-\gamma|<t)\,dt + \frac{\int \rho(dv)}{T_\gamma^s}\,,\end{aligned} \tag{8.51}$$

where we introduced $T_\gamma := |\gamma|/2$ and estimated the contribution to the second integral using the trivial bound $\rho(|v-\gamma|<t) \leq \rho(\mathbb{R})$. For any $t \geq 0$ the set $\{v \in \mathbb{R} \,|\, |v-\gamma| \leq t\}$ is an open interval centered about $\operatorname{Re}\gamma$ of width twice

$$R_{\gamma,t} := \sqrt{[t^2 - (\operatorname{Im}\gamma)^2]_+}\,, \tag{8.52}$$

where the plus indicates the positive part. We now distinguish two cases.

If $\operatorname{Im}\gamma \geq T_\gamma$, then $R_{\gamma,t} = 0$ for all $t \in [0, T_\gamma]$ and hence the first integral on the right in (8.51) is zero.

If $\operatorname{Im}\gamma \leq T_\gamma$, then $(\operatorname{Im}\gamma)^2 \leq \frac{1}{4}(\operatorname{Im}\gamma)^2 + \frac{1}{4}(\operatorname{Re}\gamma)^2$ and hence

$$|\operatorname{Re}\gamma| \geq \sqrt{3}\,T_\gamma\,. \tag{8.53}$$

For any $t \in [0, T_\gamma]$ we may then combine (8.39) and (8.40) with $q = 1$ to estimate

$$\rho(|v-\gamma|<t) \leq \frac{C}{|\operatorname{Re}\gamma| - R_{\gamma,t}}\max\{t^\tau, t\} \;\leq\; \frac{c}{T_\gamma}\max\{t^\tau, t\}\,. \tag{8.54}$$

Here the last inequality results from (8.53) and the facts that $R_{\gamma,t} \leq t \leq T_\gamma$ and that we abbreviated $c := C/(\sqrt{3}-1)$. Inserting this estimate into the first integral on the right in (8.51) we conclude the following:

$$\int_0^{T_\gamma} \frac{s}{t^{1+s}}\,\rho(|v-\gamma|<t)\,dt \leq \frac{c\,s}{\tau - s}\frac{1}{T_\gamma^{s+1-\tau}} + \frac{c\,s}{1-s}\frac{1}{T_\gamma^s}\,. \tag{8.55}$$

This concludes the proof. □

We are now ready for the proof of the main sufficiency criterion.

Proof of Theorem 8.7. We fix $s \in (0, \sigma)$. The proof rests on the fact that under the present assumption the measure

$$\rho_s(dv) := |v|^s \rho(dv) \tag{8.56}$$

enjoys the following properties:

i. ρ_s is locally uniformly τ-Hölder continuous. For a proof, we split ρ_s into a compactly supported part in the ball of radius two and a remainder $\hat{\rho}_s$. The first part is locally uniformly τ-Hölder continuous since it is uniformly τ-Hölder continuous; cf. Exercise 8.5. To establish this property for $\hat{\rho}_s$ we need to consider $|u| \geq 2$ and estimate for any $\delta \in (0, 1]$:

$$\begin{aligned} \hat{\rho}_s\left([u-\delta, u+\delta]\right) &\leq (|u|+1)^s \rho\left([u-\delta, u+\delta]\right) \\ &\leq C\,\delta^\tau\ (|u|+1)^s \rho\left([u-1, u+1]\right) \\ &\leq C\,\delta^\tau \sup_{|v|\geq 2} \left(\frac{|v|+1}{|v|-1}\right)^s \hat{\rho}_s\left([u-1, u+1]\right) . \end{aligned} \tag{8.57}$$

ii. ρ_s has a regular decay since for any $u \in \mathbb{R}$

$$\rho_s([u-1, u+1]) \leq (1+|u|)^s\ \rho([u-1, u+1]) \leq \frac{C}{1+|u|} . \tag{8.58}$$

Through Lemma 8.9 these properties ensure that

$$\limsup_{|\gamma|\to\infty} |\gamma|^s\, \psi_s(\gamma) < \infty . \tag{8.59}$$

The main assertion (8.41) will now follow from the facts that under the present assumptions the ratio

$$r_s(\gamma) := \frac{\psi_s(\gamma)}{\phi_s(\gamma)}$$

is a continuous function of $\gamma \in \mathbb{C}$ and that there is some $R_s < \infty$ such that $r_s(\gamma)$ is uniformly bounded for all $|\gamma| \geq R_s$. This is implied by (8.47) and (8.59).

The continuity of $r_s(\gamma)$ in $\gamma \in \mathbb{C}$ is established by the continuity of the numerator and denominator on $\mathbb{C}$ and the fact that $\phi_s(\gamma) \geq (\int |v|^s \rho(dv) + |\gamma|^s)^{-1} > 0$. The continuity of ϕ_s follows directly from Lemma 8.9. The latter also implies the continuity of ψ_s since ρ_s is uniformly τ-Hölder continuous.

This completes the proof of the finiteness of $D_{s,1}(\rho)$.

In case $\operatorname{supp}\rho$ is bounded, it is straightforward to check that

$$\lim_{|\gamma|\to\infty} r_s(\gamma) = \int |v|^s \rho(dv)\,. \tag{8.60}$$

The function r_s hence extends to a continuous function on the one-point compactification of $\mathbb{C}$. In order to establish that $D_{s,1}(\rho) < \sup\{|v|^s \,|\, v \in \operatorname{supp}\rho\} =: v_0^s$, we proceed by contradiction and assume equality. The continuity implies that $D_{s,1}(\rho) = r_s(\gamma_0)$ for some $\gamma_0 \in \mathbb{C}\cup\{\infty\}$. In case $\gamma_0 = \infty$ where $\gamma_0^s = r_s(\gamma_0) = \int |v|^s\rho(dv)$, the function ρ is concentrated on $|v| = v_0$ which contradicts its Hölder continuity. In case $\gamma_0 \in \mathbb{C}$, we have $0 = \int (v_0^s - |v|^s)\,|v-\gamma_0|^s\rho(dv)$, which also implies that ρ is concentrated on $|v| = v_0$. □

For future reference we also add the following comparison principle.

Lemma 8.10. *Let ρ be a locally uniformly τ-Hölder continuous, finite measure on $\mathbb{R}$ for some $\tau\in(0,1]$. Then for any $0<t<s<\tau$,*

$$C_{s,t}(\rho) := \sup_{\gamma\in\mathbb{C}} \frac{\phi_s(\gamma)}{\phi_t(\gamma)} < \infty\,. \tag{8.61}$$

Proof. We split the integral $\phi_s(\gamma)$ into

$$\int \frac{1[|v-\gamma|>1]}{|v-\gamma|^s}\rho(dv) \le \int \frac{1[|v-\gamma|>1]}{|v-\gamma|^t}\rho(dv), \tag{8.62}$$

and the remaining part, which in a computation similar to (8.49), is estimated with the help of (8.39):

$$\begin{aligned}\int \frac{1[|v-\gamma|\le 1]}{|v-\gamma|^s}\rho(dv) &= \int_0^\infty \frac{s}{t^{1+s}}\,\rho\{|v-\gamma|\le \min(t,1)\}\,dt\\ &\le \frac{\tau}{\tau-s}\,C\rho\{|v-\gamma|\le 1\}\,.\end{aligned} \tag{8.63}$$

This term should be compared to

$$\int \frac{1[|v-\gamma|\le 1]}{|v-\gamma|^t}\rho(dv) \;\ge\; \rho\{|v-\gamma|\le 1\}\,, \tag{8.64}$$

which completes the proof of (8.61). □

We may turn now to the second decoupling principle:

Proof of Theorem 8.8. The Cauchy-Schwarz inequality yields

$$\gamma_s(\alpha,\beta,\gamma) \le \sqrt{\xi_{2s}(\alpha,\beta)\,\phi_{2s}(\gamma)}. \tag{8.65}$$

Hence it is sufficient to uniformly bound the fractions

$$(\alpha,\beta)\mapsto \frac{\sqrt{\xi_{2s}(\alpha,\beta)}}{\xi_s(\alpha,\beta)}, \quad \gamma\mapsto \frac{\sqrt{\phi_{2s}(\gamma)}}{\phi_s(\gamma)}\,. \tag{8.66}$$

This is done by proving that the fractions (i) are continuous functions on $(\alpha, \beta) \in \mathbb{C} \times \mathbb{C}$ and $\gamma \in \mathbb{C}$ and (ii) are uniformly bounded for $|\alpha|$, $|\beta|$, and $|\gamma|$ large enough.

The continuity of ξ_s on $\mathbb{C} \times \mathbb{C}$ is established along the lines of the proof of Lemma 8.9 by estimating the singular part in the integrals using (8.49).

The uniform boundedness of $\sqrt{\phi_{2s}(\gamma)}/\phi_s(\gamma)$ is established using (8.47) and (8.48). In turn, the uniform boundedness of the other fraction follows from the estimate

$$0 < \liminf_{|\alpha|,|\beta| \to \infty} \frac{|\beta|^\sigma}{|\alpha|^\sigma} \xi_\sigma(\alpha, \beta) \leq \limsup_{|\alpha|,|\beta| \to \infty} \frac{|\beta|^\sigma}{|\alpha|^\sigma} \xi_\sigma(\alpha, \beta) < \infty \,, \tag{8.67}$$

valid for all $\sigma \in [0, 2s]$. This is shown using

$$|\xi_\sigma(\alpha, \beta) - |\alpha|^\sigma \phi_\sigma(\beta)| \ \leq \ \psi_\sigma(\beta) \leq D_{\sigma,1}(\rho)\, \phi_\sigma(\beta) \tag{8.68}$$

for all $\alpha, \beta \in \mathbb{C}$, which relied on Theorem 8.7. □

For example, an application of Theorem 8.8 which will play some role in a subsequent chapter is

Corollary 8.11. *Let H be a random operator of the form* (8.3) *with an iid random potential whose single-site distribution ρ is locally uniformly τ-Hölder continuous for some $\tau \in (0, 1]$ and let $s \in (0, \tau/2)$ such that*

i. $\int |v|^{2s} \rho(dv) < \infty$ *and*
ii. ρ *has a regular* $(1 + 2s)$*-decay.*

Then the conditional averages of the corresponding restricted Green functions averaged over a single potential variable $V(x)$ satisfy

$$\boxed{\begin{aligned} &\mathbb{E}\left[|G_\Lambda(x, y; z)|^s \left|G_{\Lambda'}(x', y'; z)\right|^s \,\middle|\, V_{\neq x}\right] \\ &\quad \leq D_{s,2}\, \mathbb{E}\left[|G_\Lambda(x, y; z)|^s \,\middle|\, V_{\neq x}\right] \mathbb{E}\left[\left|G_{\Lambda'}(x', y'; z)\right|^s \,\middle|\, V_{\neq x}\right] \end{aligned}} \tag{8.69}$$

for any $\Lambda, \Lambda' \subseteq \mathbb{G}$ and all $x, y \in \Lambda$ and $x', y' \in \Lambda'$.

Notes

Boole's original derivation of (8.11) concerned Stieltjes functions of the form

$$F(E) \ = \ \sum_{k=1}^{N} \frac{p_k}{u_k - E}$$

with $p_1, p_2, \ldots \geq 0$ and $u_1 \leq u_2 \leq \cdots \in \mathbb{R}$ [**63**]. The statement has the remarkable implication that the Lebesgue measures of the level sets of F

are independent of the location of the poles $\{u_k\}$ and agree with the trivial case when all $\{u_k\}$ are made to coincide.

Boole's proof proceeds as follows. The set $\{E \in \mathbb{R} : F(E) \geq t\}$ is the union of disjoint intervals $\bigcup_{k=1}^{N} [v_k, u_k)$, with $v_1 \leq \cdots \leq v_N$ the ordered collection of solutions of $F(v) = t$ for $t > 0$. Thus

$$|\{E \in \mathbb{R} : F(E) \geq t\}| = \sum_{k=1}^{N} (u_k - v_k) .$$

The points $\{v_k\}$ form the roots of the polynomial function

$$\begin{aligned} Q(E) = & [t - F(E)] \prod_{n=1}^{N} (E - u_n) = t \prod_{n} (E - u_n) + \sum_{k} p_k \prod_{n \neq k} (E - u_k) \\ = & \; t\, E^N + \left[-t \sum_{n} u_n + \sum_{k} p_k \right] E^{N-1} + \text{lower powers of } E . \end{aligned}$$

In comparison, the product representation of Q in terms of its roots gives

$$Q(E) = t \prod_{n=1}^{N} (E - v_n) = t\, E^N - t \sum_{n} v_n \, E^{N-1} + \text{lower powers of } E .$$

Equating the coefficients of E^{N-1}, one arrives at (8.11).

Yet another proof, which is related to the complex integration argument presented in Section 8.2, can be found in the Appendix of [**22**]. Some of the subject's rich history and more alternative proofs of Proposition 8.2 can be found in the notes of [**360**, Ch. 5].

The finiteness of fractional moments of boundary values of Stieltjes functions extends to the broader class of Herglotz-Pick functions. More is said on the subject in Section B.3; see also [**128**].

Generalizations and variations of the decoupling inequalities can be found in [**8, 137**].

Exercises

8.1. In the settings of Theorems 8.1 and 8.3 suppose additionally that $\mathbb{G}$ is finite. Show that the fractional moments in (8.8) and (8.26) exist for almost all $E \in \mathbb{R}$ and that they extend to continuous functions on $\mathbb{C}$.

8.2. Show how Proposition 8.2 allows deducing (8.2) for all $x, y \in \mathbb{G}$.

8.3. Give a proof of (8.36).

8.4. In the setting of Theorem 8.3 suppose additionally that $\mathbb{G}$ is finite.

a. Show that for all $E \in \mathbb{R}$ and all normalized $\psi \in P\ell^2(\mathbb{G}) := \mathrm{span}\{\delta_x, \delta_y\}$

$$\rho_2\left(\{(u,v) \in \mathbb{R}^2 \mid |\langle \psi, (H_{u,v} - E)^{-1}\psi\rangle| \geq t\}\right) \;\leq\; \mathbb{E}\left[\mathrm{tr}\, P_{(-\frac{1}{t},\frac{1}{t})}(B_{u,v}(E))\right]$$

where $B_{u,v}(E) := \begin{pmatrix} u & 0 \\ 0 & v \end{pmatrix} + A(E)$ and $A(E)$ is some hermitian 2×2-matrix.

b. Give an alternative proof of Theorem 8.3 for real energies $E \in \mathbb{R}$ (with modified constants) based on the Wegner estimate.

c. Show that the assertions of Theorem 8.3 for real energies imply the one for complex energies.
Hint: Use the fact that for any Borel-Stieltjes transform, $|F(z)|^s$ is a subhamonic function of $z \in \mathbb{C}^+$ for any $s \in (0,1)$ and hence

$$|F(z)|^s \;\leq\; \frac{1}{\pi}\int \frac{\mathrm{Im}\, z}{(\mathrm{Re}\, z - E)^2 + (\mathrm{Im}\, z)^2}\, |F(E+i0)|^s \, dE\,.$$

8.5. Let ρ be a uniformly τ-Hölder continuous probability measure on $\mathbb{R}$. Show that ρ is even locally uniformly τ-Hölder continuous in the following special cases:

a. ρ has compact support.

b. ρ is absolutely continuous with a density $\varrho \in L^\infty$ which is monotone decreasing outside some compact interval.

8.6. Let ρ be a probability measure on $\mathbb{R}$ which is uniformly τ-Hölder continuous for some $\tau \in (0,1]$ and consider $s \in (0,\tau)$.

a. Show that

$$D_{3,s}(\rho) \;:=\; \inf_{\gamma \in \mathbb{C}} \frac{\psi_s(\gamma)}{\phi_s(\gamma)} > 0\,.$$

b. In case $\int |v|^q \rho(dv) < \infty$ for some $q > s$, show that there is some increasing function $d_s : [0,\infty) \to (0,\infty)$ which depends on ρ with the property

$$\lim_{u\to\infty} \frac{d_s(u)}{u} \;=\; 1$$

such that for all $u \in \mathbb{R}$ and $\gamma \in \mathbb{C}$

$$\int \frac{|v-u|^s}{|v-\gamma|^s}\,\rho(dv) \;\geq\; d_s(|u|) \int \frac{\rho(dv)}{|v-\gamma|^s}\,.$$

Chapter 9

Strategies for Mapping Exponential Decay

In this section we focus on techniques for mapping localization, i.e., the phase diagram regimes in which the resolvent's off-diagonal elements decay exponentially. Similar challenges are encountered in models of statistical mechanics, such as independent percolation and systems of locally interacting spin variables, each with its relevant two-point function.

The analogy has been useful in guiding one to **commonly effective methods for establishing exponential decay** and also in alerting one to other phenomena. An example of the latter is the possibility of operators which in the bulk exhibit exponential decay yet in some domains may have modes propagating along the surface. These are reminiscent of **whispering gallery modes** (WGM) of which counterparts have also been noted in the behavior of correlations in certain models of statistical mechanics.

9.1. Three models with a common theme

Before we turn to the common mathematical argument let us list a few models with differently defined two-point functions, which can nevertheless be approached by similar means. A common thread is that one deals with a bounded kernel $\tau(x, y)$ on a graph $\mathbb{G}$ which is an expression of **correlations** (or connection in percolation models) arising through short-range interactions, or hopping amplitudes. In the different models this kernel appears as

$$\textbf{random operators:} \quad \tau(x,y;E) = \mathbb{E}(|G(x,y;E+i0)|^s),$$

$$\textbf{percolation:} \quad \tau(x,y;p) = \mathbb{P}_p\begin{pmatrix} x \text{ and } y \text{ are connected by} \\ \text{a path of occupied bonds} \end{pmatrix},$$

$$\textbf{spin models:} \quad \tau(x,y;\beta) = \langle \sigma_x \sigma_y \rangle_\beta .$$

In each case, τ also depends on a control parameter, which for Anderson localization is the energy E, for percolation is the single-bond occupation probability p, or the parameter β of (9.1) below, and for ferromagnetic spin models is the inverse temperature β. The dependence on it will often be omitted in our notation, especially when the discussion concerns the behavior of $\tau(x,y)$ at a fixed value of the corresponding parameter. In each case, the model formulated on an infinite graph $\mathbb{G}$ has finite-volume versions on subgraphs $\Lambda \subset \mathbb{G}$ for which we denote the corresponding function by $\tau_\Lambda(x,y)$.

Since familiarity with statistical mechanics is not assumed here, let us briefly recall the setup of the systems mentioned above.

Independent bond percolation. In independent bond percolation on a graph $\mathbb{G}$ with edge set $E_{\mathbb{G}}$ one assigns to every "bond" (which is an equivalent term to "edge") an independent Bernoulli random variable, declaring the bond to be occupied, or unoccupied, with probabilities $(p, 1-p)$, with $p \in [0,1]$ serving as the model's parameter. In a relevant generalization, the bond occupation probabilities are

$$\mathbb{P}_\beta\,(\text{the bond } [x,y] \text{ is occupied}) = 1 - e^{-\beta J(x,y)} \quad =: \; T_\beta(x,y)\,. \tag{9.1}$$

Percolation is defined as the occurrence of an infinite connected cluster of occupied bonds. The probability of this event is monotone increasing as a function of p (or β), and by Kolmogorov's zero-one law it takes only the values 0 or 1. Thus there exists p_c (or β_c) at which the probability of the occurrence of the infinite cluster changes from 0 to 1. For example, in case $\mathbb{G} = \mathbb{Z}^2$ one has $p_c = 1/2$.

Within a broad range of models throughout the subcritical regime, i.e., $p < p_c$ or $\beta < \beta_c$, the probability for a given site x to be connected to a distance R away decays exponentially fast in R [**7, 298**]. For more information, see [**181**].

Ferromagnetic system of Ising spins. A system of Ising spins, $\sigma_x \in \{\pm 1\}$, associated to the vertices of a graph, $x \in \mathbb{G}$, is characterized by an

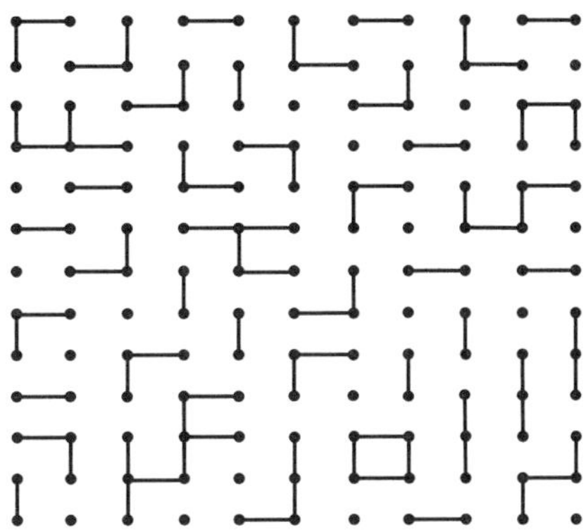

Figure 9.1. A realization of the bond percolation on $\mathbb{Z}^2$ (at $p < p_c$).

energy functional

$$H(\sigma) := - \sum_{[x,y] \in E_{\mathbb{G}}} J(x,y)\, \sigma_x\, \sigma_y\,, \tag{9.2}$$

where $J(x, y) \in \mathbb{R}$ is the coupling between the spin at x and y. The **ferromagnetic** model corresponds to $J(x,y) \geq 0$, in which case alignment of spins is favored. Associated to H is a **Gibb's measure** on the configurations $\sigma = (\sigma_x)_{x \in \mathbb{G}}$ of Ising spins. For finite graphs averages with respect to Gibbs measures are given by

$$\langle \cdot \rangle_\beta := \frac{1}{Z_\beta} \sum_{\sigma \in \{\pm 1\}^{\mathbb{G}}} e^{-\beta H(\sigma)}\, (\cdot)\,, \tag{9.3}$$

where Z_β is a normalization rendering this a probability average. The parameter $\beta \in (0, \infty)$ is physically interpreted as the inverse temperature ($\beta = 1/(kT)$, with k the Boltzmann constant). On an infinite graph, such as $\mathbb{Z}^d$, the Gibbs states are defined through suitable infinite-volume limits. E.g., the two-point function is

$$\tau(x, y; \beta) = \lim_{\Lambda \nearrow \mathbb{G}} \langle \sigma_x \sigma_y \rangle_\beta^{\Lambda,\mathrm{bc}} \tag{9.4}$$

where b.c. indicates the boundary conditions (e.g., free, or + b.c.).

For ferromagnetic models with summably decaying interactions, e.g., the nearest-neighbor case $J(x, y) = \delta_{|x-y|,1}$ on $\mathbb{Z}^d$, there is a critical inverse temperature $\beta_c \leq \infty$, below which there is a unique Gibb's measure and the correlation functions $\tau(x, y; \beta)$ decay exponentially. Above β_c, the Gibb's measure is not unique and this correlation function ceases to decay exponentially. The non-uniqueness expresses the existence of two phases, with positive and negative magnetization, respectively. For more information, see [**168, 353**].

In the following, we shall discuss some general methods for the study of the exponential decay of $\tau(x,y)$ by means of a class of bounds which can be established for each of the above models in suitable regimes of the parameter space. We shall omit here the derivation of the basic inequalities, which requires some model-specific tools, but rather focus on their implications. These include the following:

1. Explicit sufficiency criteria for uniform exponential decay of the two-point function:
$$\boxed{\tau_\Lambda(x,y) \ \le\ A\, e^{-\mu\, \mathrm{dist}(x,y)}} \tag{9.5}$$
with bounds which are uniform in $\Lambda \subset \mathbb{G}$. Examples of such bounds appeared already in a previous chapter, in the proof of exponential localization at strong disorder.
2. Finite-volume criteria for exponential decay which enable one to extend the regime where (9.5) holds, based on additional information, such as Lifshitz tail bounds on the density of states or the results of an auxiliary finite computation.
3. Criteria which allow us to establish exponential localization also for phases where such decay is guaranteed to occur only in the bulk, with WGM possibly occurring in certain subsets of $\mathbb{G}$. A modified version of the bound (9.5) which is compatible with modes with better propagation along the boundary is presented in (9.37).

9.2. Single-step condition: Subharmonicity and contraction arguments

For each of the systems mentioned above, it is not difficult to establish (references are given in the chapter's Notes) that its two-point function $\tau(x,y)$ satisfies a bound of the following form:

B1. A single-step bound: *For any $x, y \in \mathbb{G}$,*
$$\boxed{\tau(x,y) \ \le\ g(x,y) + \sum_{u\in\mathbb{G}\setminus\{x\}} T(x,u)\,\tau(u,y)} \tag{9.6}$$
with g a rapidly decaying non-negative function and T a non-negative kernel.

More specifically:

1. For the independent bond percolation the probability of sites being connected satisfies

$$\mathbb{P}_\beta(x \leftrightarrow y) \;\leq\; \sum_u T_\beta(x,u)\,\mathbb{P}_\beta(u \leftrightarrow y) \qquad \text{for all } x \neq y \tag{9.7}$$

   with $T_\beta(x,u)$ the $[x,u]$-bond-occupation probability. Thus the corresponding function $\tau(x,y)$ satisfies (9.6) with $g(x,y) = \delta_{x,y}$, expressing in this case the fact that in order for x to be connected to y, either $x = y$ or there is a path linking the two which starts with an occupied bond $[x,u]$.

2. For ferromagnetic Ising spin models the corresponding relation is [**156**]

$$\langle \sigma_x \sigma_y \rangle_\beta \;\leq\; \sum_u \tanh(\beta J(x,u))\,\langle \sigma_u \sigma_y \rangle_\beta \qquad \text{for all } x \neq y\,. \tag{9.8}$$

3. For random operators such single-step bounds are derived in the next chapter for the high-disorder regime in (10.7), and for weak disorder or the regime of extremal energies in (10.26).

The single-step bound is informative when

$$\|T\|_{\infty,\infty} \;<\; 1\,. \tag{9.9}$$

It is particularly simple to see the implications in the **nearest-neighbor case**, with $g(x,y) = \delta_{x,y}$ and T a nearest-neighbor hopping operator; i.e.,

$$T(x,y) \;=\; 0 \qquad \text{if } d(x,y) \neq 1\,. \tag{9.10}$$

That includes $T = -\Delta - D_{\mathbb{G}}$ on $\ell^2(\mathbb{G})$, for which

$$\|T\|_{\infty,\infty} = \sup_{x\in\mathbb{G}} \sum_{d(u,x)=1} |T(x,u)| \tag{9.11}$$

is the maximal degree of the graph $\mathbb{G}$. There are a number of ways to extract information from (9.6) under the above assumption. One of the simplest follows, cast in a probabilistic language and for convenience formulated for the nearest-neighbor case.

Lemma 9.1. *For a graph $\mathbb{G}$ and a site $x_0 \in \mathbb{G}$, let $f \in \ell^\infty(\mathbb{G})$ be a bounded positive function which satisfies*

$$\boxed{f(x) \;\leq\; \gamma \sum_{d(u,x)=1} |T(x,u)|\,f(u) \qquad \text{for each } x \neq x_0} \tag{9.12}$$

with T a non-negative kernel of nearest-neighbor type (9.10) *and γ such that $e^{-\mu} \;:=\; \gamma\,\|T\|_{\infty,\infty} <\; 1$. Then*

$$\boxed{f(x) \;\leq\; f(x_0)\; e^{-\mu d(x,x_0)}}\,. \tag{9.13}$$

Proof. For an iteration of the bound (9.12) applied to $f(u)$ at sites other than x_0 it is natural to organize the resulting collection of terms with the help of a simple random walk $(b_0, b_1, \ldots)$ on $\mathbb{G}$ with the transition probabilities

$$\mathbb{P}\left(b_{k+1} = u \,\middle|\, b_k = x\right) \;=\; |T(x,u)| \Big/ \sum_{\mathrm{dist}(v,x)=1} |T(x,v)| \,. \tag{9.14}$$

Let τ be the stopping time, defined as the first time $k \in \mathbb{N}$ at which b_k reaches x_0, or $+\infty$ in case it does not, and let $\tau_N := \min\{\tau, N\}$ be its finite truncation. The bound (9.12) implies that for any $N < \infty$

(9.15)

$$\begin{aligned} f(x) \;&\le \mathbb{E}\left(e^{-\mu\tau_N} f(b_{\tau_N}) \,\middle|\, b_0 = x\right) \\ &\le \mathbb{E}\left(e^{-\mu\tau} f(b_\tau) \,\middle|\, b_0 = x, \tau < N\right) \;+\; e^{-\mu N}\, \mathbb{E}\left(f(b_N) \,\middle|\, b_0 = x, \tau \ge N\right) \,. \end{aligned}$$

By the stopping rule $f(b_\tau) = f(x_0)$ if $\tau < N$, and moreover $\tau \ge d(x, x_0)$ for walks starting at $b_0 = x$. The last term in (9.12) is bounded by $e^{-\mu N}\|f\|_\infty$, which under the assumptions converges to zero. Thus we get

$$\begin{aligned} f(x) \;&\le \lim_{N\to\infty} \mathbb{E}\left(e^{-\mu\tau_N} f(b_{\tau_N}) \middle|\, b_0 = x\right) \\ &\le\; e^{-\mu\,\mathrm{dist}\,(x,x_0)}\, f(x_0)\,, \end{aligned} \tag{9.16}$$

which completes the proof. □

Under the condition $\gamma\|T\|_{\infty,\infty} < 1$ the relation (9.12) implies that the function f has the submean property, which is also referred to as subharmonicity. In the discrete case subharmonicity and the submean property are equivalent, whereas in the continuum the former is usually defined with an extra semicontinuity assumption.

A slightly more effective way to process the the single step bound (9.6), under the condition (9.9), is by building from it **iterated contractions**, as in the proof of the following statement. The improvement are (i) the analysis is not limited to the nearest-neighbor case and (ii) the result is a bound on an increasingly weighted sum, and not just a pointwise estimate. (The second point is particularly valuable in situations where exponential decay alone does not guarantee summability, e.g., on trees.)

Theorem 9.2. *Let $K : \ell^\infty(\mathbb{G}) \to \ell^\infty(\mathbb{G})$, $(Kf)(x) := \sum_{u\in\mathbb{G}} K(x,u)\, f(u)$, be an operator with non-negative kernel $K(x,u) \ge 0$, of a bounded norm:*

$$\|K\|_{\infty,\infty} := \sup_{x\in\mathbb{G}} \sum_{u\in\mathbb{G}} K(x,u) < 1\,. \tag{9.17}$$

Also let $W : \mathbb{G} \mapsto (0, \infty)$ *and* $g \in \ell^\infty(\mathbb{G})$ *be positive functions such that*

(9.18)
$$r := \sup_{y \in \mathbb{G}} \sum_{u \in \mathbb{G}} \frac{W(u)}{W(y)} K(u, y) < 1 \quad \text{and} \quad b := \sum_x W(x) g(x) < \infty .$$

Then any $f \in \ell^\infty(\mathbb{G})$ *which satisfies*

$$0 \le f(x) \le g(x) + (Kf)(x) \qquad \text{for all } x \in \mathbb{G} \tag{9.19}$$

also obeys the bound

$$\boxed{\sum_{x \in \mathbb{G}} W(x) f(x) \le \frac{b}{1-r} \qquad \text{for all } x \in \mathbb{G}} \;. \tag{9.20}$$

The leading example to keep in mind here is

$$\boxed{W(x) = \exp\left(+\mu \operatorname{dist}(x, y)\right)} \tag{9.21}$$

with $y \in \mathbb{G}$ held fixed during the discussion.

Proof. Starting from (9.19) and iterating it N times, one gets

$$f(x) \le \sum_{m=0}^{N} (K^m g)(x) \;+\; (K^{N+1} f)(x) . \tag{9.22}$$

Since $\|K\|_{\infty,\infty} < 1$, in the limit $N \to \infty$ the last term tends to zero for each $x \in \mathbb{G}$; i.e.,

$$f(x) \;\le\; \lim_{N \to \infty} \sum_{m=0}^{N} (K^m g)(x) . \tag{9.23}$$

Applying the monotone convergence theorem and then the assumption (9.18) we may deduce

$$\begin{aligned} \sum_x W(x)\, f(x) \;&\le \lim_{N \to \infty} \sum_{m=0}^{N} \sum_x W(x)\, (K^m g)(x) \\ &\le \lim_{N \to \infty} \sum_{m=0}^{N} b\, r^m \;=\; \frac{b}{1-r} . \end{aligned} \tag{9.24}$$

□

Worth noting is the following implication of (9.20). For every $x \in \mathbb{G}$,

$$f(x) \;\le\; \frac{b}{1-r} \frac{1}{W(x)} . \tag{9.25}$$

An example of such a bound with immediate relevance to our discussion is the exponential decay provided in Corollary 10.1. In the context of the other models mentioned above, the bound proves the exponential decay of $\tau(x,y)$ in the regimes (which by implication are subcritical)

$$2dp < 1, \quad \text{or} \quad \|T_\beta\|_{\infty,\infty} < 1, \qquad \text{for independent percolation}$$

$$\sum_u \tanh(\beta J(0,u)) \geq 1, \qquad \text{for ferromagnetic Ising models}.$$

9.3. Mapping the regime of exponential decay: The Hammersley stratagem

While the single-step bound **B1** under the condition (9.9) allows us to establish exponential decay through an explicit a proiri estimate, improved results can be obtained through suitable finite-volume computations. For that it is useful to have the following type of non-linear relation, which reflects the effect of a separating surface on correlations which are created through a local mechanism.

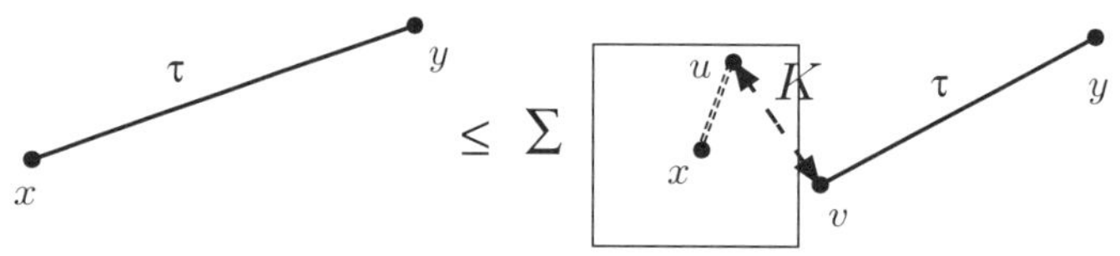

Figure 9.2. Schematic depiction of the sum in bound **B2**. It is significant here that in the factor $\tau_B(x,u)$ (i) the argument is a pair of sites which are far apart and (ii) the two-point function refers to just a finite system.

For the separating surfaces we shall take here the boundaries of balls of radius L with respect to the graph metric: $B_L(x) := \{y \in \mathbb{G} \,|\, d(x,y) \leq L\}$, and for any $S \subset \mathbb{G}$ we denote by $\tau_S(x,u)$ the two-point function τ for the systems whose interactions, or connections, are restricted to the set $S \subset \mathbb{G}$.

B2. Separating surface condition: *For a graph with metric d suppose that for some $L < \infty$ and any $x \in \mathbb{G}$ and $y \notin B_L(x)$,*

$$\boxed{\tau(x,y) \leq \sum_{\substack{u \in B_L(x) \\ v \in \mathbb{G}\setminus B_L(x)}} \tau_{B_L(x)}(x,u)\, K_{x,L}(u,v)\, \tau(v,y)} \tag{9.26}$$

with $K_{x,L}(u,v)$ a non-negative kernel.

The bound **B2** allows us to conclude exponential decay if the sum

$$\gamma_L(\mu) := \sup_{x \in \mathbb{G}} \sum_{\substack{u \in B_L(x) \\ v \in \mathbb{G}\setminus B_L(x)}} \tau_{B_L(x)}(x,u)\, K_{x,L}(u,v)\, e^{\mu\, d(x,v)} \tag{9.27}$$

satisfies the following: **(i)** $\gamma_L(0) < 1$ and **(ii)** $\gamma_L(\mu) < \infty$ for some $\mu > 0$. The full statement is presented below in Theorem 9.3.

An important observation underlying this approach to the mapping of regimes of exponential decay is that in the presence of such, one would have $\lim_{L\to\infty} \gamma_L(0) = 0$, and thus the condition $\gamma_L(0) < 1$ would be met at some $L < \infty$. In essence, up to finite corrections, the value where $\gamma_L(0) \approx e^{-1}$ should provide the length scale ξ for the exponential decay. (If desired, one may also get sharper results for ξ by calculating $\gamma_L(0)^{1/L}$ for higher values of L.)

As was done with **B1**, let us illustrate the nature of **B2** through some examples:

1. For independent bond percolation on $\mathbb{G} = \mathbb{Z}^d$, a separating surface bound (9.26) holds with

$$K_{x,L}(u,v) = \begin{cases} p\, 1_{d(u,v)=1} & \text{for the nearest-neighbor model,} \\ T_\beta(u,v) & \text{for the more general case.} \end{cases}$$

   The reason is that for x and y to be connected there needs to be an occupied bond, (u,v), linking the interior of $B_L(x)$ with its complement, and a path connecting v with y. The probability of that event can be shown to be dominated by the corresponding term in (9.26) (cf. J. M. Hammersley [**186**] and [**6, 54**]).

2. In the case of the ferromagnetic Ising model with nearest-neighbor interaction, $J(x,y) = \delta_{|x-y|,1}$ on $\mathbb{Z}^d$, the condition **B2** holds with

$$K_{x,L}(u,v) = \beta\, J(u,v)\,.$$

   This was proven by E. Lieb [**287**], who refined an inequality of B. Simon [**347**]. In that context, the relation is often referred to as the Simon-Lieb inequality.

3. For random Schrödinger operators the separating surface condition for the fractional moment of the Green function reduces to (11.16). Its derivation will be presented in detail in Chapter 11.

Through a rather direct application of Theorem 9.2, the above structure allows us to boost the single-site condition into the following finite-volume criterion.

Theorem 9.3 (A finite-volume criterion). *Let $\mathbb{G}$ be the vertex set of a graph with metric d and suppose that $\tau : \mathbb{G}\times\mathbb{G} \to [0,\infty)$ is bounded and satisfies (9.26) with some non-negative kernels τ_{B_L} and $K_{x,L}$. If for some $L < \infty$ the quantity defined by (9.27) satisfies*

$$(9.28) \qquad \boxed{\gamma_L(0) < 1} \quad \text{and for some } \mu > 0 \quad \boxed{\gamma_L(\mu) < \infty},$$

then there is $\xi \in (0, \infty)$ *such that*

$$\sup_{y \in \mathbb{G}} \sum_{x \in \mathbb{G}} \tau(x,y)\; e^{d(x,y)/\xi} \;< \infty\,. \tag{9.29}$$

In particular, for all $x, y \in \mathbb{G}$,

$$\boxed{\tau(x,y) \;\le\; C\, e^{-d(x,y)/\xi}} \tag{9.30}$$

with some $C < \infty$.

Proof. The assumed bounds (9.28) imply by continuity (using the dominated convergence theorem) that there exists some $\mu' = \xi^{-1} > 0$ for which

$$\gamma_L(\mu') \;<\; 1\,. \tag{9.31}$$

We now fix $y \in \mathbb{G}$ and consider the function given by $f(x) := \tau(x,y)$. The separating surface condition **B2** together with the uniform boundedness of τ can be summarized as

$$0 \le f(x) \;\le\; g(x) + (Kf)(x) \tag{9.32}$$

$$\text{with}\quad g(x) := 1_{B_L(y)}(x) \sup_{u,v \in \mathbb{G}} \tau(u,v) \quad\text{and}$$

$$K(x,v) \;:=\; 1\,[v \notin B_L(x)] \sum_{u \in B_L(x)} \tau_{B_L(x)}(x,u)\, K_{x,L}(u,v)\,.$$

Choosing $W(x) := e^{\mu d(x,y)}$ with μ as in (9.31), we find that the assumptions of Theorem 9.2 are satisfied. Condition (9.17) is met since $\|K\|_\infty = \gamma_L(0) < 1$ by (9.27). The triangle inequality for the metric and (9.31) imply the validity of the first condition in (9.18). The second condition also holds since it refers to a finite sum. Thus, Theorem 9.2 yields (9.29), from which (9.30) directly follows. □

The existence of exponential decay at given values of $A, \xi < \infty$ is, in principle, decidable through a finite calculation. Curiously, we do not have such a criterion for the existence of extended states.

Theorem 9.3 has additional interesting implications. Let us highlight those for the case $\mathbb{G} = \mathbb{Z}^d$, assuming also that $K_{x,L}(u,v)$ is bounded and non-zero only for $u, v \in \partial B_L(x)$:

1. If the two-point function does exhibit exponential decay, then $\gamma_L(0)$ would decay as, roughly,

$$\gamma_L(0) \;\le\; C\, L^{\#(d)}\; e^{-L/\xi} \underset{L\to\infty}{\longrightarrow} 0\,, \tag{9.33}$$

with some $\#(d) \le 2(d-1)$. Thus the strategy outlined above may in principle allow us to computationally delineate an increasing sequence of regimes in the **phase diagram** whose union is guaranteed to give exactly the full range of **exponential decay**. Taking note of its early appearance in the above-cited 1957 work on percolation [**186**], we refer to it as Hammersley's stratagem.

2. In the three classes of models mentioned above, the finite-volume quantities $\tau_{B_L(x)}(x,u)$ vary continuously within the phase space parameters. One may then conclude from the above criterion that for these models any **transition** to a regime in which $\tau(x,y)$ ceases to decay exponentially, e.g., at the mobility edge if one exists, is **continuous** in the sense that as the transition is approached the two-point function cannot remain uniformly bounded by a fixed exponentially decaying function. This may be contrasted with what is found at first-order (discontinuous) phase transitions in statistical mechanics, where the correlation length ξ stays uniformly bounded up to the transition point (such as found in the two-dimensional q-state Potts model, at q large enough) (Exercise 9.1).

3. The continuity argument also allows us to conclude from Theorem 9.3 that at the phase transition points (e.g., along the mobility edge—in the case of main interest here) for each $L < \infty$,

$$\sum_{\substack{u \in B_L(0) \\ v \in \mathbb{G} \setminus B_L(0)}} \tau_{B_L(0)}(0,u)\, K_{0,L}(u,v) \;\geq\; 1\,. \tag{9.34}$$

   From this one may readily extract a power-law lower bound on the critical values of $\tau_{B_L(0)}(0,u)$, and on the suitably defined **critical exponent**, which would depend on the kernel $K_{0,L}(u,v)$ in the **B2** condition proven for the model.

9.4. Decay rates in domains with boundary modes

The above considerations can also be adapted to situations in which the two-point function has modes with a higher mobility along the domain's boundary, or more generally along some set which would be designated as the boundary and denoted by $\partial\mathbb{G}$.

In such a situation the assumptions of Theorem 9.3 may be satisfied for a given $L < \infty$, but only under the additional restriction that $\operatorname{dist}(x, \partial\mathbb{G}) > L$. For such cases we have the following extension of Theorem 9.3.

Theorem 9.4. *For a domain $\mathbb{G}$ with a boundary set $\partial\mathbb{G}$, if for some $L < \infty$ the assumptions of Theorem* 9.3 *are satisfied, but only under the additional restriction that* $\operatorname{dist}(x, \partial\mathbb{G}) > L$, *i.e., that*

$$\hat{\gamma}_L(\mu) \;:=\; \sup_{\substack{x \in \mathbb{G} \\ \operatorname{dist}(x,\partial\mathbb{G}) > L}} \sum_{\substack{u \in B_L(x) \\ v \in \mathbb{G} \setminus B_L(x)}} \tau_{B_L(x)}(x,u)\, K_{x,L}(u,v)\, e^{\mu\, d(x,v)} \tag{9.35}$$

satisfies

$$\boxed{\hat{\gamma}_L(0) \; < \; 1} \qquad \textit{and for some } \mu > 0 \qquad \boxed{\hat{\gamma}_L(\mu) < \infty}, \tag{9.36}$$

then for all $x, y \in \mathbb{G}$,

$$\boxed{\tau(x,y) \; \leq \; A \; e^{-\mu \, \mathrm{dist}_\Lambda(x,y)}} \tag{9.37}$$

with

$$\mathrm{dist}_\Lambda(x,y) \; := \; \min\left\{ \mathrm{dist}(x,y), [\mathrm{dist}(x,\partial\Lambda) + \mathrm{dist}(y,\partial\Lambda)] \right\}. \tag{9.38}$$

The proof is left as Exercise 9.2.

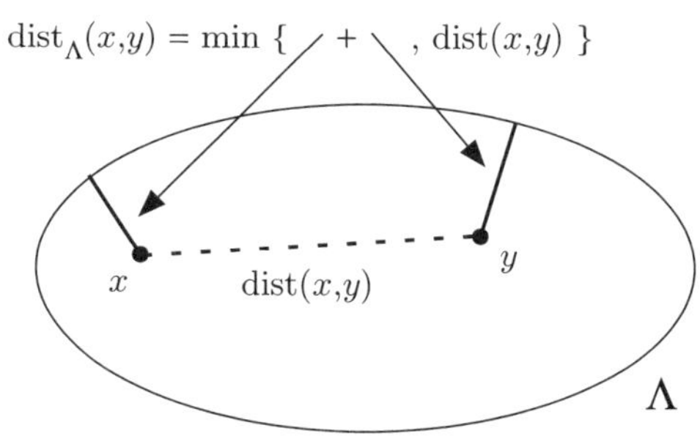

Figure 9.3. A sketch of the distance function which is relevant for the exponential decay derived in Theorem 9.4.

Comparing the conclusion of Theorem 9.4 with that of Theorem 9.3, one may note that in (9.38) the regular distance function was replaced by one for which the boundary is regarded as a point. The resulting bound implies exponential decay throughout the bulk, but it does not rule out the existence of WGM by which along the boundaries of selected subsets (e.g., hyperplane cuts) τ may have only slow decay.

WGM are a phenomenon which does not occur in percolation or ferromagnetic systems, due to the monotonicity there of $\tau_\Lambda(x,y)$ in the volume, but it is found in other more complicated statistical mechanical models and is also of relevance to optics and random Schrödinger operators. In statistical mechanics one may find models with phases for which the correlation functions decay exponentially in the bulk but exhibit long-range correlations along the surfaces of finite-volume versions of the system. An example due to Navratil is quoted by S. Shlosman [**346**], who termed it the Czech model. In the theory of random Schrödinger operators, WGM may be manifested in the presence of extended states which are supported along the boundaries of half-spaces and other subsets of $\mathbb{Z}^d$ but attenuate in the interior. Examples of this nature were discussed in [**202**] and references therein. This phenomenon has recently gained recognition due to the somewhat related manifestation of "topological states" in certain materials.

The above may serve as a reminder that, in a sense, large finite systems are richer than infinite systems since they also exhibit a boundary. It may now also be better appreciated that WGM are ruled out under the full assumption (9.28) where no restriction is placed on the position of x relative to $\partial\mathbb{G}$.

Notes

The use of bounds such as condition **B1** for the purpose discussed here, along with some of its immediate improvements, goes back to J. M. Hammersley [**186**] in the percolation context and M. E. Fisher [**156**] for Ising spins. Further related results for both models can be found, e.g., in [**347, 287, 6, 4**].

A generalized form of the strategy discussed here can be spotted in the (presumably independent) work of R. L. Dobrushin on general conditions for the decay of correlations at high temperatures in a broad class of statistical mechanical models (cf. [**122**] and references therein). The latter has inspired works of B. Simon [**347**] and others [**287, 13, 334**] on ferromagnetic Ising and $O(n)$ spin models, for which correlation inequalities yield simple relations which are strikingly similar to those used earlier in Hammersley's studies of percolation. The analysis of Anderson localization along such lines was developed in [**11**]. The topics will be discussed further in the next chapters.

Localization results for states caused by the presence of a surface are established and discussed in [**202, 203, 204, 242, 117**].

Exercises

9.1. Prove for systems on $\mathbb{Z}^d$ with a two-point function $\tau(x,y)$ which satisfies the condition **B2** that if there is a phase transition at which exponential decay is lost, then the kernel $\tau_{B_L(x)}(x,y)$ cannot satisfy a bound of the form

$$\tau_{B_L(x)}(x,y) \;\le\; A\, e^{-\operatorname{dist}(x,y)/\xi}$$

with A and ξ staying uniformly bounded up to the transition point. [As mentioned in the text, there are examples of systems with first-order phases transitions for which the two-point function itself does exhibit such a uniform decay up to the transition point.]

9.2. Prove Theorem 9.4, highlighting the points at which the argument goes beyond the derivation of the corresponding bulk statement, i.e., Theorem 9.3.

Chapter 10

Localization at High Disorder and at Extreme Energies

In Chapter 7 it was shown how spectral and dynamical localization follows from exponential decay of the Green function's fractional moments. A quick proof of that condition, under simplifying assumptions and at sufficiently high disorder, was presented in Chapter 6. We now return to the subject from the perspective of a **single-step condition** in the sense of Chapter 9, for a more complete presentation of this relatively simple method for establishing localization for at least certain regimes of the parameter space.

The analysis will be carried out in the context of random operators of the form

$$H(\omega) = T + \lambda V(\omega) \tag{10.1}$$

acting in $\ell^2(\mathbb{G})$. Unless stated otherwise it will be assumed that

1. $\mathbb{G}$ is a graph with metric denoted by $d : \mathbb{G} \times \mathbb{G} \to [0, \infty)$,
2. the operator T, which acts as

$$(T\psi)(x) := \sum_{y \in \mathbb{G}} T(x, y)\, \psi(y)\,, \tag{10.2}$$

   is bounded and self-adjoint, and $T(x, x) = 0$ for all $x \in \mathbb{G}$.
3. the random variables $V(x; \omega)$, $x \in \mathbb{G}$, have a uniformly τ-Hölder continuous distribution (in the sense of Definition 4.5) for some $\tau \in (0, 1]$.

For this class of random Schrödinger operators the following will be proven:

A. at high enough disorder exponential localization holds throughout the spectrum,

B. for the finite-dimensional graphs $\mathbb{G} = \mathbb{Z}^d$ localization is found at all levels of disorder at energies sufficiently close to the spectral extrema.

The first statement applies rather generally, both in terms of the hopping range of the operator T and the graph on which it acts. The limitation in the second statement is for a good reason: As was recently established, for tree graphs Schrödinger operators with bounded random potentials at weak disorder there is no localization at the spectral edges (where for λ small the spectrum is, somewhat surprisingly, absolutely continuous).

The discussion in the previous chapters has already provided some initiation to the fractional moment method. In Chapter 7 it was shown that a sufficient condition for strong exponential dynamical localization of a random operator in some energy regime $I \subset \mathbb{R}$ is that for some $s \in (0,1)$ the following estimate holds:

$$\boxed{\sum_{\substack{y \in \Lambda \\ d(x,y) \geq R}} \mathbb{E}\left[|G_\Lambda(x,y;E+i\eta)|^s\right] \leq C\, e^{-\mu R}} \tag{10.3}$$

for arbitrarily large subsets $\Lambda \subseteq \mathbb{G}$, with some uniform constants $C < \infty$ and $\mu > 0$ which may depend on $E \in I$ but are independent of Λ, $x \in \Lambda$, and $\eta > 0$.

10.1. Localization at high disorder

Our next goal is to provide a simple derivation of (10.3) for large λ. Of course, what large means needs to be quantified in terms of the other parameters of the model.

10.1.1. The one-step bound. Removing all matrix elements of T which connect a given site $x \in \Lambda \subseteq \mathbb{G}$ to the rest of the graph, the resolvent identity reads for $z \in \mathbb{C}\backslash\mathbb{R}$

(10.4)

$$G_\Lambda(x,y;z) = G_{\{x\}}(x,x;z)\,\delta_{x,y} - \sum_{y' \neq x} G_\Lambda(x,x;z)\, T(x,y')\, G_{\Lambda\backslash\{x\}}(y',y;z)$$

with $G_{\Lambda\backslash\{x\}}$ the Green function of the operator $H(\omega)$ restricted to the depleted domain. By Corollary 8.4 averages of fractional moments of the individual terms in (10.4) are bounded. In particular, for any $s \in (0, \tau)$,

$$\mathbb{E}\left[|G_\Lambda(x,x;z)|^s \,|V_{\neq x}\right] \le \sup_{\gamma\in\mathbb{C}} \int \frac{\rho(dv)}{|\lambda v - \gamma|^s} \le \frac{C_{s,\tau}}{|\lambda|^s}, \tag{10.5}$$

which relies on the observation that $G_\Lambda(x,x;z)$ is of simple dependence on $V(x)$, namely $(\lambda V(x) - \gamma)^{-1}$ with some $\gamma \in \mathbb{C}$ which is independent of this variable. To proceed we combine (6.18) with the expectation value to find from (10.4) that for any $x \neq y$

$$\mathbb{E}\left[|G_\Lambda(x,y;z)|^s\right] \le \sum_{y'} |T(x,y')|\, \mathbb{E}\left[|G_\Lambda(x,x;z)|^s \, |G_{\Lambda\setminus\{x\}}(y',y;z)|^s\right]. \tag{10.6}$$

We now extract from (10.6) a linear relation for the moment of the Green function's matrix elements by considering the conditional average in which $V(x)$ is integrated first, at fixed values of the other variables. The two observations that (i) the last factor in (10.6) does not depend on $V(x)$ and (ii) the conditional expectation of the pre-factor is dominated by the right side of (10.5) allow us to conclude that

$$\boxed{\mathbb{E}\left[|G_\Lambda(x,y;z)|^s\right] \le \begin{cases} \dfrac{C_{s,\tau}}{|\lambda|^s} \displaystyle\sum_{y'} |T(x,y')|^s \mathbb{E}\left[|G_{\Lambda\setminus\{x\}}(y',y;z)|^s\right], & x \neq y, \\[2ex] \dfrac{C_{s,\tau}}{|\lambda|^s}, & x = y \end{cases}}. \tag{10.7}$$

The above relation is particularly helpful in the regime where

$$\boxed{|\lambda|^s > C_{s,\tau} \, \|T\|_{\infty,\infty}} \tag{10.8}$$

with $\|T\|_{\infty,\infty}$ defined in (6.3).

In case T is a nearest-neighbor operator satisfying (10.8), the relation (10.7) is a subharmonicity statement for the function

$$f(x) = \sup_{\Lambda\subset\mathbb{G}} \mathbb{E}\left[|G_\Lambda(x,y;z)|^s\right].$$

Hence Lemma 9.1 implies the following minor extension of Theorem 6.3 to more singular distributions.

Corollary 10.1. *In the nearest-neighbor case, i.e., $T(x,y) = 0$ if $d(x,y) \neq 1$, for any $\lambda \in \mathbb{R}$ satisfying (10.8) for some $s \in (0,\tau)$, one has for any $z \in \mathbb{C}\backslash\mathbb{R}$ and any $x, y \in \mathbb{G}$,*

$$\boxed{\mathbb{E}\left[|G(x,y;z)|^s\right] \;\le\; \frac{C_{s,\tau}}{|\lambda|^s}\, e^{-\mu_{\lambda,s} d(x,y)}} \tag{10.9}$$

with $\exp\left(-\mu_{\lambda,s}\right) \;:=\; C_{\tau,s}\, \|T\|_{\infty,\infty} \,/\, |\lambda|^s$.

It should be noted that such a uniform bound automatically also extends to $G(x,y;E+i0) := \lim_{\eta\downarrow 0} G(x,y;E+i\eta)$ at almost every $E \in \mathbb{R}$ in the Lebesgue sense. For that, one may invoke Fatou's lemma and the fact that by the general properties of Herglotz functions, at each ω the limit exists for a.e. $E \in \mathbb{R}$ (Appendix B).

For graphs with subexponential growth, such as $\mathbb{Z}^d$, the bound provided by Corollary 10.1 yields upon summation the localization criterion (10.3) for all $E \in \mathbb{R}$. This allows us to conclude complete spectral and dynamical localization at sufficiently large disorder on such graphs. We do not summarize the result here since a more general statement is presented next.

For graphs with exponential rate of volume growth, (10.9) allows us to conclude (10.3) only where the inverse localization length μ exceeds the exponential growth rate of the graph. However, it turns out that the bound (10.7) may be processed in a somewhat more efficient way, whose stronger conclusion yields not just decay of the function but also decay of the tails of the relevant sum. The enabling tool was provided by Theorem 9.2, which also allows us to address operators T of unbounded range.

10.1.2. Complete localization in greater generality. The previous result can be improved and generalized beyond the case where T has only nearest-neighbor hopping terms. For the more general statement the high disorder threshold will be

$$\lambda_\rho(T) \;:=\; \inf_{s\in(0,\tau)} \left[C_{s,\tau} \inf_{\mu>0} \sup_x \sum_{u\neq x} |T(x,u)|^s\, e^{\mu d(x,u)} \right]^{1/s} . \tag{10.10}$$

In simple terms, $\lambda_\rho(T)$ is the smallest value such that for each $|\lambda| > \lambda_\rho(T)$ there is $s \in (0,\tau)$ and $\mu > 0$ with which

$$\boxed{r_{\lambda,s}(\mu) := \frac{C_{s,\tau}}{|\lambda|^s} \sup_{x\in\mathbb{G}} \sum_{u\neq x} |T(x,u)|^s\, e^{\mu d(x,u)} \;<\; 1} . \tag{10.11}$$

For the generic setting described at the beginning of this chapter we have the following result.

Theorem 10.2 (Localization at high disorder). *Let $H(\omega) = T + \lambda V(\omega)$ be a random operator in $\ell^2(\mathbb{G})$ with a uniformly τ-Hölder continuous distribution, and let $\lambda_\rho(T)$ be as defined by (10.10). Then:*

1. *For any $|\lambda| > \lambda_\rho(T)$ there are $s \in (0, \tau)$ and $\mu > 0$ such that for all $R \geq 0$ and $x \in \mathbb{G}$, $z \in \mathbb{C}\backslash\mathbb{R}$:*

$$\sum_{\substack{y \in \mathbb{G} \\ d(x,y) \geq R}} \mathbb{E}\left[|G(x, y; z)|^s\right] \leq \frac{C_{s,\tau}}{|\lambda|^s (1 - r_{\lambda,s}(\mu))} e^{-\mu R} . \tag{10.12}$$

2. *If additionally $\tau = 1$ and the single-site distribution satisfies*

$$\sup_{v \in \mathbb{R}} (1 + |v|) \, \rho(v) < \infty,$$

the operator $H(\omega)$ exhibits strong exponential dynamical localization throughout its spectrum; i.e., for any bounded open set $I \subset \mathbb{R}$ and all $x \in \mathbb{G}$:

$$\sum_{\substack{y \in \mathbb{G} \\ d(x,y) \geq R}} \mathbb{E}\left[\sup_{t \in \mathbb{R}} |\langle \delta_x, \, P_I(H) \, e^{-itH} \, \delta_y \rangle|^2\right] \leq A_I \, e^{-\mu R} \tag{10.13}$$

at some $A_I < \infty$.

Proof. We will apply Theorem 9.2 to (10.7). For that, we fix $y \in \mathbb{G}$ and set

$$f(x) = \sup_{\Lambda \subset \mathbb{G}} \mathbb{E}\left[|G_\Lambda(x, y; z)|^s\right] , \qquad W(x) = e^{\mu d(x,y)},$$

$$K(x, u) = \frac{C_{s,\tau}}{|\lambda|^s} |T(x, u)|^s , \qquad g(x) = \frac{C_{s,\tau}}{|\lambda|^s} \delta_{x,y} .$$

For each $|\lambda| > \lambda_\rho(T)$ the condition (9.17) is satisfied for each $s \in (0, \tau)$ and $\mu > 0$ chosen so that (10.11) holds. For such a choice the first requirement in (9.18) is satisfied since $W(x)/W(u) \leq \exp(\mu \, \mathrm{dist}(x, u))$ by the triangle inequality. The second requirement in (9.18) is fulfilled trivially. Applying Theorem 9.2 we conclude that

$$\begin{aligned}\sum_{\substack{y \in \mathbb{G} \\ \mathrm{dist}(x,y) \geq R}} \mathbb{E}\left[|G(x, y; z)|^s\right] e^{\mu R} &\leq \sum_{y \in \mathbb{G}} \mathbb{E}\left[|G(x, y; z)|^s\right] e^{\mu d(x,y)} \\ &\leq \frac{C_{s,\tau}}{|\lambda|^s (1 - r_{\lambda,s}(\mu))} ,\end{aligned} \tag{10.14}$$

and thus (10.12) holds.

The statement concerning dynamical localization immediately follows from Theorem 7.7 and (7.6). □

Let us comment on some **applications** and **extensions** of this general result:

1. For hopping operators with $\sup_x \sum_u |T(x,u)| e^{+\varepsilon \operatorname{dist}(x,u)} < \infty$ for some $\varepsilon > 0$, the large disorder condition in (10.11) translates to (10.8), which in (10.10) is optimized over $s \in (0, \tau)$. For large $\|T\|_{\infty,\infty}$ the optimal s is approximately $s \approx \tau - (\ln \|T\|_{\infty,\infty})^{-1}$. This yields

$$|\lambda| > C\,(\tau\, \|T\|_{\infty,\infty} \ln \|T\|_{\infty,\infty})^{1/\tau} \tag{10.15}$$

as a condition for the uniform localization statement (with $\tau = 1$ for absolutely continuous single-site distributions). Curiously, in case $T = -\Delta - D_{\mathbb{G}}$ is (up to its diagonal terms) the Laplacian of a graph of common degree (for which $\|T\|_{\infty,\infty} = \deg_{\mathbb{G}}$) and $\tau = 1$, condition (10.15) coincides in its dependence on the degree with the one in P. W. Anderson's original article [**27**].

2. The difference between Corollary 10.1 and Theorem 10.2 can be demonstrated on regular tree graphs, of degree $K+1$, with T the graph Laplacian. The volume of balls of radius R grows there exponentially as K^R, and hence to deduce the localization statement (10.3) from Corollary 10.1 it is required that $|\lambda|^s > C_{\tau,s}\,(K+1)K$. Theorem 10.2 requires only the lesser condition $|\lambda|^s > C_{\tau,s}\,(K+1)$. As we will see in Chapter 16 the latter has the correct power dependence on K.

3. Theorem 10.2 can also be applied to operators for which the hopping terms $T(x,y)$ are only of polynomial decay in $|x-y|$. For that, one may take for the graph metric on $\mathbb{Z}^d$ the function $d(x,y) = \ln(|x-y|+1)$, which also satisfies the triangle inequality. With that choice, the implied localization statement (10.12) yields only polynomial decay of the Green function, in terms of $|x-y|$.

4. The above strong disorder localization result is valid irrespective of the graph's global structure. Thus, the localization criterion $\lambda > \lambda_\rho(T)$ does not distinguish between an infinite strip of finite width and a regular tree graph, regardless of the fact that in the former case complete localization extends to all $\lambda > 0$ while in the latter case that is not true. In Chapter 11 the technique presented here is "renormalized" with the one-step bound replaced by a finite-volume computation which in principle enables us to extend the localization estimates by following the model's contours.

10.2. Localization at weak disorder and at extreme energies

At energies outside the spectrum of T the free Green function

$$G_0(x,y;z) := \langle \delta_x, (T-z)^{-1} \delta_y \rangle \tag{10.16}$$

is square summable in $d(x,y)$, and furthermore, by the Combes-Thomas bound which is presented in Section 10.3, it decays exponentially in any metric at which the kernel of T does. In particular, for energies with $\delta(E) := \mathrm{dist}(E, \sigma(T)) > 0$,

$$|G_0(x,y;E+i\eta)| \leq \frac{2}{\delta(E)}\, e^{-c\delta(E)\, d(x,y)} \tag{10.17}$$

for all $\eta \in \mathbb{R}$, with a constant $c > 0$ which is uniform for small $\delta(E)$.

When the operator is modified through the addition of the random potential, the spectrum spreads into the regime where initially the Green function $G_0(x,y;E)$ exhibits the decay (10.17). We shall next show that for any energy with $\delta(E) > 0$, exponential decay will persist for λ small enough.

For potentials whose values range over all $\mathbb{R}$ the spectrum of $H = T+\lambda V$ extends immediately to cover the entire real line, in which case the continuity result presented next implies localization at extreme energies. (More subtle analysis is required in discussing localization at the edges of the spectrum for random potentials of a bounded range, e.g., V with uniform distribution in $[-1,1]$, which is discussed in Chapter 11.)

The result will be formulated for random operators of the form $H(\omega) = T + \lambda\, V(\omega)$ on $\ell^2(\mathbb{G})$ under an additional assumption.

Definition 10.3. The joint probability distribution of $\{V(x)\}_{x\in\mathbb{G}}$ is said to be **uniformly σ-moment regular** if the conditional distributions of the variables $V(x)$ satisfy for all $s \in (0,\sigma)$

$$D_{s,1} := \sup_{\gamma\in\mathbb{C},\, x\in\mathbb{G}} \ \operatorname*{ess\,sup}_{V_{\neq x}} \int \frac{|v|^s\, \rho(dv\,|\,V_{\neq x})}{|v-\gamma|^s} \Big/ \int \frac{\rho(dv\,|\,V_{\neq x})}{|v-\gamma|^s} < \infty\,. \tag{10.18}$$

In the iid case this boils down to the requirement discussed in Definition 8.5. As is explain in Section 8.4, the class of absolutely continuous probability distributions $\rho(dv) = \varrho(v)dv$ which are 1-moment regular includes all compactly supported $\varrho \in L^\infty(\mathbb{R})$, the Gaussian case as well as the Cauchy distribution.

The following theorem identifies $\widehat{\lambda}(E) := \sup_{s\in(0,\min\{\tau,\sigma\})} \sup_{\mu>0} \widehat{\lambda}_{s,\mu}(E)$ with

$$\widehat{\lambda}_{s,\mu}(E) := \left[D_{s,1} \sup_{\eta\neq 0} \sup_{x\in\mathbb{G}} \sum_{u\in\mathbb{G}} |G_0(x,u;E+i\eta)|^s\, e^{\mu d(x,u)} \right]^{-1/s} \tag{10.19}$$

as an upper bound for the regime of localization at weak disorder and extreme energies. The condition $|\lambda| < \widehat{\lambda}(E)$ ensures that there is some

$s \in (0, \min\{\tau, \sigma\})$ and $\mu > 0$ such that

$$|\lambda|^s D_{s,1} \sum_{u \in \mathbb{G}} |G_0(x, u; E+i\eta)|^s \, e^{\mu d(x,u)} < 1 \tag{10.20}$$

for all $x \in \mathbb{G}$ and $\eta \neq 0$.

Theorem 10.4 (Localization at weak disorder/extreme energies). *Let $H(\omega) = T + \lambda V(\omega)$ be a random operator in $\ell^2(\mathbb{G})$ with a uniformly τ-Hölder continuous distribution which is uniformly σ-moment regular for some $\sigma \in (0, \tau)$. Then:*

1. *For any $(E, \lambda) \in \mathbb{R} \times \mathbb{R}$ in the regime defined by*

$$|\lambda| < \widehat{\lambda}(E)\,, \tag{10.21}$$

the following bound holds for some $s \in (0, \tau)$, $\mu > 0$, and $C < \infty$:

$$\sum_{y \in \mathbb{G}:\, d(x,y) \geq R} \mathbb{E}\left[|G(x, y; E+i\eta)|^s\right] \leq C\, e^{-\mu R} \tag{10.22}$$

for all $R \geq 0$, $x \in \mathbb{G}$, and $\eta \neq 0$.

2. *If additionally $\tau = 1$ and the density of the single-site distribution satisfies $\sup_{v \in \mathbb{R}} (1 + |v|)\, \rho(v) < \infty$, then for any bounded open $I \subset \mathbb{R}$ for which $|\lambda| < \sup_{s \in (0, \min\{\tau,\sigma\})} \sup_{\mu > 0} \inf_{E \in I} \widehat{\lambda}_{s,\mu}(E)$ the operator $H(\omega)$ exhibits strong exponential dynamical localization on I; i.e.,* (10.13) *holds.*

Proof. The proof starts from the resolvent identity

$$G(x, y; z) = G_0(x, y; z) - \sum_{y'} G_0(x, y'; z)\, \lambda\, V(y')\, G(y', y; z)\,, \tag{10.23}$$

which holds for all $z = E+i\eta \in \mathbb{C} \backslash \mathbb{R}$. Taking expectation values of fractional moments on both sides of (10.23) and using (6.18) one gets

$$\begin{aligned} &\mathbb{E}\left[|G(x, y; z)|^s\right] \\ &\quad \leq |G_0(x, y; z)|^s + |\lambda|^s \sum_{y'} \left|G_0(x, y'; z)\right|^s \mathbb{E}\left[|V(y')|^s \left|G(y', y; z)\right|^s\right]. \end{aligned} \tag{10.24}$$

This equation does not close for $\mathbb{E}\left[|G(x, y; z)|^s\right]$ since the expectation value in the second term does not factorize. This problem is overcome using the s-moment regularity bound (10.18). By first integrating over the random variable $V(y')$, this assumption yields

$$\mathbb{E}\left[|V(y')|^s \left|G(y', y; z)\right|^s\right] \leq D_{s,1}\, \mathbb{E}\left[\left|G(y', y; z)\right|^s\right]. \tag{10.25}$$

The fractional moment is hence seen to satisfy a closed inequality,

(10.26)
$$\begin{aligned}&\mathbb{E}\left[|G(x,y;z)|^s\right]\\&\qquad\leq |G_0(x,y;z)|^s + D_{s,1}\,|\lambda|^s \sum_{y'} \left|G_0(x,y';z)\right|^s\, \mathbb{E}\left[\left|G(y',y;z)\right|^s\right]\,,\end{aligned}$$

to which Theorem 9.2 is applicable. To do so, we again fix $y \in \mathbb{G}$ and set

$$\begin{aligned}f(x) &= \mathbb{E}\left[|G(x,y;z)|^s\right]\,, & W(x) &= e^{\mu d(x,y)}\,,\\ K(x,u) &= D_{s,1}\,|\lambda|^s\, |G_0(x,y;z)|^s\,, & g(x) &= |G_0(x,y;z)|^s\,.\end{aligned}$$

Then (9.17) is satisfies in case $|\lambda| < \widehat{\lambda}(E)$ and $s \in (0, \min\{\tau,\sigma\})$ and $\mu > 0$ chosen such that (10.20) applies. This also ensures the validity of the first requirement in (9.18) since $W(x)/W(u) \leq \exp\left(\mu\,\mathrm{dist}(x,u)\right)$ by the triangle inequality. The second requirement in (9.18) is then also fulfilled. The proof of the first assertion is concluded by analogous reasoning as in (10.14).

The statement concerning dynamical localization immediately follows from Theorem 7.7 and (7.6). □

The implication in Theorem 10.4 is localization within the specified ranges of energies—to the extent that there is spectrum there. One should therefore also pay attention to the extent of overlap of this range with the spectrum $\sigma(H)$. In light of this, let us comment on the **implications** and some of the **limitations** of the result, some of which are addressed in the following chapters.

1. The set of energies for which Theorem 10.4 in case $\tau = \sigma = 1$ establishes localization at weak (enough) disorder, that is,

$$W \;=\; \left\{E \in \mathbb{R} \,\middle|\, \widehat{\lambda}(E) > 0\right\}, \tag{10.27}$$

   coincides with the complement of the ℓ^1-spectrum of T, which we denote by $\sigma_1(T)$ in contrast to the ℓ^2-spectrum $\sigma(T)$. This curious fact reinforces the comment which was made in Chapter 7.

2. For the Laplacian on $\mathbb{Z}^d$, $\sigma_1(-\Delta) = \sigma(-\Delta)$, and thus W meets the spectrum. For random potentials whose range covers the full line, $\sigma(-\Delta + \lambda V) = \mathbb{R}$, and the result implies that in such a case at weak disorder localization sets in at extreme energies within a range which for $\lambda \to 0$ converges to the unperturbed spectrum $\sigma(-\Delta)$. Still, the edge of localization is not mapped correctly yet, and the theorem leaves room for improvement. The gap in the analysis is more pronounced in the case of potentials of bounded range. By the Combes-Thomas bound (10.34)

at weak disorder the regime covered by the above result is for small $\delta(E) = \text{dist}(E, \sigma(-\Delta)) > 0$ of the form

$$|\lambda| \, D_{s,1}^{1/s} < C_s \, \delta(E)^{1+\frac{d}{2s}} \tag{10.28}$$

with some $C_s < \infty$ and $s \in (0,1)$ and hence will be void of spectrum: The spectral edge moves linearly with λ at the rate of $|\lambda| V_{\max}$, while the edge of the set for which (10.21) holds recedes from $\sigma(-\Delta)$ at a faster rate. This renders the bound (10.22) irrelevant at weak disorder for the interesting example of bounded potential V with uniformly distributed in $[-1, 1]$.

3. The above gap in the argument for models with bounded potential on $\mathbb{Z}^d$ may be addressed through the finite-volume criteria which are presented next, in Chapter 11, combined with the Lifshitz tail estimates. An approach along these lines was suggested in [**11**] and carried out through detailed analysis by F. Klopp [**258**]. One may add that for $d \geq 3$ and potentials given by even random variables of finite moments the localization regime for small disorder $\lambda > 0$ is expected [**366**] to include all energies E of the form

$$-\lambda \inf \rho \;\leq\; E \;\leq -\lambda^2 \, \mathbb{E}\left[V(0)^2\right] \, G_0(0,0;0) + \mathcal{O}(\lambda^4)\,. \tag{10.29}$$

This was proven by A. Elgart [**135**] for $d = 3$, through a more detailed analysis of the Green function's renormalized perturbation series in λ.

4. The situation is different for random Schrödinger operators on regular tree graphs, of degree $K+1$, for which for the adjacency operator

$$\sigma_1(T) \;=\; [-(K+1), K+1] \quad \text{while} \quad \sigma(T) \;=\; \left[-2\sqrt{K}, 2\sqrt{K}\right]. \tag{10.30}$$

The fact that in this case the set W does not connect to the unperturbed spectrum leads to some interesting questions about the phase diagram, which have been resolved only fairly recently [**21**]. In particular, it was found (somewhat surprisingly) that at weak disorder on regular tree graphs:
   i. for the above bounded potential the states at the spectral edges are actually **delocalized**, despite the fact that the density of states decays there in a manner similar to the Lifshitz tail estimate,
   ii. for unbounded potential the range of the absolutely continuous spectrum **jumps** discontinuously from $\sigma(T)$ to the set $\sigma_1(T)$.

Thus, the gaps mentioned above do not just reflect a weakness of the general Theorem 10.4. We will further elaborate on this point in Chapter 16.

10.3. The Combes-Thomas estimate

It is a truism in quantum physics that eigenfunctions of Schrödinger operators corresponding to energies outside the spectrum of the Laplacian decay exponentially away from the support of the perturbing potential. Since for any such eigenfunction

$$\psi = -\frac{1}{-\Delta - E} V \psi,$$

such estimates and their generalization are proven with the help of bounds on the Green function of the unperturbed operator. In the case of the Laplacian, the Green function is known to decay exponentially for $E \notin \sigma(-\Delta)$ by an explicit calculation. For more general operators such decay estimates date back to J. M. Combes and L. Thomas [**98**]. The following Combes-Thomas bound is presented along the lines of [**5**, App. II].

Theorem 10.5 (Combes-Thomas). *Let $\mathbb{G}$ be a countable set with a metric $d : \mathbb{G} \times \mathbb{G} \to [0, \infty)$ and let H be a self-adjoint operator in $\ell^2(\mathbb{G})$ whose off-diagonal elements are exponentially summable for some $\alpha > 0$:*

$$S_\alpha := \sup_x \sum_y |H(x,y)| \left(e^{\alpha\, d(x,y)} - 1\right) < \infty. \tag{10.31}$$

Then for energies not in the spectrum, $\Delta := \operatorname{dist}(z, \sigma(H)) > 0$, the Green function of H is exponentially bounded:

$$|G(x,y;z)| \leq \frac{1}{\Delta - S_\mu} \exp\left(-\mu\, d(x,y)\right) \tag{10.32}$$

for any $\mu \leq \alpha$ with $S_\mu < \Delta$ and all $x, y \in \mathbb{G}$. In particular,

1. *In case $\Delta \leq 2S_\alpha$ we have*

$$|G(x,y;z)| \leq \frac{2}{\Delta} \exp\left(-\frac{\alpha\Delta}{2S_\alpha}\, d(x,y)\right). \tag{10.33}$$

2. *In case $\operatorname{Re} z < \inf \sigma(H)$ or $\operatorname{Re} z > \sup \sigma(H)$ with $\Delta_{\mathrm{Re}} := \operatorname{dist}(\operatorname{Re} z, \sigma(H)) \leq 2S_\alpha$, also*

$$|G(x,y;z)| \leq \frac{2}{\Delta_{\mathrm{Re}}} \exp\left(-\frac{\alpha\sqrt{\Delta_{\mathrm{Re}}}}{\sqrt{2S_\alpha}}\, d(x,y)\right). \tag{10.34}$$

Before giving a proof, let us discuss two examples illustrating the applicability of the theorem:

1. Any set $\mathbb{G} \subseteq \mathbb{Z}^d$ may be equipped with the Euclidean distance,

$$d(x,y) = |x - y|,$$

in which case (10.31) holds in case the kernel of H decays exponentially. In particular, this applies to any Schrödinger operator; i.e., $H(x,y) = \delta_{|x-y|,1}$ for $x \neq y$, for which $S_\alpha = 2d\, e^\alpha$ for any $\alpha > 0$.

2. Another choice of a metric for $\mathbb{G} \subseteq \mathbb{Z}^d$ is

$$d(x,y) = \ln(|x-y|+1)\,.$$

This choice fits operators H with algebraically decaying kernel, $|H(x,y)| \leq \kappa\,|x-y|^{-\alpha-\beta}$ with $\beta > d$.

The different versions (10.32)–(10.34) of the Combes-Thomas estimate address the decay of the Green function in different regimes away from the spectrum of H. Far away from $\sigma(H)$, the decay follows that of the kernel of H. Near $\sigma(H)$ the localization length diverges. More specifically, approaching $\sigma(H)$ within the upper or lower complex plane from above or below, the localization length is proportional to Δ^{-1} as $\Delta \to 0$. This length only diverges proportionately to $\Delta^{-1/2}$ as $\Delta \to 0$ in case the approach is on the real axis below or above $\sigma(H)$.

In the proof we shall make use of the following estimates. The first is an example of the interpolation methods.

Proposition 10.6. *Let $\mathbb{G}$ be a countable set and let B be an operator on $\ell^2(\mathbb{G})$ with kernel $B(x,y)$ satisfying*

$$\|B\|_{1,1} := \sup_y \sum_x |B(x,y)| < \infty\,,$$

$$\|B\|_{\infty,\infty} := \sup_x \sum_y |B(x,y)| < \infty\,.$$

Then B is bounded with $\|B\| \leq \sqrt{\|B\|_{1,1}\|B\|_{\infty,\infty}}$.

The proof is left as Exercise 10.3.

The next lemma is a particular case of the general fact that the spectrum of two operators can only expand into the convex hull [**217**]. In case A is a self-adjoint operator on some Hilbert space with $z \notin \sigma(A)$ and B is a bounded operator with $\|B\| < \operatorname{dist}(\sigma(A), z)$, it is elementary to see that $z \notin \sigma(A+B)$ and

$$\|(A+B-z)^{-1}\| \leq \frac{1}{\operatorname{dist}(\sigma(A),z) - \|B\|}\,. \tag{10.35}$$

The following estimate refines this bound. Through it, one may prove the bound (10.34).

Lemma 10.7. *Let A be a positive, self-adjoint operator with* $\operatorname{dist}(\sigma(A),0) > 0$ *and let B be a bounded, self-adjoint operator. Then zero belongs to the resolvent set of $A + iB$ and*

$$\|(A+iB)^{-1}\| \;\leq\; \frac{1}{\operatorname{dist}(\sigma(A),0)}\,. \tag{10.36}$$

Proof. We may factorize the resolvent as

$$(A+iB)^{-1} = \frac{1}{\sqrt{A}}\left(1+\frac{1}{\sqrt{A}}iB\frac{1}{\sqrt{A}}\right)^{-1}\frac{1}{\sqrt{A}}\,. \tag{10.37}$$

For the middle factor, we note that $\frac{1}{\sqrt{A}}B\frac{1}{\sqrt{A}}$ is self-adjoint, and hence using its spectral representation,

$$\left\|\left(1+\frac{1}{\sqrt{A}}iB\frac{1}{\sqrt{A}}\right)^{-1}\right\| \leq \sup_{\eta\in\mathbb{R}}|1+i\eta|^{-1} \leq 1\,. \tag{10.38}$$

Thus $\|(A+iB)^{-1}\| \leq \|(\sqrt{A})^{-1}\|^2 \leq \operatorname{dist}(\sigma(A),0)^{-1}$. □

We are now ready to give a

Proof of Theorem 10.5. For any fixed $y\in\mathbb{G}$ and any cutoff $R>0$, the multiplication operator

$$M = \exp\left(\mu\,\min\{d(\cdot,y),R\}\right) \tag{10.39}$$

is bounded and invertible on $\ell^2(\mathbb{G})$ for any $\mu\in(0,\infty)$. The Combes-Thomas estimate is based on the observation that for any x with $d(x,y)\leq R$ and any $z\notin\sigma(H)\cup\sigma(MHM^{-1})$,

$$\begin{aligned} G(x,y;z)\,e^{\mu\,d(x,y)} &= \left\langle\delta_x\,,M\,(H-z)^{-1}\,M^{-1}\,\delta_y\right\rangle \\ &= \left\langle\delta_x\,,\left(MHM^{-1}-z\right)^{-1}\,\delta_y\right\rangle \\ &= \left\langle\delta_x\,,(H+B-z)^{-1}\,\delta_y\right\rangle\,. \end{aligned} \tag{10.40}$$

The idea is to control the shift of the spectrum of H introduced by

$$B := MHM^{-1} - H\,. \tag{10.41}$$

For a proof of (10.32) and (10.33), this will be done by estimating the operator norm of B. Its kernel is bounded independently of R with the help of the triangle inequality according to

$$|B(x,x')| \leq |H(x,x')|\left(e^{\mu\,d(x,x')}-1\right)\,. \tag{10.42}$$

Together with (10.31) and Proposition 10.6, this yields the following upper bound on the operator norm:

$$\|B\| \leq \sqrt{\|B\|_{1,1}\|B\|_{\infty,\infty}} \leq S_\mu \leq S_\alpha < \infty \tag{10.43}$$

for all $\mu\leq\alpha$. The spectrum of $B+H$ has distance at most $\|B\|$ from the spectrum of H. Hence $B+H-z$ is invertible as long as $\Delta>\|B\|$, for which

$$\left\|(B+H-z)^{-1}\right\| \leq \frac{1}{\Delta-\|B\|} \leq \frac{1}{\Delta-S_\mu}; \tag{10.44}$$

cf. (10.35). In view of (10.40) and the fact that R was arbitrary, we have thus proved (10.32).

The proof of (10.33) follows from (10.32) with the help of the inequality

$$\frac{e^{\mu d}-1}{e^{\alpha d}-1} \leq \frac{\mu}{\alpha} \tag{10.45}$$

valid for all $\mu \leq \alpha$, which implies $S_\mu \leq \frac{\mu}{\alpha} S_\alpha$. Choosing $\mu = \alpha\,\Delta\,(2S_\alpha)^{-1}$, we hence obtain (10.33).

In case $\operatorname{Re} z < \inf \sigma(H)$ or $\operatorname{Re} z > \sup \sigma(H)$, the above estimates can be improved. We split $B = C - iD$ into two operators, both defined in terms of their kernels:

$$\begin{aligned} C(x,x') &:= H(x,x')\left(e^{\mu s(x,x')} - \mu\, s(x,x') - 1\right), \\ D(x,x') &:= i\, H(x,x')\, \mu\, s(x,x') \end{aligned} \tag{10.46}$$

where we abbreviated $s(x,x') := \min\{d(x',y), R\} - \min\{d(x,y), R\}$. The kernel of C can be estimated with the help of the elementary inequality $|e^x - 1 - x| \leq x^2\, e^{|x|}/2$ together with the triangle inequality:

$$\begin{aligned} \left|C(x,x')\right| &\leq \frac{\mu^2}{2}\,|H(x,x')|\, d(x,x')^2\, e^{\mu\, d(x,x')} \\ &= \frac{\mu^2}{2}\,|H(x,x')|\, d(x,x')^2\, \frac{e^{\mu\, d(x,x')}}{e^{\alpha\, d(x,x')}-1}\left(e^{\alpha\, d(x,x')}-1\right) \\ &\leq \frac{\mu^2}{\alpha^2}\,|H(x,x')|\left(e^{\alpha\, d(x,x')}-1\right) \end{aligned} \tag{10.47}$$

for all $\mu \leq \alpha$, and consequently

$$\|C\| \leq \sqrt{\|C\|_{1,1}\|C\|_{\infty,\infty}} \leq S_\alpha\, \frac{\mu^2}{\alpha^2}\,. \tag{10.48}$$

As long as $\Delta_{\mathrm{Re}} > \|C\|$, the distance $\operatorname{dist}(\operatorname{Re} z, \sigma(H+C)) \geq \operatorname{dist}(\operatorname{Re} z, \sigma(H)) - \|C\|$ stays strictly positive. By Lemma 10.7 and (10.40), we obtain

$$|G(x,y;z)| \leq \frac{1}{\Delta_{\mathrm{Re}} - S_\alpha\, \frac{\mu^2}{\alpha^2}}\, e^{-\mu\, d(x,y)}\,, \tag{10.49}$$

as long as the denominator is positive, which for example applies to the choice $\mu = \alpha\,\sqrt{\Delta_{\mathrm{Re}}/(2S_\alpha)}$. This concludes the proof of (10.34). □

Notes

The localization proofs presented here go back to [**8**, **5**]. The fractional moment method also leaves room for other ways to organize the argument for high disorder or extreme energies. In particular, for unbounded potentials one may also proceed as laid out in Exercise 10.2. The latter addresses the case of the so-called Llyod model, which was initially predicted in [**290**] (incorrectly) to have no localization regime since its mean Green function is analytic in the disorder parameter.

Exercises

10.1. Give the second part of the proof of Theorem 10.2 and Theorem 10.4.

10.2. Let $H(\omega) = T + \lambda V(\omega)$ be a random operator in $\ell^2(\mathbb{G})$ with an iid potential with uniformly τ-Hölder continuous single-site distribution ρ for which for some $q > s$

$$\int |v|^q \rho(dv) \ < \infty \, .$$

a. Show that (10.22) remains valid for all $E \in \mathbb{R}$, $|\lambda| > 0$, and $\mu > 0$ satisfying

$$|\lambda|^s \, d_s \left(\frac{|E|}{|\lambda|} \right) \ < \ \sup_{x \in \mathbb{G}} \sum_{u \in \mathbb{G}} |T(x,u)|^s e^{\mu d(x,u)} \, ,$$

where d_s is the function introduced in Exercise 8.6.

b. Consider the case with $|T(x,u)| \neq 0$ if only if $d(x,u) = 1$ and $\rho(dv) = \pi^{-1}(v^2+1)^{-1} dv$, i.e., the **Llyod model**. Determine explicitly a regime $(E, \lambda) \in I \times \mathbb{R}$ in which this model exhibits strong dynamical localization in the sense that (10.13) holds for some $\mu > 0$.

10.3. Provide a proof of Proposition 10.6.

Chapter 11

Constructive Criteria for Anderson Localization

The simple and rather direct methods for establishing localization presented in Chapter 10 have a few drawbacks—some of which we already addressed there. First and foremost, the range they cover is limited to the regime of high disorder or extreme energies. Pursuing the ideas described in Chapter 9 we show here how to reach beyond those limitations (at least in principle) through a sequence of finite-volume localization criteria. These, in effect, renormalize the one-step estimate on which Theorem 10.2 is based.

11.1. Finite-volume localization criteria

To reduce the complexity of the presentation the discussion in this chapter is restricted to random operators with just nearest-neighbor hopping terms,

$$\begin{aligned} H(\omega) &= T + \lambda V(\omega) \quad \text{in } \ell^2(\mathbb{Z}^d) \\ &\qquad \text{with } |T(x,y)| = \delta_{|x-y|,1}\,, \end{aligned} \tag{11.1}$$

and iid random potentials $V(\omega)$ with a uniformly τ-Hölder continuous single-site probability distribution ρ. This setting includes the original Anderson model and the model with a constant magnetic field (which is of relevance for the integer quantum Hall effect (IQHE) which is discussed in Chapter 14).

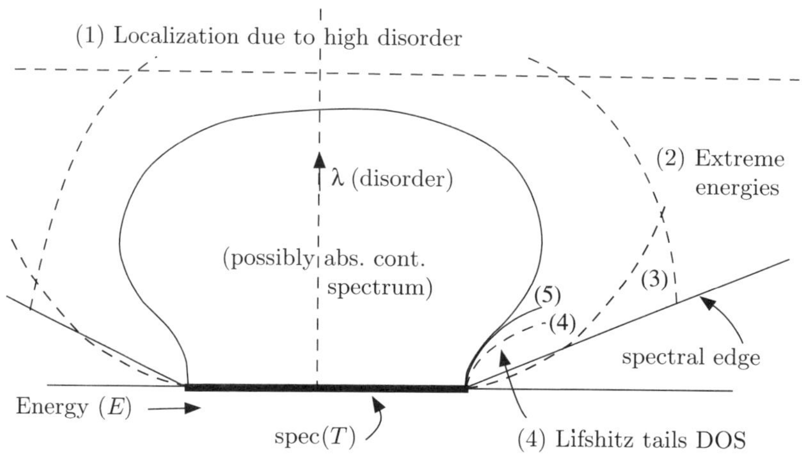

Figure 11.1. Sketch of the expected phase diagram for the Anderson localization in $d > 2$ dimensions and the regimes covered by different estimates: (1) high disorder (Corollary 10.1/Theorem 10.2), (2) extreme energy (Theorem 10.4), (3) weak disorder (Theorem 10.4), (4) weak disorder + Lifshitz tail DOS estimate (Corollary 11.6), and (5) the perturbative analysis of [**135**].

In the discussion of finite-volume criteria we denote, for subsets $\Lambda \subset \mathbb{Z}^d$,

$$\begin{aligned} \Lambda^+ &:= \left\{ x \in \mathbb{Z}^d \mid \operatorname{dist}(x, \Lambda) \le 1 \right\} , \\ \partial\Lambda &:= \{ (x,y) \in \Lambda \times \Lambda^c \mid |x-y| = 1 \} , \end{aligned}$$

which form the one-step fattening of $\Lambda \subset \mathbb{Z}^d$ and its set of (oriented) boundary edges.

Theorem 11.1 (Finite-volume criterion I). *For an operator of the form (11.1), with V an iid random potential of uniformly τ-Hölder continuous single-site probability distribution, suppose that for some finite set $\Lambda \subset \mathbb{Z}^d$ which contains the origin,*

$$r(\Lambda; z, s) := \max_{W \subseteq \Lambda} \left\{ |\partial W^+| \, \frac{4 C_{s,\tau}}{|\lambda|^s} \sum_{(u,v) \in \partial\Lambda} \mathbb{E}\left[|G_W(0, u; z)|^s \right] \right\} < 1 \tag{11.2}$$

at some $z \in \mathbb{C}$ and $s \in (0, \tau)$. Then for any $\Omega \subseteq \mathbb{Z}^d$ and any $x, y \in \Omega$,

$$\mathbb{E}\left[|G_\Omega(x, y; z)|^s \right] \le C \, e^{-\mu |x-y|} \tag{11.3}$$

with some $\mu > 0$ and $C < \infty$ (which do not vary with Ω).

The maximum in (11.2) is taken over all non-empty subsets W of Λ and terms in the summation over $(u,v) \in \partial\Lambda$ with $u \notin W$ are understood to be zero. A proof of Theorem 11.1 will be given in Section 11.3.

Since (11.3) addresses the decay of the Green function in any finite-volume $\Omega \subset \mathbb{Z}^d$, it not only rules out the presence of delocalized states in the bulk but also along the edge of Ω. In this sense it is a stronger statement than Theorem 11.2 below.

To compare this theorem to the previous single-step bound, Corollary 10.1, one may consider the localization condition that it yields with $\Lambda = \{0\}$. In that case, equation (11.2) yields the following sufficient condition for localization:

$$\left[(2d-1)2d\,\frac{4C_{s,\tau}}{|\lambda|^s}\right] \times 2d\frac{C_{s,\tau}}{|\lambda|^s} \;<\; 1 \tag{11.4}$$

(using the fractional moment bound (8.34) and the value $|\partial\{0\}^+| = 2d(2d-1)$). This is not as efficient as the less stringent high-disorder criterion (10.8). However, Theorem 11.1 allows us to move beyond the single-step bound, taking $\Lambda_L = [-L, L]^d \cap \mathbb{Z}^d$ and incorporating in the analysis the results of finite-volume calculations of the Green function.

11.2. Localization in the bulk

Since (11.2) involves the Green function for subsets of Λ the computation is somewhat cumbersome. The following alternative criterion gets by with just the Green function G_Λ of the basic domain Λ.

Theorem 11.2 (Finite-volume criterion II). *Let $H(\omega)$ be an operator of the form (11.1), with V an iid random potential of uniformly τ-Hölder continuous single-site probability distribution, which for some $s \in (0, \tau/2)$ also satisfies*

(i) $\int |v|^{2s}\rho(dv) < \infty$ *and* (ii) *ρ has a regular $(1+2s)$-decay.*

If for some $z \in \mathbb{C}$ and $\Lambda \subset \mathbb{Z}^d$ which contains the origin, at the above value of $s \in (0, \tau/2)$

$$\boxed{\left(1 + \frac{\kappa_s}{|\lambda|^s}\,|\partial\Lambda|\right)^2 \sum_{(u,v)\in\partial\Lambda} \mathbb{E}\left[|G_\Lambda(0,u;z)|^s\right] \;< 1} \tag{11.5}$$

with $\kappa_s := 4C_{s,\tau}\,D_{s,2}^2$ [and $D_{s,2}$ the constant in (8.69)], then there exists some $\mu > 0$ and $C < \infty$ with which the operator's Green function satisfies, for all $x, y \in \mathbb{Z}^d$

$$\boxed{\mathbb{E}\left[|G(x,y;z)|^s\right] \;\le\; C\,e^{-\mu|x-y|}} \tag{11.6}$$

and for all other subsets $\Omega \subset \mathbb{Z}^d$,

$$\boxed{\mathbb{E}\left[|G_\Omega(x,y;z)|^s\right] \;\leq\; C\, e^{-\mu\, \mathrm{dist}_\Omega(x,y)}} \tag{11.7}$$

with $\mathrm{dist}_\Omega(x,y)$ *the distance defined in* (9.38) *(for which the boundary* $\partial\Omega$ *is regarded as one point).*

In comparison to Theorem 11.1, the bound (11.7) allows us to establish localization in the bulk even in regimes where **extended boundary states** may be present in some geometry. The proof of Theorem 11.2 follows by the method presented in Chapter 9, based on the bounds which will be established next.

As a tool for the phase diagram's rigorous exploration, Theorem 11.2 may enable one to systematically cover the regime of localization (though at the cost of increasing complex calculations):

Theorem 11.3. *For a random Schrödinger operation of the structure described in Theorem* 11.1, *if the Green function of* $H(\omega)$ *on* $\ell^2(\mathbb{Z}^d)$ *decays exponentially, i.e.,* (11.6) *holds for all* $x, y \in \mathbb{Z}^d$ *at* $z \in \mathbb{C}\backslash\mathbb{R}$ *with some* $C < \infty$ *and* $\mu > 0$, *then for all sufficiently large* L *the finite-volume condition* (11.5) *is satisfied for* $\Lambda = (-L, L)^d \cap \mathbb{Z}^d$.

The proof of Theorem 11.3 relies on techniques used in the proof of Theorem 11.1. It is therefore left as Exercise 11.1. We shall discuss some further implications of the above finite-volume conditions in Section 11.4.

11.3. Derivation of the finite-volume criteria

The proofs of the finite-volume criteria rest on applications of the resolvent equation in which we deplete the region $\Omega \subseteq \mathbb{Z}^d$. Given a finite collection of oriented bonds $\Gamma \subset \mathbb{Z}^d \times \mathbb{Z}^d$,

$$T^{(\Gamma)}(x,y) \;:=\begin{cases} T(x,y) & \text{if } (x,y) \in \Gamma, \\ 0 & \text{otherwise} \end{cases} \tag{11.8}$$

gives rise to a self-adjoint operator $T_\Omega^{(\Gamma)}$ on $\ell^2(\Omega)$, which describes the hopping over the bonds Γ. Correspondingly,

$$H_\Omega =: \; H_\Omega^{(\Gamma)} + T_\Omega^{(\Gamma)} \tag{11.9}$$

defines the depleted operator $H_\Omega^{(\Gamma)}$ on $\ell^2(\Omega)$. For two sets of oriented bonds $\Gamma_1, \Gamma_2 \subset \mathbb{Z}^d \times \mathbb{Z}^d$, a twofold application of the resolvent equation yields

(11.10)
$$\begin{aligned}(H_\Omega - z)^{-1} &= (H_\Omega^{(\Gamma_1)} - z)^{-1} - (H_\Omega^{(\Gamma_1)} - z)^{-1} T_\Omega^{(\Gamma_1)} (H_\Omega - z)^{-1} \\ &= (H_\Omega^{(\Gamma_1)} - z)^{-1} - (H_\Omega^{(\Gamma_1)} - z)^{-1} T_\Omega^{(\Gamma_1)} (H_\Omega^{(\Gamma_2)} - z)^{-1} \\ &\quad + (H_\Omega^{(\Gamma_1)} - z)^{-1} T_\Omega^{(\Gamma_1)} (H_\Omega - z)^{-1} T_\Omega^{(\Gamma_2)} (H_\Omega^{(\Gamma_2)} - z)^{-1} .\end{aligned}$$

This formula will be exploited in different ways for the proof of the finite-volume criteria.

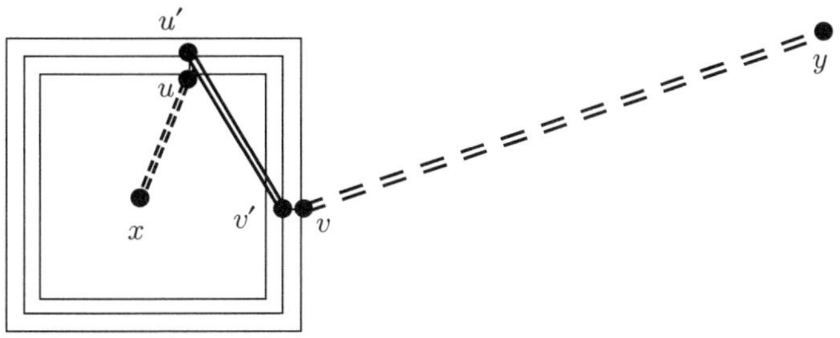

Figure 11.2. A graphic depiction of the expression for $G_\Omega(x, y; z)$ in formula (11.12). The solid double line indicates the full Green function G_Ω, whereas the dotted lines correspond to the Green functions for either the interior (when linked with x) or the exterior (when linked with y) operator.

Proof of Theorem 11.1. We fix $x \in \Omega$, and we set $\Lambda(x) := \Lambda + x$ and $W := \Omega \cap \Lambda(x)$. We will employ the depletion formula (11.10) with the following choices:

$$\begin{aligned}\Gamma_1 &:= \{(u, v) \in \Omega \times \Omega \,|\, \text{either } (u, v) \in \partial W \text{ or } (v, u) \in \partial W\} , \\ \Gamma_2 &:= \{(u, v) \in \Omega \times \Omega \,|\, \text{either } (u, v) \in \partial W^+ \text{ or } (v, u) \in \partial W^+\} . \end{aligned} \tag{11.11}$$

Using the convention that terms corresponding to indices not belonging to Ω are assumed to be zero, the depletion formula (11.10) yields for any $y \notin W^+$,

(11.12)
$$G_\Omega(x, y; z) = \sum_{\substack{(u,u')\in\partial W \\ (v',v)\in\partial W^+}} G_W(x, u; z)\, T(u, u')\, G_\Omega(u', v'; z)\, T(v', v)\, G_{\Omega\setminus W^+}(v, y; z) .$$

A diagrammatic illustration of this identity can be found in Figure 11.2. Taking fractional moments and using (6.18), we are therefore left estimating the following expectation value:

$$
\begin{aligned}
(11.13)\quad & \mathbb{E}\left[|G_W(x,u;z)|^s \, |G_\Omega(u',v';z)|^s \, |G_{\Omega\setminus W^+}(v,y;z)|^s\right] \\
&= \mathbb{E}\left[|G_W(x,u;z)|^s \, |G_{\Omega\setminus W^+}(v,y;z)|^s \, \mathbb{E}_{u',v'}\left[|G_\Omega(u',v';z)|^s\right]\right] \\
&\le \frac{4C_{s,\tau}}{|\lambda|^s}\, \mathbb{E}\left[|G_W(x,u;z)|^s\right] \, \mathbb{E}\left[|G_{\Omega\setminus W^+}(v,y;z)|^s\right] .
\end{aligned}
$$

Here the first step results from first conditioning on all random variables aside from those associated to u' and v' and we chose to abbreviate $\mathbb{E}_x[\cdot] := \mathbb{E}[\cdot \,|\, V_{\neq x}]$. Note that these random variables do not enter the Green function associated with W and $\Omega\setminus W^+$. The inequality results from Corollary 8.4 and the independence of G_W and $G_{\Omega\setminus W^+}$.

Combining (11.12) and (11.13) we thus arrive at

$$
(11.14)\quad \boxed{\mathbb{E}\left[|G_\Omega(x,y;z)|^s\right] \le \frac{r(\Lambda;z,s)}{|\partial W^+|} \sum_{(v,v')\in\partial W^+} \mathbb{E}\left[|G_{\Omega\setminus W^+}(v,y;z)|^s\right]}
$$

with $r(\Lambda;z,s)$ as defined in (11.2). In the regime where $r(\Lambda;z,s) < 1$ this turns into a subharmonicity statement. Upon iteration or through the more generally applicable Theorem 9.3 this yields the claimed exponential decay (11.3). □

Proof of Theorem 11.2. Let $\Omega \subseteq \mathbb{Z}^d$ and fix $x \in \Omega$ and $\Lambda(x) := \Lambda + x$. Picking $W = \Omega \cap \Lambda(x)$ and Γ_1 as in (11.11) in the resolvent equation (11.10), we obtain for any $y \notin W$

$$
\begin{aligned}
G_\Omega(x,y;z) &= - \sum_{(v',v)\in\partial W} G_W(x,v';z)\, T(v',v)\, G_\Omega(v,y;z) \\
&= - \sum_{(v',v)\in\partial W} G_W(x,v';z)\, T(v',v)\, G_{\Omega\setminus W}(v,y;z) \\
&\quad + \sum_{\substack{(v',v)\in\partial W \\ (u',u)\in\partial W}} G_W(x,v';z)\, T(v',v)\, G_\Omega(v,u';z)\, T(u',u)\, G_{\Omega\setminus W}(u,y;z) ,
\end{aligned}
$$

where the last equality results from iterating (11.11) with the same Γ_1 and we again use the convention that terms corresponding to indices not in Ω are assumed to be zero. Taking expectation values and using (6.18) we thus arrive at

$$
\begin{aligned}
\mathbb{E}\left[|G_\Omega(x,y;z)|^s\right] &\le \sum_{(v',v)\in\partial W} \mathbb{E}\left[|G_W(x,v';z)|^s \, |G_{\Omega\setminus W}(v,y;z)|^s\right] \\
&\quad + \sum_{\substack{(v',v)\in\partial W \\ (u',u)\in\partial W}} \mathbb{E}\left[|G_W(x,v';z)|^s \, |G_\Omega(v,u';z)|^s \, |G_{\Omega\setminus W}(u,y;z)|^s\right] .
\end{aligned}
$$

The first expectation value on the right side factorizes. However, in comparison to the situation in (11.13), due to the lack of a separating layer, the situation is more complicated for the second expectation. We cannot proceed by simply integrating over the random variables associated to v and u' since the random variable $V(v)$ enters the last two factors and $V(u')$ enters the first two factors. However, the decoupling inequality (8.69) ensures that this expectation value can at least be bounded from above according to

(11.15)

$$\begin{aligned}
&\mathbb{E}\left[|G_W(x,v';z)|^s\, |G_\Omega(v,u';z)|^s\, |G_{\Omega\setminus W}(u,y;z)|^s\right] \\
&\leq D_{s,2}^2\, \mathbb{E}\left[\mathbb{E}_{u'}\left[|G_W(x,v';z)|^s\right] \mathbb{E}_{v,u'}\left[|G_\Omega(v,u';z)|^s\right] \mathbb{E}_v\left[|G_{\Omega\setminus W}(u,y;z)|^s\right]\right] \\
&\leq D_{s,2}^2\, \frac{4C_{s,\tau}}{|\lambda|^s}\, \mathbb{E}\left[|G_W(x,v';z)|^s\right] \mathbb{E}\left[|G_{\Omega\setminus W}(u,y;z)|^s\right] .
\end{aligned}$$

Here the last line results from the a priori fractional moment (FM) estimate of Corollary 8.4. In order to arrive at an inequality of Simon-Lieb type for $\mathbb{E}\left[|G_\Omega(x,y;z)|^s\right]$, we use Lemma 11.4, presented below, to replace $G_{\Omega\setminus W}$ in the last expression with terms involving G_Ω. Using it we arrive at

(11.16)

$$\boxed{\mathbb{E}\left[|G_\Omega(x,y;z)|^s\right] \leq \sum_{v\in W,\, u\in\Omega} \mathbb{E}\left[|G_W(x,v;z)|^s\right] k_{x,\Lambda}(v,u)\, \mathbb{E}\left[|G_\Omega(u,y;z)|^s\right]}$$

with

(11.17)

$$k_{x,\Lambda}(v,u) :=$$

$$\sum_{v'\in\Omega} 1_{(v,v')\in\partial W} \left(\delta_{u,v'} + \sum_{u'\in\Omega} \frac{\kappa_s}{|\lambda|^s} \left(1_{(u,u')\in\partial W} + \left(1+\frac{\kappa_s}{|\lambda|^s}\right) 1_{(u',u)\in\partial W}\right)\right) .$$

In case $\Omega = \mathbb{Z}^d$, the inequality fits the framework described in Theorem 9.3. Since

$$\sum_{u\in\Omega} k_{x,\Lambda}(v,u) \leq \left(1+\frac{\kappa_s}{|\lambda|^s}|\partial W|\right)^2 \sum_{v'\in\Omega} 1_{(v,v')\in\partial W} , \tag{11.18}$$

the condition (11.5) implies the assumption (9.27), which by that theorem implies (11.6).

The more general law (11.7) was the subject of Exercise 9.2 and follows by iteration of (11.16) and its adjoint equation. This can be carried out only as long as the arguments of the Green function stay at distance $L := \sup\{|u| \,|\, u \in \Lambda^+\}$ not only from each other but also from the boundary $\partial\Omega$ such that W is just a translate of Λ. The relevant observation is that for every pair of sites $x,y\in\Omega$ there is a pair of integers n,m such that (i) $n+m = \operatorname{dist}_\Omega(x,y)$ and (ii) the ball of radius n centered at x and the ball of radius m centered at y form a pair of disjoint subsets of Ω. We then iterate (11.16) n/L times from the left, and its adjoint m/L times from the right. $\square$

The proof of Theorem 11.2 relied on the following procedure for restoring the FM of the depleted Green function into terms involving the full one.

Lemma 11.4. *In the situation of Theorem* 11.2, *for any* $z \in \mathbb{C}\backslash\mathbb{R}$, *any subset* $W \subset \Omega \subseteq \mathbb{Z}^d$, *and any* $u, y \in \Omega\backslash W$,

$$\mathbb{E}\left[\left|G_{\Omega\backslash W}(u,y;z)\right|^s\right] \leq \mathbb{E}\left[|G_\Omega(u,y;z)|^s\right] + \frac{4C_{s,\tau}D_{s,2}^2}{|\lambda|^s} \sum_{(v',v)\in\partial W} \mathbb{E}\left[|G_\Omega(v,y;z)|^s\right]. \tag{11.19}$$

Proof. The resolvent equation (11.10) with Γ_1 as in (11.11) yields

$$G_{\Omega\backslash W}(u,y;z) = G_\Omega(u,y;z) + \sum_{(v',v)\in\widetilde{\Gamma}_1} G_{\Omega\backslash W}(u,v;z)\, T(v,v')\, G_\Omega(v,y;z). \tag{11.20}$$

Taking expectation values and using (6.18) we therefore arrive at

$$\begin{aligned}\mathbb{E}\left[\left|G_{\Omega\backslash W}(u,y;z)\right|^s\right] &\leq \mathbb{E}\left[|G_\Omega(u,y;z)|^s\right] \\ &\quad + \sum_{(v',v)\in\partial W} \mathbb{E}\left[\left|G_{\Omega\backslash W}(u,v';z)\right|^s \; |G_\Omega(v,y;z)|^s\right].\end{aligned} \tag{11.21}$$

We first integrate over the random variables associated with u and v' and employ the decoupling inequality (8.69) twice. We hence end up estimating $\mathbb{E}_{u,v'}\left[\left|G_{\Omega\backslash W}(u,v';z)\right|^s\right] \leq 4C_{s,\tau}|\lambda|^{-s}$ by Corollary 8.4. This completes the proof of (11.19). $\square$

11.4. Additional implications

Let us call attention to three additional implications of the finite-volume criteria, the first of which is in line with the discussion in Section 9.3.

Fast power decay implies exponential localization: As is generally the case when Hammersley-Simon-Lieb type inequalities apply, Theorem 11.2 can be used to establish fast power law decay of the FM of the Green function as a criterion for exponential decay.

Corollary 11.5 (Fast power law implies exponential decay). *In the situation of Theorem* 11.2 *there is some* $A, L_0 < \infty$ *(which depend on* λ*) such that if for* $E \in \mathbb{R}$ *and some* $L \geq L_0$,

$$\sup_{(x,y)\in\partial\Lambda_L} \mathbb{E}\left[\left|\langle\delta_0\,, (H_{\Lambda_L}-E)^{-1}\,\delta_x\rangle\right|^s\right] \;\leq\; A\, L^{3(1-d)}, \tag{11.22}$$

then there exists some complex neighborhood $U(E) \subset \mathbb{C}$ containing E and $\mu > 0$, $C < \infty$ such that for all $x, y \in \mathbb{Z}^d$,

$$\sup_{z \in U(E)\setminus\mathbb{R}} \mathbb{E}\left[|G(x,y;z)|^s\right] \;\le\; C\, e^{-\mu|x-y|} . \tag{11.23}$$

Proof. Since $|\partial\Lambda_L| = \mathcal{O}(L^{d-1})$, the finite-volume condition (11.5) with $\Lambda = \Lambda_L$ is implied by (11.22) for a given E at some $L < \infty$.

The extension of the (locally uniform) bound to a complex neighborhood follows from the continuity of the function $z \mapsto \mathbb{E}\big[|\langle\delta_0\,, (H_{\Lambda_L} - z)^{-1}\,\delta_x\rangle|^s\big]$ in a finite volume $L < \infty$; cf. Exercise 8.1. □

Explaining complete localization in one dimension: Corollary 11.5 is particularly effective in $d = 1$, for which it implies that in the presence of homogeneous disorder the two-point function (of an operator with short-range matrix elements) has to decay exponentially fast if it decays at all. As slow decay is not difficult to establish, this statement can be used for a proof of complete localization in one dimension at arbitrarily small $\lambda > 0$ (a topic to which we turn next, in Chapter 12).

Localization at the edges of the spectrum: Localization in the regime of Lifshitz tails is another consequence of Corollary 11.5. By Theorem 4.18 the subsequent assumption (11.25) is indeed tailored to imply localization for the operator (11.1) in a regime of the from

$$\{(E,\lambda) \in \mathbb{R}\times\mathbb{R} \,|\, E \le -2d + \lambda\,\mathbb{E}[V(0)] - c\lambda^{1+q}\}\,, \tag{11.24}$$

with some c finite and $0 < q < \frac{d}{d+4}$; cf. Figure 4.1.

Corollary 11.6 (Localization via Lifshitz tails)**.** *In the situation of Theorem 11.2 suppose that for $E \in \mathbb{R}$ and some $\beta \in (0,1)$, $C_1 > 0$, $C_2 < \infty$, $\alpha > 3(d-1)$, and $L_0 > 0$, one has that for all $L \ge L_0$,*

$$\mathbb{P}\left\{\operatorname{dist}(\sigma(H_{\Lambda_L}), E) \le C_1\, L^{-\beta}\right\} \;\le\; C_2 L^{-\alpha} . \tag{11.25}$$

Then there exists some complex neighborhood $U(E) \subset \mathbb{C}$ containing E and $\mu > 0$, $C < \infty$ such that (11.23) holds for all $x, y \in \mathbb{Z}^d$.

Proof. We will show that the condition (11.22) is satisfied at the given $E \in \mathbb{R}$.

For fixed $L \in (1,\infty)$ we set $\delta_L := C_1\, L^{-\beta}$ and define as the good set of realizations

$$\Omega_G := \{\omega \,|\, \operatorname{dist}(\sigma(H_{\Lambda_L}), E) > \delta_L\}\;, \tag{11.26}$$

on which the Green functions decay exponentially,

$$|G_{\Lambda_L}(0,x;E;\omega)| \le \frac{2}{\delta_L} e^{-c\,\delta_L\,|x|}, \quad \omega \in \Omega_G, \tag{11.27}$$

according to the Combes-Thomas estimate (Theorem 10.5) with some constant $c > 0$ (which depends on the dimension d). The probability of the bad realizations $\Omega_B := \Omega \backslash \Omega_G$ is controlled by (11.25). We may hence estimate

$$\begin{aligned}
\mathbb{E}\left[|G_{\Lambda_L}(0,x;E)|^s\right] &= \mathbb{E}\left[|G_{\Lambda_L}(0,x;E)|^s\, 1_{\Omega_G}\right] + \mathbb{E}\left[|G_{\Lambda_L}(0,x;E)|^s\, 1_{\Omega_B}\right] \\
&\le 2^s\,\delta_L^{-s}\, e^{-sc\,\delta_L\,|x|} + \left(\mathbb{E}\left[|G_{\Lambda_L}(0,x;E)|^{sp}\right]\right)^{\frac{1}{p}} \left(\mathbb{P}\left(\Omega_B\right)\right)^{1-\frac{1}{p}} \\
&\le 2^s\,\delta_L^{-s}\, e^{-sc\,\delta_L\,|x|} + |\lambda|^{-s}\left(C_{sp,\tau}\right)^{\frac{1}{p}} \left(C_2 L^{-\alpha}\right)^{1-\frac{1}{p}} \qquad (11.28)
\end{aligned}$$

where $p \in (1, \tau/s)$ is any number such that the fractional moments are finite. The required bound (11.22) is satisfied if one chooses $s \in (0,\tau)$ small enough and hence p large enough such that $\alpha > 3(d-1)\,\frac{p}{p-1}$. If $L \ge L_0$ is large enough, then both terms on the right side of (11.28) for x such that $(x,y) \in \partial\Lambda_L$ are bounded by $AL^{3(1-d)}/2$:

$$\begin{aligned}
2^s\, C_1^s\, L^{3(d-1)+s\beta}\, e^{-s\,c\,C_1\,L^{1-\beta}} &\le \tfrac{1}{2}A, \\
|\lambda|^{-s}\left(C_{sp,\tau}\right)^{\frac{1}{p}} \left(C_2 L^{3(d-1)\frac{p}{p-1}-\alpha}\right)^{\frac{p-1}{p}} &\le \tfrac{1}{2}A. \qquad (11.29)
\end{aligned}$$

This completes the proof of (11.22) and hence of the result. □

Notes

The material in this chapter draws on the results which were presented in [**11**], with minor modification.

Exercises

11.1. Prove Theorem 11.3.

Chapter 12

Complete Localization in One Dimension

Disorder effects are particularly striking in one-dimensional systems. In this chapter we present a number of results which demonstrate this in the context of random Schrödinger operators on $\ell^2(\mathbb{Z})$:

$$(H(\omega)\,\psi)\,(x) \;=\; -\sum_{y:|x-y|=1} \psi(y) + V(x;\omega)\,\psi(x). \tag{12.1}$$

Some of the statements presented here will also be of value in discussing disorder effects in other situations, including for the mobility edge of random operators on tree graphs. Following is an informal summary of this chapter's highlights.

1. **Dynamical system perspective:** Similar to other one-dimensional systems such as those mentioned in Chapter 9, whose correlations, connections, or transmission result from a local mechanism, the Schrödinger operator's Green function can be studied through an iterative relation, or a transfer matrix formalism. This allows us to cast the question of existence of ac spectrum in terms of a **Lyapunov exponent** $\mathcal{L}(E) \geq 0$.

   The main results here are the theorems of K. Ishii and L. Pastur [**200, 322**] and S. Kotani and B. Simon [**263, 349**] which when combined state that for ergodic Schrödinger operators (12.1) (satisfying a mild assumption) at Lebesgue-almost every $E \in \mathbb{R}$,

   $$\mathcal{L}(E) \;>\; 0 \quad\Longleftrightarrow\quad \operatorname{Im} G(x,x;E+i0) = 0 \text{ almost surely for all } x \in \mathbb{Z}.$$

   Equivalently stated, the operator's ac spectrum coincides with the Lebesgue-essential closure of the set $\{E \in \mathbb{R} \,|\, \mathcal{L}(E) = 0\}$.

2. **Restriction of ac spectrum to deterministic potentials:** Kotani's result comes with a catch, which is remarkable on its own. For ergodic Schrödinger operators

$$|\{E \in \mathbb{R} \,|\, \mathcal{L}(E) \;=\; 0\}| \;>\; 0 \Longrightarrow \{V(x)\} \text{ forms a } \textbf{deterministic} \text{ sequence.}$$

Among the deterministic potentials one finds the periodic ones, whose spectrum is indeed ac, and quasi-periodic potentials whose spectrum need not be so. However, the random Schrödinger operator with iid potential is certainly not in that class.

3. **Complete dynamical localization:** For iid random potentials the above results imply that at any non-zero disorder $\mathcal{L}(E) > 0$ for almost all $E \in \mathbb{R}$. From this, the tools presented in earlier chapters allow us to conclude full exponential localization, in the spectral sense and the dynamical sense. In particular, the fractional moment finite-volume sufficiency criterion of Chapter 11 implies exponential decay for the eigenfunction correlator from just the knowledge that $\mathbb{E}\left(|G(0,L;E+i0)|^s\right) \to 0$ as $L \to \infty$ for some $s > 0$.

   A convenient tool for carrying out the above argument is found in the function

$$\varphi(s;z) \;=\; \lim_{|x|\to\infty} \frac{\ln \mathbb{E}\left[|G(0,x;z)|^s\right]}{|x|}\,, \tag{12.2}$$

which shares some features with the free-energy in statistical mechanics, and the **moment-generating function** in the theory of large deviations. Among its properties proven here is convexity in s and the relation

$$\frac{\partial \varphi}{\partial s}(0;E) \;=\; -\mathcal{L}(E)\,. \tag{12.3}$$

The function φ will also play an essential role in the analysis of the mobility edge on tree graphs.

4. **Scattering perspective:** One-dimensional systems may be viewed as quantum wires. Looking beyond the class of disordered potentials, into operators which do allow conduction in the sense discussed in Section 2.5, the Kotani theory yields another interesting implication: For shift covariant operators, conduction (ac spectrum and $\mathcal{L}(E) = 0$) requires the transmission through the wire to be **reflectionless**. Thus considerations of scattering suggest yet another perspective on the strong localization effects of disorder in one dimension.

12.1. Weyl functions and recursion relations

In one dimension and more generally on tree graphs the Schrödinger operator's Green functions have a product structure. The values of $G(x, y; z)$ can be determined from the **Weyl functions**,

$$\begin{aligned} \Gamma_+(x;z) &:= \langle \delta_x, \left(H_x^+ - z\right)^{-1} \delta_x \rangle, \\ \Gamma_-(x;z) &:= \langle \delta_x, \left(H_x^- - z\right)^{-1} \delta_x \rangle, \end{aligned} \tag{12.4}$$

which are the Green functions of the operator's restrictions $H_x^{\pm}$ to the half-lines $[x, \infty) \cap \mathbb{Z}$ and, correspondingly, $(-\infty, x] \cap \mathbb{Z}$.

Following is a summary of the basic relations. They have an intuitive explanation for those comfortable with such terms through the random walk expansion presented in Section 6.4 but can also be derived through resolvent identities.

Proposition 12.1. *The Weyl functions of the one-dimensional Schrödinger operator $H = -A + V$ satisfy the* **discrete Riccati equation**

$$\boxed{\frac{1}{\Gamma_\pm(x;z)} = V(x) - z - \Gamma_\pm(x \pm 1; z)}\,. \tag{12.5}$$

Moreover:

(i) *They also yield the full Green function through*

$$G(x, x; z) = \frac{1}{V(x) - z - \Gamma_-(x-1;z) - \Gamma_+(x+1;z)}, \tag{12.6}$$

and for $x, y \in \mathbb{Z}$ and $z \in \mathbb{C}^+$,

$$G(x, y; z) = G(x, x; z) \prod_{x<u\leq y} \Gamma_+(u; z) = G(y, y; z) \prod_{x\leq u<y} \Gamma_-(u; z)\,. \tag{12.7}$$

(ii) *For all $x < y$ and $z \in \mathbb{C}^+$,*

$$\begin{aligned} G_+(x, y; z) &:= \langle \delta_x, \left(H_x^+ - z\right)^{-1} \delta_y \rangle = \prod_{x\leq u\leq y} \Gamma_+(u; z)\,, \\ G_-(y, x; z) &:= \langle \delta_y, \left(H_x^- - z\right)^{-1} \delta_x \rangle = \prod_{y\leq u\leq x} \Gamma_-(u; z)\,. \end{aligned} \tag{12.8}$$

Proof. The relations (12.7) and (12.8) form particular cases of (6.26).

Using the factorization (12.8), the Riccati equation (12.5) emerges as an equivalent form of the equation satisfied by the Green function $G_+(x_0, x; z)$ at $x > x_0$ and $z \in \mathbb{C}^+$,

$$(12.9) \qquad - \sum_{y:\,|y-x|=1} G_+(x_0, y; z) + [V(x) - z]\; G_+(x_0, x; z) \;= 0\,,$$

and similarly for $G_-(x, x_0; z)$ at $x < x_0$.

The derivation of (12.6) is left as Exercise 12.1. □

For completeness, let us add that the Riccati equation (12.5) is just an equivalent way of writing the one-dimensional Schrödinger equation

$$(12.10) \qquad - \sum_{y:\,|y-x|=1} \psi(y; z) + (V(x) - z)\; \psi(x; z) \;= 0\,, \quad x \in \mathbb{Z}\,,$$

expressed in terms of the ratios

$$(12.11) \qquad \Gamma_{\pm}(x; z) \;=\; \frac{\psi(x)}{\psi(x \mp 1)}\,.$$

The Weyl functions $\Gamma_{\pm}(x; z)$ correspond to the two special solutions of (12.10) which are ℓ^2-summable towards $\pm\infty$ correspondingly. For each $z \in \mathbb{C}^+$ this summability condition characterizes the functions $\psi_{\pm}(x)$ uniquely up to an overall multiplicative factor, and thus the Weyl functions $\Gamma_{\pm}$ encode $\psi_{\pm}$'s unique projective images.

12.2. Lyapunov exponent and Thouless relation

For ergodic operators $\Gamma_{\pm}(x; z, \omega)$ is a shift-covariant function of ω:

$$(12.12) \qquad \Gamma_{\pm}(x; z, \omega) = \Gamma_{\pm}(0; z, T_{-x}\omega)\,,$$

and hence the factorization (12.7) and Birkhoff's ergodic theorem (Proposition 3.5) imply that for any ergodic Schrödinger operator, at almost every ω:

(12.13)

$$\begin{aligned} \lim_{L\to\infty} \frac{1}{L} \ln |G(0, L; z, \omega)| &= \lim_{L\to\infty} \frac{1}{L} \ln |G_+(0, L; z, \omega)| \\ &= \lim_{L\to\infty} \frac{1}{L} \sum_{n=1}^{L} \ln |\Gamma_{\pm}(0; z, T_{-n}\,\omega)| = \mathbb{E}\left(\ln |\Gamma_{\pm}(0; z)|\right). \end{aligned}$$

This motivates the following definition.

Definition 12.2. For any ergodic Schrödinger operator (12.1) the **Lyapunov exponent** is for $z \in \mathbb{C}^+$

$$\boxed{\mathcal{L}(z) := -\mathbb{E}\left[\ln|\Gamma_+(0;z)|\right]} \tag{12.14}$$

and for $E \in \mathbb{R}$ at which the limit exists

$$\mathcal{L}(E) := \mathcal{L}(E+i0) = \lim_{\eta\downarrow 0} \mathcal{L}(E+i\eta). \tag{12.15}$$

Following are some of its immediate properties and notable points:

1. By (12.7), the decay or growth rate of products of Γ_+ is the same as of products of Γ_-. Hence, also
$$\mathcal{L}(z) = -\mathbb{E}\left[\ln|\Gamma_-(0;z)|\right].$$
2. The Lyapunov exponent forms the negative real part of the Herglotz-Pick function $\mathbb{C}^+ \ni z \mapsto \mathbb{E}\left[\ln \Gamma_+(0;z)\right]$. By implication, the limit (12.15) exists for Lebesgue-almost all $E \in \mathbb{R}$.
3. The recursion relation (12.5) implies, for any $z \in \mathbb{C}^+$,
$$\operatorname{Im}\Gamma_\pm(0;z) \;=\; |\Gamma_\pm(0;z)|^2 \;\left[\operatorname{Im}\Gamma_\pm(\pm 1;z) + \operatorname{Im} z\right]. \tag{12.16}$$
Translation invariance hence yields
$$\mathbb{E}\left[\ln \operatorname{Im}\Gamma_+(0;z)\right] \;=\; -2\,\mathcal{L}(z) + \mathbb{E}\left[\ln\left(\operatorname{Im}\Gamma_+(0;z) + \operatorname{Im} z\right)\right] \tag{12.17}$$
and therefore for all $E \in \mathbb{R}$ and $\eta \geq 0$,
$$\boxed{\mathcal{L}(E+i\eta) \geq 0}. \tag{12.18}$$

The Lyapunov exponent has a remarkably useful relation to the density of states, expressed in the following formula of D. J. Thouless [**381**] (with precursors and special cases due to D. C. Herbert and R. Jones [**192**] and T. Hirota and K. Ishii [**193**]).

Theorem 12.3 (Thouless formula)**.** *Let $H(\omega)$ be a standard ergodic Schrödinger operator in $\ell^2(\mathbb{Z})$, with a random potential satisfying $\mathbb{E}\left[\ln\left(1+|V(0)|\right)\right] < \infty$. Then for all $z \in \mathbb{C}^+$,*

$$\boxed{\mathcal{L}(z) = \int \ln|u-z|\,\nu(du)}. \tag{12.19}$$

Proof. The result is based on the merger of two observations:

1. By Cramer's rule, the Green function of the restriction $H_L(\omega)$ of the operator $H(\omega)$ to the finite segment $[1,L]\cap\mathbb{Z}$,
$$G_L(1,L;z) := \langle \delta_1, (H_L - z)^{-1}\delta_L\rangle,$$

can be expression as the ratio

$$G_L(1,L;z) = \frac{1}{\det(H_L - z)} = \exp\left(\operatorname{tr}\ln(H_L - z)^{-1}\right). \tag{12.20}$$

2. For the Green function, $G_+(1,L;z)$ of the half-infinite system $[1,\infty)\cap\mathbb{Z}$, the factorization (12.8) yields

$$\lim_{L\to\infty}\frac{1}{L}\mathbb{E}\left[\ln|G_+(1,L;z)|\right] \;=\; -\mathcal{L}(z) \tag{12.21}$$

for all $z\in\mathbb{C}^+$. The quantity averaged here is bounded since

$$\frac{1}{|V(x)|+|z|+2(\operatorname{Im} z)^{-1}} \le |G_\#(x,x;z)| \le (\operatorname{Im} z)^{-1}, \quad \# = L,+. \tag{12.22}$$

To merge the two observations, one may note that the corresponding pairs differ by just the ratios of two diagonal elements:

$$\frac{G_+(1,L;z)}{G_L(1,L;z)} = \frac{G_+(L,L;z)}{G_L(L,L;z)}. \tag{12.23}$$

This, along with (12.22), allows us to conclude from (12.21) that for any $z\in\mathbb{C}^+$ also

$$\mathcal{L}(z) = -\lim_{L\to\infty}\frac{1}{L}\mathbb{E}\left[\ln|G_L(1,L;z)|\right] \;=\; \lim_{L\to\infty}\frac{1}{L}\mathbb{E}\left[\operatorname{tr}\ln|H_L - z|\right], \tag{12.24}$$

with the last equality holding due to (12.20).

To relate the limit to the density of states, we note that for any fixed $z\in\mathbb{C}^+$ the function $f_z(u) := \ln|u-z|$ is continuous in u and bounded on compact sets, however not at infinity. We therefore split the integral

$$\frac{1}{L}\mathbb{E}\left[\operatorname{tr}\ln|H_L - z|\right] =: \int f_z(u)\,\nu_L(du) \tag{12.25}$$

into the contribution from u with $|u-z|\le K$ and the remainder. For the most part, the function f_z is bounded and continuous, and by the convergence of the finite-volume DOS measure (Theorem 3.15) its mean converges to the integral with respect to the limiting measure ν. To control the contribution from $|u-z|>K$ we invoke the bound, derived in Exercise 12.2:

$$\begin{aligned}\int_{|u-z|\ge K} f_z(u)\,\nu_L(du) \;&\le \int \ln\left(1+\frac{|u-z|^2}{K}\right)\nu_L(du)\\ &\le C\,\mathbb{E}\left[\ln\left(1+\frac{c}{\sqrt{K}}\,(1+|V(0)|+z)\right)\right],\end{aligned} \tag{12.26}$$

which is valid for all $K\ge 1$ with some numerical constants $c, C\in(0,\infty)$.

In the limit $K\to\infty$ the last term's contribution vanishes. Hence the integral (12.25) can be approximated arbitrarily closely by quantities for which the limit $L\to\infty$ can be taken, and the claimed relation (12.19) follows. □

The Thouless relation has been found useful in a myriad of applications, including the Kotani theory presented below. In [**104, 103**] it is used to conclude the log-Hölder continuity of the DOS measure, and it provides a convenient starting point for estimates on $\mathcal{L}(E)$ and asymptotic expansions in the strength of the disorder [**324**].

12.3. The Lyapunov exponent criterion for ac spectrum

At energies $E \in \mathbb{R}$ at which the limits defining $\Gamma_{\pm}(x; E + i0)$ exist for all $x \in \mathbb{R}$ one has by (12.6) that

$$\text{(12.27)} \quad \operatorname{Im} G(x, x; E + i0) = 0 \quad \text{if and only if} \quad \operatorname{Im}\Gamma_{-}(x-1; E+i0) = \operatorname{Im}\Gamma_{+}(x+1; E+i0) = 0$$

and by (12.16) that

$$\text{(12.28)} \qquad \operatorname{Im}\Gamma_{+}(0; E+i0) \;=\; |\Gamma_{+}(0; E+i0)|^2 \operatorname{Im}\Gamma_{+}(x+1; E+i0)\,,$$

with the reflected relation for $\operatorname{Im}\Gamma_{-}(0; E + i0)$. It readily follows that $\operatorname{Im}\Gamma_{+}(0; E + i0) \neq 0$ if and only if $\operatorname{Im}\Gamma_{+}(x; E + i0) \neq 0$ for an infinite collection of $x \geq 0$. Since the latter event is shift invariant, one may conclude that for ergodic operators the following dichotomy holds:

$$\text{(12.29)} \qquad \mathbb{P}\left(\operatorname{Im}\Gamma_{\pm}(0; E+i0) = 0\right) \quad \text{is either } 0 \text{ or } 1$$

with the possible exception of a null set of E, at which the condition subsumed above could fail.

If $\mathcal{L}(E) > 0$, the recursion relation (12.17) (taken at $z = E+i0$) suggests that the distribution of the imaginary part decreases towards zero under iteration. The theorem asserting that this is indeed the case goes back to K. Ishii [**200**] and L. Pastur [**322**].

Theorem 12.4 (Ishii-Pastur). *Let $H(\omega)$ be an ergodic Schrödinger operator in $\ell^2(\mathbb{Z})$ with a random potential for which $\mathbb{E}\left[\ln\left(1 + |V(0)|\right)\right] < \infty$. If for a Borel set $I \subset \mathbb{R}$*

$$\text{(12.30)} \qquad \mathcal{L}(E) > 0 \quad \textit{for Lebesgue-almost all } E \in I,$$

then $\sigma^{\mathrm{ac}}(H) \cap I = \emptyset$ almost surely.

Proof. The proof is guided by the observation that $\mathcal{L}(E) > 0$ implies that the quantity $\mathbb{E}\left(\ln \operatorname{Im}\Gamma_{+}(x; E + i0)\right)$ is monotone decreasing under the iteration $x \mapsto (x - 1)$, and hence it may serve as a *Lyapunov function(al)* for the evolution under iterations of the probability distribution of $\operatorname{Im}\Gamma_{+}(x; E+i0)$. The argument is, however, carried out at $\eta > 0$, and that counters the monotonicity at small values of this random variable. Nevertheless, (12.17) implies, using the stationarity of the distribution of $\Gamma_+(x; E+i0)$, that for any $E \in \mathbb{R}$ and $\eta > 0$,

$$2\,\mathcal{L}(E+i\eta) \;=\; \mathbb{E}\left[\ln\left(1+\frac{\eta}{\operatorname{Im}\Gamma_\pm(0;E+i\eta)}\right)\right]. \tag{12.31}$$

The integrand can be estimated uniformly in $\eta \in (0,1]$ using

$$\begin{aligned}\frac{\eta}{\operatorname{Im}\Gamma_\pm(0;E+i\eta)} &= \left(\int \frac{\mu^\pm_{\delta_0}(du)}{(u-E)^2+\eta^2}\right)^{-1} \\ &\le \int \left[(u-E)^2+\eta^2\right]\mu^\pm_{\delta_0}(du) = \langle \delta_0, (H_0^\pm - E)^2\delta_0\rangle + 1\,,\end{aligned} \tag{12.32}$$

which is Jensen's inequality for the average with respect to the spectral measure $\mu^\pm_{\delta_0}$ associated with δ_0 and $H_0^\pm$, the operators restriction to the half-lines. The first term on the right side is just $(V(0)-E)^2+1$.

If $\mathbb{P}\left(\operatorname{Im}\Gamma_\pm(0;E+i0)=0\right) = 0$, the integrand in (12.31) almost surely tends to 0 as $\eta \downarrow 0$, and the above bound combined with the dominated convergence principle allows us to conclude that $\mathcal{L}(E) = 0$, which by assumption is not true for almost all $E \in I$. Thus, given the dichotomy (12.29), we conclude that for almost all $E \in I$,

$$\mathbb{P}\left(\operatorname{Im}\Gamma_+(0;E+i0)=0\right) = 1\,, \tag{12.33}$$

and similarly for Γ_-. Through (12.27) this implies the absence of ac spectrum. □

The condition $\mathcal{L}(E) = 0$ is not only necessary for the presence of absolutely continuous spectrum but it is also sufficient. That is one of two linked results of the Kotani theory:

Theorem 12.5 (Kotani-Simon I). *Under the assumptions of Theorem* 12.4, *if for a Borel set $I \subset \mathbb{R}$*

$$\mathcal{L}(E) = 0 \quad \textit{for Lebesgue-almost all } E \in I, \tag{12.34}$$

then $\sigma^{\mathrm{ac}}(H) \cap I = \emptyset$ almost surely.

This result and Theorem 12.6 below were first presented by S. Kotani for the continuous case [**263**]. The argument was adapted to the discrete setup by B. Simon [**349**] whose derivation guides our presentation.

Proof. From the relation (12.31), applying the bound $\ln(1+x) \ge x/(1+\frac{1}{2}x)$ (for $x > 0$) one gets

$$2\mathcal{L}(E+i\eta) \ge \mathbb{E}\left[\frac{\eta}{\operatorname{Im}\Gamma_\pm(0;E+i\eta)+\eta/2}\right]. \tag{12.35}$$

Since the denominator on the right side converges to $\operatorname{Im}\Gamma_\pm(0;E+i0)$ almost surely and for almost all $E\in\mathbb{R}$, we conclude using Fatou's lemma that

$$\lim_{\eta\downarrow 0}\frac{\mathcal{L}(E+i\eta)}{\eta}\geq\frac{1}{2}\mathbb{E}\left[\frac{1}{\operatorname{Im}\Gamma_\pm(0;E+i0)}\right]. \tag{12.36}$$

For energies $E\in\mathbb{R}$ at which $\mathcal{L}(E)=0$, the limit on the left side can be written as

$$\lim_{\eta\downarrow 0}\frac{\mathcal{L}(E+i\eta)}{\eta}=\lim_{\eta\downarrow 0}\frac{\mathcal{L}(E+i\eta)-\mathcal{L}(E+i0)}{\eta}=\lim_{\eta\downarrow 0}\frac{\partial\mathcal{L}(E+i\eta)}{\partial\eta}, \tag{12.37}$$

with the last equality holding provided the derivative $\frac{\partial\mathcal{L}}{\partial\eta}(E+i\eta)$ is continuous from above at $\eta=0+$. Such continuity is implied by the Thouless relation (12.19) since it gives

$$\frac{\partial\mathcal{L}(E+i\eta)}{\partial\eta}=\operatorname{Re}\int\frac{\partial\ln(E+i\eta-u)}{\partial\eta}\,\nu(du)=\operatorname{Im}\int\frac{\nu(du)}{u-E-i\eta} \tag{12.38}$$

and the integral on the right defines a function in the Herglotz-Pick class. By the de la Valleé-Poussin theorem for such functions the limit $\eta\downarrow 0$ exists and is finite at Lebesgue-almost all $E\in\mathbb{R}$. Consequently, for almost every $E\in\mathbb{R}$ at which $\mathcal{L}(E)=0$ the expression on the right in (12.36) is finite.

It follows that for almost all $E\in\mathbb{R}$ at which $\mathcal{L}(E)=0$

$$\mathbb{P}\left(\operatorname{Im}\Gamma_\pm(0;E+i0)>0\right)=1, \tag{12.39}$$

which through (12.27) implies the claim. □

12.4. Kotani theory

Kotani continued his version of Theorem 12.5 with the observation that for ergodic operators absolutely continuous spectrum requires the potential to be deterministic.

Theorem 12.6 (Kotani-Simon II). *Under the assumptions of Theorem* 12.4, *if* (12.34) *holds for a set $I\subset\mathbb{R}$ of positive Lebesgue measure, then:*

1. *$V(\omega)$ is deterministic in the sense that $V(1;\omega)$ is a measurable function of $\{V(x;\omega)\,|\,x\leq 0\}$.*
2. *The system is reflectionless throughout I in the sense that for almost all $E\in I$ and almost surely*

$$\overline{\Gamma_+(1;E+i0)}=\left(\Gamma_-(0;E+i0)\right)^{-1}. \tag{12.40}$$

It may be added that a deterministic potential is fully determined by its tail; i.e., $\{V(x;\omega)\,|\,x\leq -L\}$ for arbitrarily large $L<\infty$ up to events of measure zero. Among the examples of this is the almost-Mathieu potential in (3.15), as well as a number of other curiously deterministic potentials of the form $V(x;\omega)=\sum_{n\in\mathbb{Z}}\omega(n)\,f(x-n)$ which can be found in [**236**] (see also [**105**]). The prime example of a non-deterministic ergodic random potential is the iid case with any non-trivial single-site distribution.

The key step in the proof of Theorem 12.6 is the following result. It will be used to control the fluctuations in the imaginary parts of the Weyl functions

$$n_\pm(E;\eta) := \operatorname{Im}\Gamma_\pm(\pm 1; E+i\eta) + \eta/2 \tag{12.41}$$

associated to the left and right half-lines.

Lemma 12.7. *For any $E \in \mathbb{R}$ and $\eta > 0$*

$$\begin{aligned} &\frac{\mathcal{L}(E+i\eta)}{\eta} - \frac{\partial \mathcal{L}(E+i\eta)}{\partial \eta} \\ &\quad \geq \frac{1}{2}\,\mathbb{E}\left[\frac{1}{2}\left(\frac{1}{n_+(E;\eta)} + \frac{1}{n_-(E;\eta)}\right) - \frac{1}{\frac{1}{2}(n_+(E;\eta)+n_-(E;\eta))}\right] \\ &\quad = \mathbb{E}\left(\frac{(n_+(E;\eta)-n_-(E;\eta))^2}{4\,[n_+(E;\eta)+n_-(E;\eta)]\; n_+(E;\eta)\, n_-(E;\eta)}\right). \end{aligned} \tag{12.42}$$

As with Theorem 12.5, the first inequality is based on the Thouless relation. The quantity in the middle is non-negative by Jensen inequality, and the last expression makes this positivity manifest.

Proof. The inequality is established by combining the lower bound (12.35), rewritten as

$$\frac{\mathcal{L}(E+i\eta)}{\eta} \geq \mathbb{E}\left[\frac{1}{2\,n_\pm(E;\eta)}\right], \tag{12.43}$$

with the Thouless relation. The latter upon differentiation yields (12.38) and hence

$$\frac{\partial \mathcal{L}(E+i\eta)}{\partial \eta} = \mathbb{E}\left[\operatorname{Im} G(0,0;E+i\eta)\right] \leq \mathbb{E}\left[\frac{1}{n_+(E;\eta)+n_-(E;\eta)}\right], \tag{12.44}$$

with the inequality based on (12.6). The equality claimed in (12.42) follows by a simple computation. □

Proof of Theorem 12.6. We start by proving the theorem's second assertion. For that, let us further transform the bound (12.42) into

$$\frac{\mathcal{L}(E+i\eta)}{\eta} - \frac{\partial \mathcal{L}(E+i\eta)}{\partial \eta} \geq \mathbb{E}\left(\frac{(n_+(E;\eta)-n_-(E;\eta))^2}{4\,[n_+(E;\eta)+n_-(E;\eta)]^3}\right). \tag{12.45}$$

By Theorem 12.5, in the limit $\eta \downarrow 0$ for almost all $E \in I$ and almost surely

$$\lim_{\eta\downarrow 0} n_\pm(E;\eta) \;=\; n_\pm(E;0) \;=\; \operatorname{Im}\Gamma_\pm(\pm 1; E+i0) \in (0,\infty)\,, \tag{12.46}$$

while by (12.37) the left side in (12.45) tends to zero for almost all $E \in I$ for $\eta \downarrow 0$. Invoking Fatou's lemma this allows us to conclude from (12.45) the almost-sure equality

$$n_+(E;0) \;=\; n_-(E;0)\,. \tag{12.47}$$

This in turn implies

$$\lim_{\eta\downarrow 0} \frac{\partial \mathcal{L}(E+i\eta)}{\partial\eta} = \lim_{\eta\downarrow 0} \frac{\mathcal{L}(E+i\eta)}{\eta} \;=\; \lim_{\eta\downarrow 0} \mathbb{E}\left[\frac{1}{2\,n_\pm(E;\eta)}\right]. \tag{12.48}$$

Returning to the upper bound (12.44), we conclude that the inequality is sharp in the limit $\eta \downarrow 0$ from which it also easily follows that

$$V(0) - E = \operatorname{Re}\Gamma_+(1;E+i0) + \operatorname{Re}\Gamma_-(-1;E+i0) \tag{12.49}$$

almost surely. The two equalities (12.47) and (12.49) constitute the equality (12.40). This concludes the proof of the second assertion of the theorem.

The deduction that the potential is deterministic proceeds along the following steps:

1. By its definition, $\Gamma_-(0, E+i0;\omega)$ is a measurable function of $\{V(x;\omega) \mid x \leq 0\}$.
2. Through (12.40), $\Gamma_+(1, E+i0;\omega)$ is determined for Lebesgue-almost all $E \in I$ through $\Gamma_-(0, E+i0;\omega)$. Since both are Herglotz functions, this implies that the relation (12.40) holds for Lebesgue-almost all $E \in \mathbb{R}$; cf. Proposition B.5.
3. $V(1;\omega) = \langle \delta_1, H_1^+(\omega)\,\delta_1\rangle + 2$, where $H_1^+(\omega)$ denotes the operator's restriction to the half-line $[1,\infty)\cap\mathbb{Z}$. It can therefore be determined from the spectral measure associated with the vector δ_1. This measure is in turn uniquely determined by the boundary values of $\Gamma_+(1, E+i0;\omega)$, which form the measure's Borel-Stieltjes transform. □

12.5.* Implications for quantum wires

The one-dimensional Schrödinger equation (12.10) may be viewed as describing a quantum wire. To explain why under the condition (12.40) the potential is called reflectionless, let us first recall some basic concepts.

Associated with any quantum state function $\psi : \mathbb{Z} \mapsto \mathbb{C}$ is the **current functional** $J[\psi] = (J_{x,x+1}[\psi])_{x\in\mathbb{Z}}$, which is defined on the bonds $[x, x+1]$

by

$$J_{x,x+1}[\psi] \;:=\; \frac{1}{i}\left(\overline{\psi}(x)\,\psi(x+1) - \psi(x)\,\overline{\psi}(x+1)\right) . \tag{12.50}$$

Under the Schrödinger dynamics $\psi_t(x) = (e^{-itH}\psi)(x)$ the following *continuity equation* holds:

$$\operatorname{div} J[\psi_t](x) \;+\; \frac{\partial}{\partial t}\,|\psi_t(x)|^2 \;=\; 0\,, \tag{12.51}$$

where $\operatorname{div} J[\psi_t](x) := J_{x,x+1}[\psi] - J_{x-1,x}[\psi]$. The Schrödinger operator H has the time reversal symmetry, implemented by the antiunitary mapping $\psi \mapsto \overline{\psi}$, under which $J_{x,x+1}[\overline{\psi}] = -J_{x,x+1}[\psi]$.

Solutions of (12.10) describe stationary states for which the current is constant along the line. In particular, the solutions $\psi_\pm$ which are related to the Weyl functions through

$$\begin{aligned}\psi_+(x;E+i0) &= \Gamma_+(x;E+i0)\,\psi_+(x-1;E+i0)\,,\\ \psi_-(x;E+i0) &= \Gamma_-(x;E+i0)\,\psi_-(x+1;E+i0)\end{aligned} \tag{12.52}$$

may be viewed as the steady-state quantum amplitudes for particles in a situation in which current is injected into the wire at one of its ends: $-\infty$ in the case of ψ_+ and $+\infty$ in the case of ψ_-. This interpretation can be made somewhat more transparent by first considering finite segments of the wire, say restricted to $[-L,L]$, under the dissipative dynamics generated by $H - i\eta$, and then taking the limits in the following order: $\lim_{\eta\downarrow 0}\lim_{L\to\infty}$. In this limit, non-zero conduction of particles of energy E requires $\overline{\psi}_\pm \neq \psi_\pm$, i.e., $E \in \sigma^{\mathrm{ac}}(H)$. In that case, any solution of (12.10) can be presented as a linear combination of ψ_+ and $\overline{\psi}_+$, or of ψ_- and $\overline{\psi}_-$, which we normalize by the current, so that

$$J_{x,x+1}[\psi_\pm] \;=\; \pm 1\,. \tag{12.53}$$

This, along with the unambiguously defined $\Gamma_\pm$, determines $\psi_\pm$ up to a unitary factor.

The relation between the two bases states yields a natural notion of the reflection and transmission coefficients, $|r|^2$ and $|t|^2$, defined through the relation

$$\psi_+(x,E+i0) \;+\; \overline{r}\,\overline{\psi}_+(x,E+i0) \;=\; \overline{t}\,\overline{\psi}_-(x,E+i0)\,. \tag{12.54}$$

The arguments of r and t depend on the phase choice for $\psi_\pm$, but regardless of that, $1 - |r|^2 = |t|^2$ by a simple comparison of the currents. Combining (12.54) with (12.52), one gets

$$r \;=\; e^{-2i\theta(x)}\,\frac{(\Gamma_-(x;E+i0))^{-1} - \overline{\Gamma_+(x+1;E+i0)}}{(\Gamma_-(x;E+i0))^{-1} - \Gamma_+(x+1;E+i0)} \tag{12.55}$$

with $\theta(x) := \operatorname{Arg} \psi_+(x, E + i0)$. Note that r does not vary with x for the given choice of phase for $\psi_\pm$ (which in general would not be a shift-covariant function of ω).

The above coefficient r can be alternatively regarded [**72, 208**] as describing the scattering effect of the *quantum dot* at x when connected to the pair of the semi-infinite wires represented by $(-\infty, x-1] \cap \mathbb{Z}$ and $[x+1, \infty) \cap \mathbb{Z}$.

The relation (12.40) in Kotani's theorem may hence indeed be interpreted as the statement that for any ergodic operator **the reflection coefficient vanishes** (i.e., $r = 0$) within the ac spectrum!

This principle seems to have an intuitive explanation: In the ergodic setup, if reflection is possible, then it will have occurred infinitely often and the transmission rate from infinity would vanish. This also explains the point, which may nevertheless still be worth stressing, that the conclusion that $r = 0$ is not implied by just the existence of *ac* spectrum. Indeed it would not remain true under a (generic) local modification of an initially shift-covariant potential. A simple example of that is seen in the effect of the addition of a single delta-potential to the shift-invariant Laplacian.

An interesting extension of these ideas to general one-dimensional Schrödinger operators was formulated by C. Remling [**332**]. His oracle theorem limits the possibility of ac spectrum to potential for which the collection of the asymptotically recurring patterns at $\pm\infty$ (in the sense of accumulation points under shifts) includes only sequences in the class identified by Kotani, that is, potentials which are deterministic and reflectionless.

12.6. A moment-generating function

Beyond the Lyapunov exponent, further information about the decay of the Green functions is encoded in its large-deviation or **moment-generating function**:

$$\boxed{\varphi(s; z) \; := \; \lim_{|x| \to \infty} \frac{\ln \mathbb{E}\left[|G(0, x; z)|^s\right]}{|x|}} \,. \tag{12.56}$$

The existence of the limit will be established for the following choice of parameters:

At *complex energies*, $z \in \mathbb{C}^+$: for all $s \in [-\varsigma, \infty)$.

At *real energies*, $z = E + i0$: for all $s \in [-\varsigma, \tau)$, with E restricted to the full measure set of energies for which the boundary values of the Green function exist for all $x \in \mathbb{Z}$.

Also included in the following result is the relation of the moment-generating function to the Lyapunov exponent. This would allow us to establish complete dynamical localization from the Kotani theorem.

Theorem 12.8 (Generating function and Lyapunov exponent). *Let $V(\omega)$ be and iid random potential on $\mathbb{Z}$ whose single-site distribution ρ satisfies the following:*

i. *ρ is locally uniformly τ-Hölder continuous for some $\tau \in (0,1]$.*
ii. *$\int |v|^{\varsigma}\rho(dv) < \infty$ for some $\varsigma \in (0,\tau)$.*

Then, for the one-dimensional random Schrödinger operator $H(\omega)$ of (12.1)*:*

1. *Over the range $(s,z) \in [-\varsigma,\infty) \times \mathbb{C}^+$ the limit* (12.56) *exists and the function it defines satisfies the following:*
   (i) *$\varphi(s;z)$ is convex and non-increasing in $s \in [-\varsigma,\infty)$.*
   (ii) *The following relation to the Lyapunov exponent holds:*

$$\boxed{\frac{\partial\varphi}{\partial s}(0;z) \quad = \quad -\mathcal{L}(z)}\,. \tag{12.57}$$

2. *For Lebesgue-almost all $E \in \mathbb{R}$ the limit* (12.56) *with $z = E + i0$ exists for any $s \in [-\varsigma,\tau)$, and the function*

$$\varphi(s;E) := \lim_{|x|\to\infty}\lim_{\eta\downarrow 0}\frac{\ln \mathbb{E}\left[|G(0,x;E+i\eta)|^s\right]}{|x|} \tag{12.58}$$

   shares the above properties (i) *and* (ii) *and satisfies*

$$\varphi(s;E) = \lim_{\eta\downarrow 0}\varphi(s;E+i\eta)$$

   (i.e., the limits in (12.58) *are interchanged).*

3. *For any $z \in \mathbb{C}^+ \cup \mathbb{R}$ there are $c_s(z), C_s(z) \in (0,\infty)$ such that for all $x \in \mathbb{Z}$*

$$\boxed{c_s(z)^{-1}\,e^{\varphi(s;z)\,|x|} \;\le\; \mathbb{E}\left[|G(0,x;z)|^s\right] \;\le\; C_s(z)\,e^{\varphi(s;z)\,|x|}}\,. \tag{12.59}$$

   Moreover:
   (i) *The constants can be chosen locally uniformly bounded; i.e.,*

$$\sup_{s\in S}\;\sup_{z\in K+i(0,1]}\;\max\{c_s(z),C_s(z)\} < \infty \tag{12.60}$$

   for any compact subset $K \subset \mathbb{R}$ and $S \subset [-\varsigma,\tau)$.
   (ii) *The same bounds also apply to the Green function $\widehat{G}(x,y;z)$ of the operator's restriction to $[x,y] \cap \mathbb{Z}$.*

In the proof we shall employ the following super/subadditivity principle of M. Fekete [**150**]. Some readers may recognize it as the handy tool which is often used in statistical mechanics in existence proofs of the free energy per volume.

Proposition 12.9 (Fekete). *Let (a_n) be a positive sequence for which there is some $C \in (0,\infty)$ such that all $n, m \in \mathbb{N}$ either*

A. **supermultiplicative case:** $\quad a_{n+m} \ \geq\ C\, a_n\, a_m \qquad$ *or*

B. **submultiplicative case:** $\quad a_{n+m} \ \leq\ C\, a_n\, a_m\,.$

Then the limit

$$\lim_{n\to\infty} \frac{\ln a_n}{n} \quad =: \alpha \tag{12.61}$$

exists and for any $n \in \mathbb{N}$,

$$a_n \begin{cases} \leq C^{-1} e^{\alpha n} & \textit{case A}, \\ \geq C^{-1} e^{\alpha n} & \textit{case B}. \end{cases} \tag{12.62}$$

The proof is left as Exercise 12.3.

We shall apply this principle to the sequence

$$a_n(z,s) := \mathbb{E}\left[|G(0,n;z)|^s\right] \tag{12.63}$$

and the related sequence

$$\hat{a}_n(z,s) := \mathbb{E}\left[\left|\widehat{G}(1,n-1;z)\right|^s\right], \qquad n \geq 2\,, \tag{12.64}$$

with $\widehat{G}(x,y;z) = G_{[x,y]}(x,y;z)$ the Green function of the operator restricted to the segment $[x,y] \cap \mathbb{Z}$. The two sequences are comparable and are both super- and submultiplicative:

Lemma 12.10. *Under the assumptions of Theorem 12.8:*

1. *There are $C_\pm(z,s) < \infty$ such that for all $n, m \geq 2$,*

$$C_-^{-1}(z,s)\; a_n(z,s) \leq \hat{a}_n(z,s) \leq\ C_+(z,s)\; a_n(z,s)\,. \tag{12.65}$$

2. *The sequence $(\hat{a}_n)_{n\geq 2}$ is both super- and submultiplicative.*

Furthermore, for any compact $K \subset \mathbb{R}$ and $S \subset [-\varsigma, \tau)$, all arising constants are uniformly bounded for $(z,s) \in (K + i(0,1]) \times \in S$ (with $K + i(0,1]$ the set sum).

Since the proof of the first part of this lemma involves techniques similar to those employed in the second part, we will leave its proof as Exercise 12.4. The proof of the super- or submultiplicativity of $(\hat{a}_n)$ rests on the factorization property of the Green function. Abbreviating

$$Q_{n,m} := G_{[1,n+m-1]}(n,n;z) = \Big(V(n) - z - \hat{\Gamma}_-(n-1;z) - \hat{\Gamma}_+(n+1;z)\Big)^{-1}, \tag{12.66}$$

with

$$\hat{\Gamma}_-(n-1;z) := G_{[1,n-1]}(n-1,n-1;z)$$

and

$$\hat{\Gamma}_+(n+1;z) := G_{[n+1,n+m-1]}(n+1,n+1;z),$$

we may write

$$\begin{aligned}(12.67)\qquad \hat{a}_{n+m} &= \mathbb{E}\left[|\widehat{G}(1,n-1;z)|^s\,|Q_{n,m}|^s\,|\widehat{G}(n+1,n+m-1;z)|^s\right]\\ &=: \mathrm{Av}\left(|Q_{n,m}|^s\right)\,\hat{a}_n\,\hat{a}_m\,.\end{aligned}$$

The last line defines an average, whose dependence on s, z, and $n,m \geq 2$ we drop from the notation for convenience. The bounds,

$$(12.68)\qquad 0 < \inf_{n,m\geq 2} \mathrm{Av}\left(|Q_{n,m}|^s\right) \leq \sup_{n,m\geq 2} \mathrm{Av}\left(|Q_{n,m}|^s\right) < \infty\,,$$

immediately establish the second claim in Lemma 12.10, and a closer inspection of the subsequent proof establishes the claimed local uniformity of the involved constant. It thus remains to give the proof of these bounds, which is somewhat technical.

Proof of the bounds (12.68)**.** We distinguish the three cases:

Case $s \geq 0$**:** In this case we take $z \in \mathbb{C}^+$ and use

$$(12.69)\qquad |G(y,y;z)| \leq (\mathrm{Im}\, z)^{-1}$$

for the upper bound. For the lower bound Jensen's inequality for averages ensures that for any $\varepsilon \in (0,s]$,

$$(12.70)\qquad \mathrm{Av}\left(|Q_{n,m}|^s\right) \geq \mathrm{Av}\left(|Q_{n,m}|^{-\varepsilon}\right)^{-\frac{s}{\varepsilon}}\,.$$

The triangle inequality applied to the sum in the dominator in (12.66) shows that it is sufficient to bound the average of the ε-moment of each of the terms on the right side. Clearly, in case $\varepsilon \in (0,\varsigma]$,

$$(12.71)\qquad \mathrm{Av}(|V(n)|^\varepsilon) = \int |v|^\varepsilon \varrho(v)\,dv < \infty\,.$$

The other terms are estimated trivially using (12.69) again.

Case $s \in [0,\tau)$**:** For the upper bound, we note that the conditional average, conditioned on all the potential's values aside from $V(n)$, coincides with

$$(12.72)\qquad \mathrm{Av}\left(|Q_{n,m}|^s\,|V_{\neq n}\right) = \mathbb{E}\left[|Q_{n,m}|^s\middle|V_{\neq n}\right] \leq C_{s,\tau}\,,$$

where the inequality is the basic fractional moment estimate (8.8).

For the lower bound, we again start with Jensen's inequality (12.70) and the subsequent estimate (12.71). To bound the terms in (12.66) note that

$$\text{(12.73)} \qquad \operatorname{Av}\left(|\hat{\Gamma}_-(n-1;z)|^{\varepsilon}\right) = \frac{\mathbb{E}\left[|\widehat{G}(1,n-1;z)|^{s}\,|\hat{\Gamma}_-(n-1;z)|^{\varepsilon}\right]}{\mathbb{E}\left[|\widehat{G}(1,n-1;z)|^{s}\right]}$$

and similarly for Γ_+. Conditioning on all values of the potential aside from $V(n-1)$ and using the product structure of the Green function $\widehat{G}(1,n-1;z) = \widehat{G}(1,n-2;z)\,\hat{\Gamma}_-(n-1;z)$, we are hence left with estimating

(12.74)

$$\begin{aligned}\mathbb{E}\left[|\hat{\Gamma}_-(n-1;z)|^{s+\varepsilon}\,|\,V_{\neq n-1}\right] &= \int \frac{\rho(dv)}{|v-\hat{\Gamma}_-(n-2;z)|^{s+\varepsilon}}\\ \leq C_{s+\varepsilon}(\rho)\int \frac{\rho(dv)}{|v-\hat{\Gamma}_-(n-2;z)|^{s}} &= C_{s+\varepsilon}(\rho)\,\mathbb{E}\left[|\hat{\Gamma}_-(n-1;z)|^{s}\,|\,V_{\neq n-1}\right],\end{aligned}$$

where we used (8.61) with $C_{s+\varepsilon}(\rho) < \infty$ finite and independent of $z \in \mathbb{C}^+$ as long as $0 < s+\varepsilon < \tau$. The right side in (12.73) is thus upper bounded by $C_{s+\varepsilon}(\rho)$. Proceeding similarly with Γ_+ concludes the proof of the lower bound in (12.68) in this case.

Case $s \in [-\varsigma, 0)$**:** Through (12.66) the proof of the upper bound mirrors the lower bound in case $s \in (0,1)$. For the lower bound we use Jensen's inequality

$$\operatorname{Av}\left(|Q_{n,m}|^{s}\right) \geq \frac{1}{\operatorname{Av}\left(|Q_{n,m}|^{-s}\right)}.$$

The proof hence follows the lines of reasoning of the upper bound in case $s \in (0,\tau)$.

This completes the proof of (12.68). □

Proof of Theorem 12.8. Following is the list of the assertions of Theorem 12.8 with their proofs.

Existence of the limit and the bounds (12.59)**:** This is an immediate consequence of Fekete's result (Proposition 12.9) and the properties established in Lemma 12.10. The claimed uniformity of the constants in (12.59) in case $s \in [-\varsigma, 1)$ can be read from the explicit upper and lower bounds in the proof of (12.68).

Interchangeability of the limits: The bounds (12.59) allow us to conclude that for any $s \in [-\varsigma, \tau)$ the convergence (12.56) is in fact locally uniform; i.e.,

$$\lim_{|x|\to\infty} \sup_{s\in S} \sup_{z\in K+i(0,1]} \left| \varphi(s;z) - \frac{\ln \mathbb{E}\left[|G(0,x;z)|^s\right]}{|x|} \right| = 0 \tag{12.75}$$

for any compact $K \subset \mathbb{R}$, $S \subset [-\varsigma, \tau)$. Since the fractional moments $\mathbb{E}\left[|G(0,x;E+i\eta)|^s\right]$ are continuous in η at $\eta = 0$ (cf. (8.36) and Exercise 8.3), this implies the interchangeability of the limits $|x| \to \infty$ and $\eta \downarrow 0$ in (12.58).

Convexity and monotonicity: Since the prelimits are convex functions of s, the limit (12.56) is convex. Since for any $\varepsilon \geq 0$

$$\mathbb{E}\left[|G(0,x;z)|^{s+\varepsilon}\right] \leq (\operatorname{Im} z)^{-\varepsilon}\, \mathbb{E}\left[|G(0,x;z)|^s\right], \tag{12.76}$$

the limit is non-increasing in s for all $z \in \mathbb{C}^+$. These properties are inherited by the limit $\lim_{\eta\downarrow 0} \varphi(s;E+i\eta)$.

Relation to the Lyapunov exponent: We will give a simultaneous argument for $z \in \mathbb{C}^+$ and $z = E + i0$. Homogeneity and the factorization property of the Green function imply that for any $x \geq 1$,

$$\mathcal{L}(z) = -\frac{1}{|x|}\, \mathbb{E}\left[\ln |G_+(1,x;z)|\right]. \tag{12.77}$$

Since $\varphi(s;z) = \lim_{|x|\to\infty} |x|^{-1} \ln \mathbb{E}\left[|G_+(1,x;z)|^s\right]$, Jensen's inequality shows that the negative Lyapunov exponent is a lower bound on the slope of the moment-generating function; i.e., for any $s \in [-\varsigma, \tau)$ and $x \geq 1$,

$$\begin{aligned} 0 \leq\ & \varphi(s;z) + s\mathcal{L}(z) \\ \leq\ & \frac{1}{|x|}\left(\ln \mathbb{E}\left[|G_+(1,x;z)|^s\right] - \mathbb{E}\left[\ln |G_+(1,x;z)|^s\right] + \ln[c(z)]\right). \end{aligned} \tag{12.78}$$

Here the last inequality used the lower bound in (12.59) and the fact that one may truncate the first vertex from the Green function at the cost of a finite factor; cf. (12.65). The constant $c(z) < \infty$ can be taken uniformly in $\operatorname{Im} z \in (0,1]$ and in s in a sufficiently small interval about zero. To estimate the first two terms we employ the bound

$$0 \leq \ln \mathbb{E}\left[|Y|\right] - \mathbb{E}\left[\ln |Y|\right] \leq \frac{1}{2}\, \mathbb{E}\left[(1+|Y|)\,(\ln |Y|)^2\right], \tag{12.79}$$

which derives from the elementary estimates $e^\alpha \leq 1 + \alpha + \alpha^2(e^\alpha + 1)/2$ and $1 + \beta \leq e^\beta$ valid for all $\alpha, \beta \in \mathbb{R}$. Applying this bound to $Y = |G_+(1,x;z)|^s$, the right side of (12.79) reads

$$\frac{s^2}{2}\,\mathbb{E}\left[(1+|G_+(1,x;z)|^s)\,(\ln|G_+(1,x;z)|)^2\right] \le \frac{s^2}{2}\,|x|\sum_{n=1}^{|x|}\mathbb{E}\left[(1+|G_+(1,x;z)|^s)\,(\ln|\Gamma_+(n;z)|)^2\right], \tag{12.80}$$

where use of the product structure of the Green function was made. By Exercise 12.5 for any $p > 0$

$$\sup_{\operatorname{Im} z\in(0,1]} \mathbb{E}\left[(\ln|\Gamma_+(n;z)|)^p\right] < \infty\,, \tag{12.81}$$

and the right side is bounded by some constant times $s^2|x|^2$. For any sequence $s \to 0$, we may therefore choose a divergent sequence $|x(s)| \to \infty$ such that $s|x(s)| \to 0$. Then (12.78) together with (12.80) implies that

$$\lim_{s\to 0}\frac{\varphi(s;z)}{s} = -\mathcal{L}(z)\,, \tag{12.82}$$

which completes the proof of Theorem 12.8. □

12.7. Complete dynamical localization

Since any iid random potential $V(\omega)$ with a non-trivial distribution is non-deterministic, the Kotani-Simon theorem (Theorem 12.6) immediately yields that for almost all energies $E \in \mathbb{R}$,

$$\mathcal{L}(E) > 0\,. \tag{12.83}$$

Through the Ishii-Pastur theorem (Theorem 12.4) this in turn implies that the corresponding random operator $H(\omega)$ has only singular spectrum. Some further analysis is required for a proof that the spectrum is only pure point and that dynamical localization holds.

Historically, the first proof of complete spectral localization was accomplished by I. Goldsheid, S. Molchanov, and L. Pastur [**176**] in a special one-dimensional continuum model. Complete dynamical localization in the situation described by the following theorem was shown by H. Kunz and B. Souillard [**274**] (the argument given below is different, though not unrelated to theirs).

Theorem 12.11 (Complete dynamical localization). *Let $H(\omega)$ be a random Schrödinger operator on $\mathbb{Z}$ satisfying the assumptions of Theorem* 12.8. *Then for almost all $E \in \mathbb{R}$ there is some complex neighborhood $U(E) \subset \mathbb{C}$ and some $s \in (0,\tau)$ and $C,\mu \in (0,\infty)$ such that for all $x,y \in \mathbb{Z}$*

$$\sup_{z\in U(E)\setminus\mathbb{R}} \mathbb{E}\left[|G(x,y;z)|^s\right] \le C\, e^{-\mu|x-y|}\,. \tag{12.84}$$

If additionally $\tau = 1$, then the operator $H(\omega)$ exhibits strong exponential dynamical localization on $I = \operatorname{Re} U(E)$; i.e., (10.13) *holds.*

Proof. With a minor refinement, the finite-volume criterion of Corollary 11.5 allows us to conclude that for one-dimensional operators (12.84) holds throughout some complex neighborhood of E from just the knowledge that

$$\mathbb{E}\left[\left|G_{[0,L]}(0, L; E)\right|^s\right] \underset{L\to\infty}{\longrightarrow} 0 \,. \tag{12.85}$$

The proof of this assertion is left as Exercise 12.6.

Condition (12.85) can be inferred for the iid potentials considered here, from the positivity of the Lyapunov exponent, $\mathcal{L}(E) > 0$, and the proven properties of the moment-generating function. By convexity and since $-\mathcal{L}(E) = \partial\varphi(0; E)/\partial s$ (cf. (12.57)), there is some $s(E) \in (0, \tau)$ such that for all $s \in (0, s(E)]$,

$$\varphi(s; E) \;\leq\; -\frac{s}{2}\mathcal{L}(E) \,. \tag{12.86}$$

The bounds (12.59), applied to the Green function of the operator's restriction to $[0, L] \cap \mathbb{Z}$, then imply the exponential decay

$$\mathbb{E}\left[\left|G_{[0,L]}(0, L; E)\right|^s\right] \;\leq C_s(E) \exp\left(-\frac{s}{2}\,\mathcal{L}(E)\,L\right) \,, \tag{12.87}$$

thereby proving that (12.85) is satisfied. □

Notes

One-dimensional random Schrödinger operators have been extensively studied, and by a variety of techniques which were not covered in this chapter's discussion. Following are pointers to some of them.

An argument that for a range of energies the eigenstates of a one-dimensional Schrödinger operator with a (particular) random potential are localized was given by R. E. Borland [**64**], who studied the scattering through a random chain of delta functions. An interesting, though still early, account of the developments of ideas may be found in [**307**].

Since the Schrödinger equation is a linear second-order difference equation, its solutions span a two-dimensional space. This is evident by rewriting (12.10) as an evolution of the vectors $\Psi(x) := \begin{pmatrix} \psi(x) \\ \psi(x-1) \end{pmatrix}$ by means of a transfer matrix:

$$\Psi(x+1) \;=\; A(x; z)\,\Psi(x) \,, \quad \text{with} \quad A(x; z) := \begin{pmatrix} V(x) - z & -1 \\ 1 & 0 \end{pmatrix} . \tag{12.88}$$

This brings in a dynamical system perspective, in the light of which solutions of the one-dimensional Schrödinger equation have been studied through products of random matrices (of determinant one). Of particular relevance has been the theorem of Furstenberg [**166**], which together with Oseledec's theorem [**317**] provides information on the growth rates of the solutions, through the relevant theory of Lyapunov exponents. Ishii and Pastur's proof of Theorem 12.4 in fact relied on these fundamental results. An extensive account of the transfer-matrix approach to random operators through the Furstenberg theorem can be found in [**82**] and in the excellent textbook [**66**]; see also [**358**, Sect. 10.6].

Extending the reach of the Ishii-Pastur theorem, B. Simon [**359**] showed that within the set $\{E \in \mathbb{R} \,|\, \mathcal{L}(E) > 0\}$ the spectral measure of an ergodic one-dimensional Schrödinger operator is supported on a subset of capacity zero, and hence not only of Lebesgue-measure zero but also of zero Hausdorff dimension.

H. Kunz and B. Souillard [**274**] proved complete dynamical localization by studying the distribution of Γ_+ through implications of the recursion relation (12.5). This proof was later simplified by Delyon, Kunz, and Souillard [**113**] and can also be found in [**105**]. A related proof (whose main focus is a localization result for tree graphs) is contained in [**163**].

B. Simon and T. Spencer [**361**] noted that in one dimension there is no absolutely continuous spectrum for any potential which assumes arbitrarily large values as $x \to \pm\infty$. Proof of exponential decay of fractional moments (as claimed in Theorem 12.11) via the positivity of the Lyapunov exponents was presented in the appendix of [**301**] and extended to the continuum in [**190**].

Spectral localization for iid potentials with Bernoulli distribution, which is the most singular case and regularly omitted in our discussions, was established in [**81**]; see also [**107**] for the extension to the one-dimensional continuum. For general dimensions $d \geq 1$ localization at band-edges or large disorder is still an open problem in the discrete case. For operators over the continuum this was established by J. Bourgian and C. Kenig [**69**] through an extension of the multi-scale method and considerations of the unique continuation principle.

In discussing the Lyapunov exponent we have approached the subject by first considering $\mathcal{L}(E+i\eta)$ for which $\eta > 0$ serves as a regularizing cutoff.

The technical issues which needed to be addressed in order to study $\mathcal{L}(E)$ directly (at $\operatorname{Im} z = 0$) have been tackled in [**37, 104**].

More on the spectral theory of one-dimensional (not necessarily random) Schrödinger operators on $\ell^2(\mathbb{Z})$ can be found in [**379**]. Extensions of the theory of random operators on $\mathbb{Z}$ have been pursued in a number of directions:

- replacing $\mathbb{Z}$ by a sequence of strips, such as $\mathbb{Z} \times [0, \dots, k]$,
- replacement of the nearest-neighbor kinetic term by one with finite-range hopping,
- decorating the line through the attachment of additional finite structures, as suggested by Figure 12.1.

The extension to strips, or band matrices, leads to interesting questions concerning the dependence of the localization length on the width [**83, 84**]. Partial results were presented by J. Schenker [**340**] through a fluctuation-based analysis.

Interesting results on the crossover from one dimension to random matrix behavior have been presented in [**271, 338, 388, 389**].

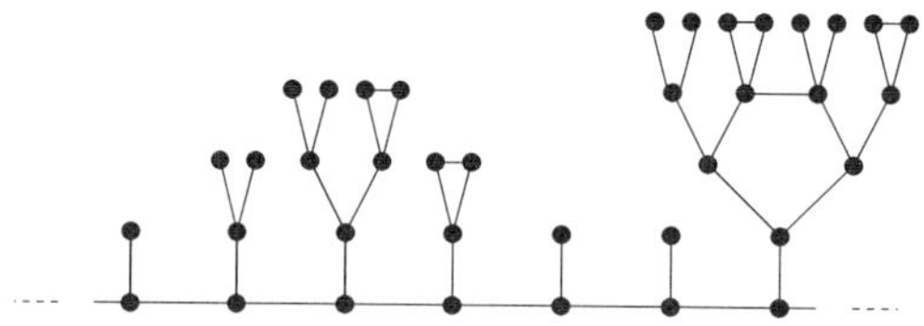

Figure 12.1. On graphs where each point may be separated from infinity by the removal of not more than a fixed number of sites, such as the one indicated in the figure, the spectra of random operators with iid potentials have no absolutely continuos components [**113, 16**].

An essential feature of the decorated line is that any finite region on the resulting graph can be disconnected from infinity by cutting the graph at not more than a fixed number of points. In such cases one may conclude that in the presence of an iid random potential the operator has no absolutely continuous spectrum (which is a generalization along the lines of [**16**] of the one-dimensional result of Delyon, Kunz, and Souillard [**113**]). An example to which this applies is the canopy graph, depicted in Figure 17.3(b), which is of relevance for the discussion of the spectral statistics on tree graphs. It may be added that there are examples of graphs in this class with *singular continuous* spectrum [**16**].

Other, and earlier, constructions of *singular continuous* spectra for Schrödinger operators were presented by D. Pearson [**325**], and a surprising topological robustness of the *sc* spectrum, as an alternative to dense pure-point spectrum, was discovered by B. Simon [**355**].

Exercises

12.1. Prove the relations (12.6), by which the full Green function can be determined from the Weyl functions, and the recursion relation (12.5).

12.2. Prove for the Schrödinger operator H_L in $\ell^2([1,L]\cap\mathbb{Z})$ that for all $\alpha \geq 0$,

$$\operatorname{tr}\left(\ln(1+\alpha\, H_L^2)\right) \leq \sum_{x=1}^{L} \ln\left(1+\alpha\, 2(4^2+V(x)^2)\right) .$$

Show how this may be used to complete the proof of (12.26) for Schrödinger operators with potentials which are ergodic and satisfy

$$\mathbb{E}\left[\ln(1+|V(0)|)\right] < \infty.$$

12.3. Prove Fekete's lemma (Lemma 12.9), also showing that in the super-multiplicative case the limit equals

$$\alpha = \sup_{n\in\mathbb{N}} \frac{b_n}{n} \quad \text{with } b_n := \ln(C a_n).$$

What is the corresponding statement for the submultiplicative case? Hint: It is convenient to think of the sequences $n = mL + r$ with L chosen suitably large.

12.4. Prove part (1) of Lemma 12.10.

12.5. Establish the moment condition expressed by (12.81). [Hint: $\ln|x| \leq |x| + |x|^{-1}$.]

12.6. Consider a one-dimensional Schrödinger operator with a random potential whose conditional single-site distribution ρ is uniformly τ-Hölder continuous with some $\tau \in (0,1]$. Show that if (12.85) is satisfied for some $E \in \mathbb{R}$ and $s \in (0,\tau)$, then there exists a complex neighborhood $U(E) \subset \mathbb{C}$ containing E throughout which (12.84) holds for all $x, y \in \mathbb{Z}$ with some uniform constants $\mu > 0$, $C < \infty$.

Chapter 13

Diffusion Hypothesis and the Green-Kubo-Streda Formula

In this chapter we present the Kubo formula for electrical conductivity in an independent electron gas model of condensed matter. We shall not discuss its full justification, and/or limitations, but will show how the formula is arrived at within the linear response approximation. Not surprisingly, the formula shows the vanishing of the direct conductance at energies within the localization regime. Also presented is a proof of the asymptotic vanishing of the off-diagonal elements of the one-particle density function for the corresponding system of non-interacting fermions if the Fermi energy falls within the localization regime.

We start, however, with the general diffusion conjecture for disordered systems, where the diffusion constant is linked via an Einstein relation to the *dc* conductivity.

13.1. The diffusion hypothesis

Some of the relevant information on the dynamics generated by a self-adjoint operator H in the Hilbert space over a graph with metric d may be encoded in the rate of growth, in time, of the position moments ($\alpha > 0$) in generic states $\psi \in \ell^2(\mathbb{G})$; i.e.,

$$M_\psi(\alpha, t) \;:=\; \sum_{x \in \mathbb{G}} d(x,0)^\alpha \; |\psi(x;t)|^2 \,. \tag{13.1}$$

The corresponding Abel average over times of order $t \approx 1/\eta$ is given by

$$\begin{aligned} \widehat{M}_\psi(\alpha, \eta) \;&:=\; 2\eta \int_0^\infty e^{-2\eta t}\, M_\psi(\alpha, t)\, dt \\ &= \frac{\eta}{\pi} \sum_{x \in \mathbb{G}} d(x,0)^\alpha \int \left| \left((H - E - i\eta)^{-1} \psi \right)(x) \right|^2 dE \,. \end{aligned} \tag{13.2}$$

The RAGE theorem, or more precisely (2.37), implies that $\widehat{M}_\psi(\alpha, \eta) \to \infty$ as $\eta \downarrow 0$ for any $\psi \in \mathcal{H}^c$. The asymptotic growth rates can indicate qualitative differences between systems. Particularly worthy of mentioning are the two cases

$$M_\psi(\alpha, \eta)^{1/\alpha} \;\approx\; \begin{cases} C_\psi\, \eta^{-1}, & \text{ballistic growth,} \\ D_\psi\, \eta^{-1/2}, & \text{diffusive growth.} \end{cases} \tag{13.3}$$

More generally, one refers to the growth rate $M_\psi(\alpha, \eta)^{1/\alpha} \approx C_\psi\, \eta^{-r}$ as subdiffusive if $r < 1/2$ and superdiffusive if $r > 1/2$. Following are some general observations:

1. The "free motion" generated by the Laplacian on $\mathbb{Z}^d$ is ballistic for generic initial states $\psi \in \ell^2(\mathbb{Z}^d)$ and all values of $\alpha > 0$.
2. For Hamiltonians with tempered hopping amplitudes, such as Schrödinger operators (2.2), the growth rate is at most ballistic (i.e., $r \le 1$); cf. Exercise 2.2.
3. Lower bounds on the moments $M_\psi(\alpha, t)$ for $\psi \in \mathcal{H}^c$ can be obtained by bounding above the probability of lingering: $\mathrm{Prob}_{\psi(t)}(|x| < b\, t^r)$. At $r = 0$ these tend to zero for any $b > 0$. More explicit bounds can be derived based on the finer distinction among spectral types along the lines of Definition 2.2. Such bounds were due originally to I. Guarneri [**182**] with generalization found in [**91, 283, 243**]. They are presented in Exercise 2.3.
4. For disordered systems it is expected that the motion is generally diffusive, provided the generator's continuous spectrum is not empty. Thus, starting in the localized state $\psi = \delta_0$ the growth rate of the second moment is expected to be $M_{\delta_0}(2, t) \approx 2tD_{\delta_0}$ with a finite diffusion constant D_{δ_0}.

To expand on the last point, if indeed for $\psi = \delta_0$

$$|\psi(x;t)|^2 \;\approx\; \langle \delta_x \,,\, e^{tD_{\delta_0}\Delta} \delta_{x0} \rangle \,, \qquad \widehat{M}_\psi(\alpha, \eta) \;\approx\; D_{\delta_0} \eta^{-1}, \tag{13.4}$$

then under the time average one is led to the following **diffusion hypothesis** for the Green function:

$$\text{(13.5)}\quad \frac{\eta}{\pi}\left|\langle\delta_x\,,(H-E-i\eta)^{-1}\delta_0\rangle\right|^2 \sim 2\eta\,\langle\delta_x\,,(-D_{\delta_0}(E)\,\Delta+2\eta)^{-1}\delta_0\rangle\,.$$

For the total diffusion constant this yields $D_{\delta_0} = \int \nu(E)D_{\delta_0}(E)dE$, with $\nu(E)$ the density of states (defined in Chapter 5).

The above considerations lead to the **Kubo-Greenwood formula** for the *conjectured relation* between the diffusion constant and the Green function:

$$\text{(13.6)}\quad \boxed{\nu(E)\,D_{\delta_0}(E) = \lim_{\eta\downarrow 0}\frac{\eta^2}{\pi}\sum_x d(x,0)^2\left|\langle\delta_x\,,(H-E-i\eta)^{-1}\delta_0\rangle\right|^2}\,.$$

We will see later in this chapter that the right side here coincides with the *dc* conductance in linear response theory.

Establishing the diffusion hypothesis for a random Schrödinger operator on any finite-dimensional configuration space is a major open challenge. Existing partial results in this direction include the following:

i. a proof of diffusive/ballistic behavior on tree graphs, where diffusion is ballistic (discussed in Chapter 16),
ii. proofs of diffusive behavior up to a bounded times scale which depends on the disorder [**140, 141, 142**],
iii. results in this vein for certain supersymmetric models [**120**]. The latter go beyond the framework of operators discussed here, yet an analog of (13.5) formulated in terms of the relevant correlation function (substituting the Green function) provides there the defining feature of diffusive behavior.

Related to the discussion in the next chapter is the proof of the existence of some transport in the quantum Hall situation $\mathbb{R}^2$ with a perpendicular magnetic field, where the constancy of the Hall conductance over regimes of localization together with the fact that the Hall conductance jumps over the so-called Landau levels is used for a proof [**171, 172**].

13.2. Heuristic linear response theory

The Kubo formula expresses the electrical conductivity within the framework of linear response theory. Here one considers the effects of an electric field $\mathbf{E} \in \mathbb{R}^d$ which is adiabatically switched on. For non-interacting particles with unit charge $q \in \mathbb{R}$ the time evolution is thus governed by a time-dependent operator on some Hilbert space of the form

$$H(t) = H - q\,\mathbf{E}\cdot\mathbf{x}\,e^{\eta t} \qquad (-\infty < t \leq 0\,,\ \eta \downarrow 0)\,.$$

The **adiabatic time evolution** of the system which is described by a density matrix W is given in terms of the following initial-value problem of the **Liouville equation**:

$$\text{(13.7)} \qquad \frac{d}{dt}W(t) \;=\; -i\,[H(t), W(t)]\,, \qquad \lim_{t\to-\infty} e^{itH}\,W(t)\,e^{-itH} = W\,.$$

The state at $t = -\infty$ is thereby chosen as the one which would generate W at $t = 0$ under the time evolution generated by H if started at $t = -\infty$. Simplifications arise if the initial state W is invariant under the time evolution of H. A physically relevant example is the thermal equlilibrium state of a non-interacting system of fermions at inverse temperature $\beta \in (0, \infty]$; i.e.,

$$\text{(13.8)} \quad W = \theta^{\beta}_{E_F}(H)\,, \qquad \theta^{\beta}_{E_F}(\xi) := \begin{cases} \left(1 + e^{\beta(\xi - E_F)}\right)^{-1}, & \beta \in (0,\infty), \\ 1_{(-\infty, E_F]}(\xi)\,, & \beta = \infty\,, \end{cases}$$

where $E_F \in \mathbb{R}$ is the Fermi energy. Then, to first order in $\mathbf{E}$, the approximate solution of (13.7) is

$$\text{(13.9)} \qquad W(0) - W \;=\; iq \int_{-\infty}^{0} e^{itH}\,[\mathbf{E}\cdot\mathbf{x}, W]\,e^{-itH}\,e^{\eta t}\,dt\,.$$

This equation constitutes the **linear response ansatz**.

The **current density** caused by the application of the electric field in a system with elementary charges $q \in \mathbb{R}$ which is translation invariant on average is given by

$$\mathbf{j} := q\,\mathrm{Tr}\,(\mathbf{v}\,(W(0) - W))$$

where $\mathbf{v} := i\,[H, \mathbf{x}]$ is the **velocity operator** and

$$\mathrm{Tr}\,(\cdot) \;:=\; \lim_{L\to\infty} \frac{1}{|\Lambda_L|}\;\mathrm{tr}\,1_{\Lambda_L}(\cdot)$$

denotes the **trace per unit volume**. In the ergodic setup, the trace per unit volume can be replaced by an average over the disorder,

$$\mathrm{Tr}\,(\cdot) = \mathbb{E}\left[\langle \delta_0, (\cdot)\,\delta_0\rangle\right]\,.$$

If H is a magnetic Schrödinger operator $-\Delta^{(A)} + V$ on $\ell^2(\mathbb{Z}^d)$, the components of the velocity operator are given by

$$\text{(13.10)} \qquad (v_\nu \psi)\,(x) \;=\; i \sum_{|y-x|=1} e^{-iqA(x,y)}\,(x-y)_\nu\,\psi(y)\,.$$

In case $A = 0$, they coincide up to a factor of $-i$ with the components of the discrete derivative.

Within the linear response ansatz (13.9), the relation between the components of the current density and the electric field is hence linear, $j_\nu = \sum_\mu \sigma_{\nu,\mu} E_\mu$, as in Ohm's law with

$$\sigma_{\nu,\mu} = \lim_{\eta \downarrow 0} iq^2 \operatorname{Tr}\Big(\int_{-\infty}^{0} v_\nu \, e^{itH} \, [x_\mu, W] \, e^{-itH} \, e^{\eta t} dt \Big) \tag{13.11}$$

defining the **Kubo-Greenwood conductivity tensor**. The above arguments constitute the customary heuristic motivation for the definition of the conductivity tensor. From a mathematical point of view, several questions arise:

1. In what sense can one give meaning to the solutions of the Liouville equation in the ergodic setup?
2. Can one justify the linear response ansatz (13.9) as a first-order approximation to the exact solution of the Liouville equation?
3. Is there a mathematical meaning to the quantities on the right side of (13.11) and does the limit exist?

In the following, we will sketch an answer to the first question and provide the answer to the third question. Answers to these questions were pioneered by works of J. Bellissard, A. van Elst, and H. Schulz-Baldes [**49**]. We skip the justification of the linear response ansatz and refer the reader to the remarks in the Notes.

13.3. The Green-Kubo-Streda formulas

As discussed in Chapter 3, disordered systems are modeled in terms of random, translation-covariant operators. Let us restrict our discussion to the lattice case $\mathbb{Z}^d$ and, in order to incorporate the effects of a constant magnetic field B, implement the group of translations on $\ell^2(\mathbb{Z}^d)$ in terms of magnetic translations given by $U_x^{(A)}\psi(y) = e^{-iBq(y_1x_2 - x_1y_2)}\psi(y-x)$. In such a setup, all operators entering Liouville's equation (13.7) may be taken to be bounded and linear operators $M(\omega)$ on $\ell^2(\mathbb{Z}^d)$. In their dependence on ω, they are required to be weakly measurable and satisfy

$$U_x^{(A)} \, M(\omega) \, \big(U_x^{(A)}\big)^{-1} = M(T_x\omega) \, . \tag{13.12}$$

Such operators constitute the linear space $\mathcal{K}_{\mathrm{mc}}$ of **measurable covariant operators**. Of particular interest is the subspace

$$\mathcal{K}_2 \; := \; \big\{ M \in \mathcal{K}_{\mathrm{mc}} \, \big| \, \mathbb{E}\left[\langle M\delta_0 \, , M\delta_0 \rangle \right] < \infty \big\} \tag{13.13}$$

which together with the scalar product

$$\langle\!\langle \, M_1, M_2 \, \rangle\!\rangle \; := \; \mathbb{E}\left[\langle M_1 \, \delta_0 \, , \, M_2 \, \delta_0 \rangle \right] \tag{13.14}$$

forms a Hilbert space. Solutions to the Liouville equation can now be discussed on this space. Namely, given a bounded self-adjoint $H \in \mathcal{K}_{\mathrm{mc}}$ with

$$\operatorname*{ess\,sup}_{\omega \in \Omega} \|H(\omega)\| < \infty,$$

the linear operator

$$\begin{aligned} \mathcal{L}_H : \mathcal{K}_2 &\to \mathcal{K}_2 \,, \\ M &\mapsto [H, M] \,, \end{aligned} \tag{13.15}$$

called the **Liouville operator**, is self-adjoint on $\mathcal{K}_2$. It therefore generates a unitary group $e^{it\mathcal{L}_H}$ on $\mathcal{K}_2$ and its resolvent at $\eta > 0$ is a bounded operator given by

$$(\mathcal{L}_H - i\eta)^{-1} \;=\; i \int_{-\infty}^{0} e^{it\mathcal{L}_H}\, e^{\eta t}\, dt \,. \tag{13.16}$$

This sketches the answer to the first question.

For future reference, let us mention that, as expected for a trace, the trace per unit volume as given by $\mathbb{E}\left[\langle \delta_0, (\cdot)\delta_0\rangle\right]$ is cyclic; i.e., for any $M_1, M_2 \in \mathcal{K}_2$,

$$\begin{aligned} \mathbb{E}\left[\langle \delta_0, M_1\, M_2 \delta_0\rangle\right] \;&=\; \sum_{x \in \mathbb{Z}^d} \mathbb{E}\left[\langle \delta_0, M_1\, \delta_x\rangle \langle \delta_x, M_2 \delta_0\rangle\right] \\ &=\; \sum_{x \in \mathbb{Z}^d} \mathbb{E}\left[\langle \delta_{-x}, M_1\, \delta_0\rangle \langle \delta_0, M_2 \delta_{-x}\rangle\right] \\ &=\; \mathbb{E}\left[\langle \delta_0, M_2\, M_1 \delta_0\rangle\right] \,. \end{aligned} \tag{13.17}$$

Here the first and third equalities use Fubini's theorem together with the completeness of $(\delta_x)_{x\in\mathbb{Z}^d}$. The second equality is based on the covariance relation (13.12).

As preparation for an answer to the third question in the list, we note that within the above framework the conductivity tensor (13.11) can formally be re-expressed as

$$\begin{aligned} \sigma_{\nu,\mu} \;&=\; \lim_{\eta \downarrow 0} i q^2 \int_{-\infty}^{0} \operatorname{Tr}\left(v_\nu\, e^{it\mathcal{L}_H} \left([x_\mu, W]\right)\right) e^{\eta t}\, dt \\ &=\; q^2 \lim_{\eta \downarrow 0} \operatorname{Tr}\left(v_\nu\, \left(\mathcal{L}_H - i\eta\right)^{-1} \left([x_\mu, W]\right)\right) \\ &=\; q^2 \lim_{\eta \downarrow 0} \langle\!\langle\, v_\nu \,,\, \left(\mathcal{L}_H - i\eta\right)^{-1} \left([x_\mu, W]\right) \rangle\!\rangle \,. \end{aligned} \tag{13.18}$$

Clearly, in the standard situation (13.10) the components of the velocity operator v_ν are elements of $\mathcal{K}_2$. However, to provide meaning to the expression on the right side we also require

$$[x_\mu, W] \in \mathcal{K}_2 \,. \tag{13.19}$$

Even if this applies, the existence of the limit cannot be taken for granted. To further discuss these issues we restrict ourselves to the following situation:

> The Hamiltonian is given by the Laplacian corresponding to a constant magnetic field plus a bounded random potential,
> $$H(\omega) = -\Delta^{(A)} + V(\omega)\,, \quad \text{on } \ell^2(\mathbb{Z}^d),$$
> with $\operatorname{ess\,sup}_{\omega\in\Omega} \|V(\omega)\| < \infty$. The initial state is given by the thermal equilibrium, $W = \theta^{\beta}_{E_F}(H)$; cf. (13.8).

We now distinguish the cases of positive temperature, $\beta \in (0, \infty)$, and zero temperature, $\beta = \infty$.

13.3.1. Zero temperature limit. In the zero temperature limit, the thermal equilibrium state is a projection, $\theta^{\infty}_{E_F}(H) = P_{(-\infty,E_F]}(H) \;=: P$ on $\ell^2(\mathbb{Z}^d)$. The requirement (13.19) then translates to

$$\boxed{\mathbb{E}\left[\|[x_\nu, P]\,\delta_0\|^2\right] \;=\; \sum_{x\in\mathbb{Z}^d} |x_\nu|^2\, \mathbb{E}\left[|\langle\delta_0, P\delta_x\rangle|^2\right] \;<\infty} \tag{13.20}$$

for all $\nu \in \{1, \ldots, d\}$. This is a **localization condition on the two-point function** in the ground-state of non-interacting fermions whose Fermi level is at E_F. In Section 13.4 we will show that the exponential decay of a fractional moment of the Green function at E_F is sufficient for the validity of (13.20).

The condition (13.20) ensures that the prelimit in (13.18) is well-defined. This makes the representation (13.18) of the conductivity available. The remaining question about the existence of the limit $\eta \downarrow 0$ is resolved in the following theorem. Under somewhat more restrictive conditions it was first presented in [**49**]. The present version and its proof are taken from [**10**].

Theorem 13.1 (Kubo-Streda formula). *Under assumption* (13.20) *the conductivity tensor for a system of fermions with unit charge $q \in \mathbb{R}$ at zero temperature as given by the right side in* (13.18) *is well-defined and can be expressed as*

$$\sigma_{\nu,\mu} \;=\; iq^2\, \mathbb{E}\left[\langle\delta_0\,,\, P\,[[x_\nu, P], [x_\mu, P]]\;\delta_0\rangle\right]\,. \tag{13.21}$$

As a consequence, the direct conductivity vanishes; i.e., $\sigma_{\nu,\nu} = 0$ for any $\nu \in \{1, \ldots, d\}$.

The evaluation of the limit $\eta \downarrow 0$ in (13.18) will be based on the formula

$$\lim_{\eta\downarrow 0}\, (\mathcal{L}_H + i\eta)^{-1}\, \mathcal{L}_H\,(M) \;=\; M \tag{13.22}$$

valid for any $M \in (\ker \mathcal{L}_H)^\perp$, the orthogonal complement of the kernel of the Liouville operator. Since the latter is self-adjoint, this identity follows from spectral calculus. We will apply it to $M = [[x_\nu, P], P]$ and hence we need the following:

Lemma 13.2. *Under assumption* (13.20), *one has* $[[x_\nu, P], P] \in (\ker \mathcal{L}_H)^\perp$.

Proof. For any $M \in \ker \mathcal{L}_H$ we have

$$\begin{aligned}\langle\!\langle\, M\,,\, [[x_\nu, P], P]\,\rangle\!\rangle &= \mathbb{E}\Big[\langle \delta_0\,, M^\dagger [x_\nu, P]\, P\delta_0\rangle\Big] - \mathbb{E}\Big[\langle \delta_0\,, M^\dagger P\,[x_\nu, P]\,\delta_0\rangle\Big]\\ &= \mathbb{E}\Big[\langle \delta_0\,, [P, M^\dagger]\,[x_\nu, P]\,\delta_0\rangle\Big]\\ &= \langle\!\langle [M, P]\,,\, [x_\nu, P]\,\rangle\!\rangle \;=\; 0\,. \qquad (13.23)\end{aligned}$$

Here the second equality uses the cyclicity of the trace per unit volume (13.17). The vanishing of the right side follows from $M \in \ker \mathcal{L}_H$, which implies $[M, P] = 0$. □

In order to establish the existence of the limit in (13.18), it is thus useful to rewrite its right side. This will be done in the subsequent

Proof of Theorem 13.1. Denoting $P^\perp := 1 - P$, we have

$$\begin{aligned}[x_\mu, P] &= P[x_\mu, P]P^\perp + P^\perp[x_\mu, P]P,\\ i\mathcal{L}_H\,([[x_\nu, P], P]) &= i\,P^\perp[H, x_\nu]P + i\,P[H, x_\nu]P^\perp\\ &= P^\perp v_\nu P + P v_\nu P^\perp = v_\nu\,.\end{aligned}$$

The cyclic invariance of the trace per unit volume hence yields

$$\begin{aligned}&\langle\!\langle\, v_\nu\,, (\mathcal{L}_H - i\eta)^{-1}\,([x_\mu, P])\,\rangle\!\rangle\\ &\quad= \langle\!\langle\, (P^\perp v_\nu\, P + P v_\nu P^\perp), (\mathcal{L}_H - i\eta)^{-1}\,([x_\mu, P])\,\rangle\!\rangle\\ &\quad= -i\,\langle\!\langle\, \mathcal{L}_H\,([[x_\nu, P], P])\,, (\mathcal{L}_H - i\eta)^{-1}\,([x_\mu, P])\,\rangle\!\rangle\\ &\quad= -i\,\langle\!\langle\, (\mathcal{L}_H + i\eta)^{-1}\,\mathcal{L}_H\,([[x_\nu, P], P])\,, [x_\mu, P]\,\rangle\!\rangle\,. \qquad (13.24)\end{aligned}$$

Since $[[x_\nu, P], P] \in (\ker \mathcal{L}_H)^\perp$ by Lemma 13.2, the spectral calculus of $\mathcal{L}_H$ implies the validity of (13.22) in case $M = [[x_\nu, P], P]$. Inserting the above results into (13.18), we hence obtain

$$\sigma_{\nu,\mu} = -iq^2\,\langle\!\langle\, [[x_\nu, P], P]\,,\, [x_\mu, P]\,\rangle\!\rangle = iq^2\,\mathbb{E}\,[\langle \delta_0\,,\, P\,[[x_\nu, P], [x_\mu, P]]\,\delta_0\rangle]\,. \qquad (13.25)$$

This finishes the proof of the Kubo-Streda formula (13.21). The fact that the direct conductivity vanishes is an immediate consequence thereof. □

In the special case $d = 2$ with a constant perpendicular magnetic field, the transversal, i.e., Hall, conductance as given by the off-diagonals in (13.21) turns out to be an integer divided by 2π. There is a deeper topological reason, which applies more generally than to the quantum Hall situation mentioned, that the number given by the off-diagonals in (13.21) (modulo 2π) turns out to be an integer. As has been noticed in [**386, 49, 38**] in certain situations, these off-diagonals can be identified with an index or Chern number ([**42**]). A notable property of indices is that they appear to vanish under careless application of standard principles, such as commutativity of the trace. Yet they come to life if one is careful to pay attention to convergence issues. A related observation applies to the above argument. A lesson here is that nature tends not to waste beautiful mathematical principles. We will come back to this in the next chapter.

Having established conditions for the existence of the Kubo-Streda formula and the vanishing of the direct conductivity in the zero temperature limit, we now turn to the Kubo formula at positive temperatures.

13.3.2. Positive temperatures. At positive temperature, it is reasonable to rewrite (13.18) using the spectral representation for H:

(13.26)

$$\begin{aligned}\sigma_{\nu,\,\mu} &= q^2 \lim_{\eta\downarrow 0} \iint \frac{1}{E-E'-i\eta}\, \mathbb{E}\left[\langle \delta_0\,,\, v_\nu\, P_{dE}(H)\,[x_\mu, \theta^\beta_{E_F}(H)]\, P_{dE'}(H)\delta_0\rangle\right] \\ &= q^2 \lim_{\eta\downarrow 0} \iint \frac{i}{E-E'-i\eta}\, \frac{\theta^\beta_{E_F}(E)-\theta^\beta_{E_F}(E')}{E-E'}\, m_{\nu,\,\mu}(dE\, dE')\,.\end{aligned}$$

The measure appearing on the right side is the velocity correlation measure associated with H.

Definition 13.3. The **velocity correlation measure** of $H(\omega) = -\Delta^{(A)} + V(\omega)$ on $\ell^2(\mathbb{Z}^d)$ is the finite, complex-valued Borel measure on $\mathbb{R}^2$ which is defined through

$$\int\!\!\int f(E)\, g(E)\, m_{\nu,\mu}(dE\, dE') \;:=\mathbb{E}\left[\langle \delta_0\,,\, v_\nu\, f(H)\, v_\mu\, g(H)\, \delta_0\rangle\right]$$

for all $f,\, g \in C_0(\mathbb{R})$ and $\nu, \mu \in \{1, \ldots, d\}$.

The measure is well-defined through the Riesz-Markov representation theorem. There is a wealth of additional information on this measure available in the literature. For example, using the techniques of Section 3.4, one may show that the velocity-correlation measure is self-averaging in the sense that it coincides almost surely with the infinite-volume limit of finite-volume counterparts. We refer the interested reader to the Notes at the end of this chapter.

It is elementary to check that $\overline{m_{\nu,\mu}(I\times J)} = m_{\mu,\nu}(I\times J) = m_{\nu,\mu}(J\times I)$ for any Borel set $I, J \subset \mathbb{R}$. In the special case $H(\omega) = -\Delta + V(\omega)$, the velocity correlation measure is real; i.e., $m_{\nu,\mu} = \overline{m_{\nu,\mu}}$. Consequently, the Kubo conductivity tensor is symmetric and real, $\sigma_{\nu,\mu} = \overline{\sigma_{\mu,\nu}} \in \mathbb{R}$, and the left side in (13.26) can be simplified:

$$\begin{aligned} \sigma^{\text{sym}}_{\nu,\mu} &:= \frac{1}{2}\left(\sigma_{\nu,\mu} + \overline{\sigma_{\mu,\nu}}\right) \\ &= -\pi\, q^2 \lim_{\eta\downarrow 0} \int\!\!\int \delta_\eta(E-E')\, \frac{\theta^\beta_{E_F}(E) - \theta^\beta_{E_F}(E')}{E-E'}\, m_{\nu,\,\mu}(dE\, dE')\,, \end{aligned} \tag{13.27}$$

where $\delta_\eta(u) := \eta\,\pi^{-1}(u^2+\eta^2)^{-1}$ is an approximate identity. The above expression for $\sigma^{\text{sym}}_{\nu,\mu}$ corresponds to the familiar form of the Kubo formula. It is the starting point for another customary formula for the symmetric part of the conductivity involving the pointwise limit

$$\boxed{\widetilde{\sigma}_{\nu,\mu}(E) := \lim_{\eta\downarrow 0} \frac{\eta^2}{\pi} \sum_{x\in\mathbb{Z}^d} x_\mu x_\nu\, \mathbb{E}\left[|G(0,x;E+i\eta)|^2\right]}\,. \tag{13.28}$$

For energies E in the localization regime in which

$$\sup_{\eta>0} \sum_{x\in\mathbb{Z}^d} |x|^2\, \mathbb{E}\left[|G(0,x;E+i\eta)|^s\right] < \infty \tag{13.29}$$

for some $s \in (0,1)$, this limit is in fact zero; i.e.,

$$\widetilde{\sigma}_{\nu,\mu}(E) \;\le\; \lim_{\eta\downarrow 0} \frac{\eta^s}{\pi} \sum_{x\in\mathbb{Z}^d} x_\mu x_\nu\, \mathbb{E}\left[|G(0,x;E+i\eta)|^s\right] \;=\; 0$$

where the inequality results from $|G(0,x;E+i\eta)| \le \eta^{-1}$. Since $\widetilde{\sigma}_{\nu,\mu}$ is related to the diffusion tensor of the system (cf. (13.6)), the limit should exist in the presence of disorder regardless of localization.

Theorem 13.4 (Existence of the Kubo conductivity). *Suppose that*

$$\sup_{z\in\mathbb{C}^+} \sum_{x\in\mathbb{Z}^d} |x|^2\, (\operatorname{Im} z)^2\, \mathbb{E}\left[|G(0,x;z)|^2\right] \;<\; \infty \tag{13.30}$$

and that the limit (13.28) *exists for all $E \in \mathbb{R}$. Then for any $\beta \in (0,\infty)$ and $E_F \in \mathbb{R}$,*

$$\sigma^{\text{sym}}_{\nu,\mu} \;=\; -q^2 \int \widetilde{\sigma}_{\nu,\mu}(E)\, \frac{d\theta^\beta_{E_F}(E)}{dE}\, dE\,. \tag{13.31}$$

Proof. We start by rewriting the prelimit on the right side of (13.28):

(13.32)

$$\begin{aligned}
&\frac{\eta^2}{\pi} \sum_{x\in\mathbb{Z}^d} x_\mu x_\nu \mathbb{E}\left[|G(0,x;E+i\eta)|^2\right] \\
&\quad = -\frac{\eta^2}{\pi}\,\mathbb{E}\left[\langle \delta_0\,,\left[x_\nu,(H-E-i\eta)^{-1}\right]\left[x_\mu,(H-E+i\eta)^{-1}\right]\delta_0\rangle\right] \\
&\quad = \frac{\eta^2}{\pi}\,\mathbb{E}\left[\langle \delta_0\,,(H-E-i\eta)^{-1}v_\nu|H-E-i\eta|^{-2}v_\mu(H-E+i\eta)^{-1}\delta_0\rangle\right] \\
&\quad = \pi\int\!\!\int \delta_\eta(E-u)\,\delta_\eta(E-u')\,m_{\nu,\mu}(du\,du'),
\end{aligned}$$

where $\delta_\eta(u) = \pi^{-1}\eta(u^2+\eta^2)^{-1}$. The second equality is based on the equality

$$\left[x_\nu,(H-z)^{-1}\right] = (H-z)^{-1}[H,x_\nu]\,(H-z)^{-1}$$

for all $z\in\mathbb{C}^+$, and the last equality relies on the cyclic invariance of the trace per unit volume and the spectral theorem.

The dominated convergence theorem, which is applicable thanks to (13.30), guarantees that the limit $\eta\downarrow 0$ may be interchanged with the E-integration in (13.31). We hence conclude from (13.27) and (13.32) that

(13.33)

$$\sigma^{\mathrm{sym}}_{\nu,\mu} + q^2\int \widetilde{\sigma}_{\nu,\mu}(E)\,\frac{d\theta^\beta_{E_F}(E)}{dE}\,dE \;=\; \pi\,q^2\lim_{\eta\downarrow 0}\int\!\!\int g_\eta(u,u')\,m_{\nu,\mu}(du\,du'),$$

with

$$\begin{aligned}
&g_\eta(u,u') \;=\\
&\int \delta_1(x)\left(\frac{d\theta^\beta_{E_F}(u+\eta x)}{du}\delta_\eta(u-u'+\eta x) - \frac{\theta^\beta_{E_F}(u)-\theta^\beta_{E_F}(u')}{u-u'}\,\delta_\eta(u-u')\right)dx\,.
\end{aligned}$$

Using the fact that $\theta^\beta_{E_F}$ has bounded derivatives (to arbitrary order) for $\beta\in(0,\infty)$, it is a tedious but elementary exercise to show that (i) $g_\eta(u,u')$ is bounded uniformly in (u,u') and (ii) $\lim_{\eta\to 0} g_\eta(u,u') = 0$ for all $(u,u')\in\mathbb{R}^2$. Since the velocity correlation measure $m_{\nu,\mu}$ is finite, the dominated convergence theorem yields

$$\lim_{\eta\to 0}\int\!\!\int g_\eta(u,u')\,m_{\nu,\mu}(du\,du') = 0\,, \tag{13.34}$$

which is equivalent to the assertion. □

It is not hard to see that the low-temperature limit $\beta\to\infty$ of the right side of (13.31) exists and is given by

$$\lim_{\beta\to\infty}\sigma^{\mathrm{sym}}_{\nu,\mu} \;=\; q^2\,\widetilde{\sigma}_{\nu,\mu}(E_F)\,. \tag{13.35}$$

In fact, an inspection of the previous proof shows that this identity for the double limit of first taking $\eta \downarrow 0$ and then $\beta \to \infty$ in (13.27) may be established under the weaker assumption that the limit (13.28) exists for a neighborhood of the Fermi energy E_F.

13.4. Localization and decay of the two-point function

We will now come back to the question of whether localization implies condition (13.20) on the decay of the kernel of the Fermi projection. This kernel represents the two-point function in the ground-state (corresponding to $\beta = \infty$) of non-interacting fermions. More generally, the two-point function of a system of non-interacting Fermions in the thermal equilibrium state corresponding to inverse temperature $\beta \in [0, \infty]$ and Fermi level at E_F is

$$\langle \psi^\dagger(y)\psi(x)\rangle_\beta \;=\; \langle \delta_x, \theta^\beta_{E_F}(H)\delta_y\rangle \tag{13.36}$$

using the notation $\psi(x)$ for the Fock-space fermionic annihilation operator at x.

At positive temperature $\beta < \infty$, this kernel always exhibits exponential decay. This follows from the fact that the function $\theta^\beta_{E_F}$ is meromorphic with poles at $E_F + i(2n+1)\pi/\beta$, $n \in \mathbb{Z}$. It is hence analytic and bounded within the strip $|\operatorname{Im} z| < \pi/\beta$ about the real axis. For any such function the decay of the kernel is exponential.

Proposition 13.5. *For a function F which is analytic and bounded in the strip $|\operatorname{Im} z| < \eta$ and a bounded self-adjoint operator H on $\ell^2(\mathbb{Z}^d)$ with the property $\sup_x \sum_y |H(x,y)| \left(e^{\mu|x-y|} - 1\right) \le \eta/2$ for some $\mu > 0$, there is some $C > 0$ such that for all $x, y \in \mathbb{Z}^d$,*

$$|\langle \delta_x, F(H)\delta_y\rangle| \;\le\; C\, e^{-\mu|x-y|}\,. \tag{13.37}$$

Proof. By the analyticty of F in the strip, we may represent $F(E)$ for any $E \in \sigma(H)$ as a contour integral with a contour $\mathcal{C}$ which encircles the spectrum counterclockwise at a distance $\eta/2$ from $\sigma(H)$. By the spectral theorem this yields

$$\langle \delta_x, F(H)\delta_y\rangle = \frac{1}{2\pi i}\oint_{\mathcal{C}} \langle \delta_x, (z-H)^{-1}\delta_y\rangle\; F(z)\, dz\,. \tag{13.38}$$

The asserted bound is then a simple consequence of the Combes-Thomas estimate (Theorem 10.5), the fact that F is bounded on $\mathcal{C}$, and the boundedness of the length of the contour. □

The above proposition may be generalized to unbounded self-adjoint operators. This requires a more careful estimate on the contour integral which then consists of two infinite paths above and below the real line. The key here is a variant of the Combes-Thomas estimate, which also leads to an improved upper bound in the bounded case [**10**].

In case $F = \theta^\beta_{E_F}$ and H is a bounded random Schrödinger operator of the form

$$H(\omega) = -\Delta^{(A)} + V(\omega) \quad \text{on } \ell^2(\mathbb{Z}^d),$$

Proposition 13.5 implies that the two-point function decays exponentially with exponent μ proportional to β^{-1}. This bound does not survive the zero temperature limit $\beta = \infty$. However, if the spectrum of H below the Fermi level consists only of localized states and the eigenfunction correlator decays exponentially, the bound $|\langle \delta_x, P_{(-\infty,E_F]}(H)\delta_y\rangle| \le Q(x, y; (-\infty, E_F])$ implies the decay of the two-point function. At zero temperature in a Fermi system the physical intuition is that only states close to the Fermi energy participate in the dynamics. This suggests that exponential localization of states with energies in a neighborhood of E_F (instead of the full half-line $(-\infty, E_F]$) should be enough to guarantee the exponential decay of the two-point function.

Theorem 13.6. *For a random Schrödinger operator $H(\omega) - \Delta^{(A)} + V(\omega)$ on $\ell^2(\mathbb{Z}^d)$ with $\operatorname{ess\,sup}_{\omega\in\Omega} \|V(\omega)\| < \infty$, if for some $s \in (0, 1)$ and $C, \xi < \infty$ and all $x, y \in \mathbb{Z}^d$*

$$\sup_{\eta\neq 0}\ \mathbb{E}\left[|G(x, y; E_F + i\eta)|^s\right] \ \le\ C\, e^{-|x-y|/\xi},$$

then there is some $A < \infty$ such that for all $x, y \in \mathbb{Z}^d$

$$\mathbb{E}\left[|\langle \delta_x\,,\ P_{(-\infty,E_F)}(H)\delta_y\rangle|\right] \ \le\ A\, e^{-|x-y|/\xi}\,. \tag{13.39}$$

Proof. We represent the Fermi projection in terms of a contour integral with a contour $\mathcal{C}_{E_F}$ which consists of the segments connecting $E_F - i\alpha$, $E_F + i\alpha$, $E_F - \gamma + i\alpha$, and $E_F - \gamma + i\alpha$ where $\gamma > 0$ is chosen such that $\operatorname{dist}(E_F - \gamma, \sigma(H(\omega))) \ge \alpha$ with $\alpha > 0$ chosen later; cf. Figure 13.1. Splitting the integral and noting that the assumption implies that with probability one E_F is not an eigenvalue, we obtain

$$P_{(-\infty,E_F)}(H) \ =\ Q_1(H - E_F) + Q_2(H - E_F) + Q_3(H - E_F) \tag{13.40}$$

where

$$\begin{aligned} Q_1(x) &= \frac{1}{2\pi}\int_{-\alpha}^{\alpha} \frac{d\eta}{i\eta - x}, \\ Q_2(x) &= \frac{1}{2\pi i}\int_{-\gamma}^{0} \left(\frac{1}{u - i\alpha - x} - \frac{1}{u + i\alpha - x}\right) du, \\ Q_3(x) &= -Q_1(x + \gamma)\,. \end{aligned} \tag{13.41}$$

The contribution of the second and third terms is estimated using the

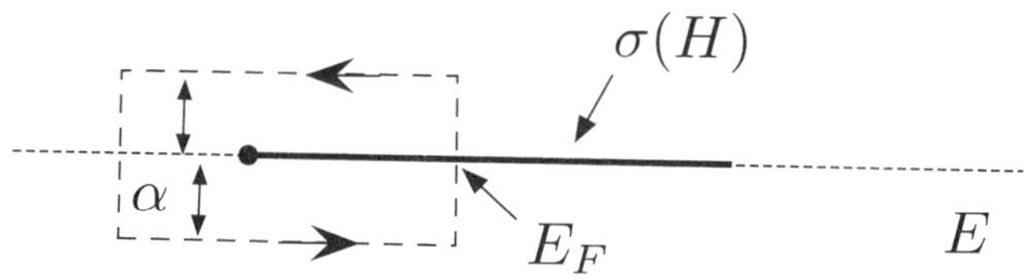

Figure 13.1. The choice of the contour for the representation of the Fermi projection in case the Hamiltonian is bounded from below.

Combes-Thomas estimate (Theorem 10.5) which for $\mathrm{dist}(z, \sigma(H(\omega))) \geq \alpha$ reads as follows:

$$\left|\langle \delta_x \,, (H(\omega) - z)^{-1} \delta_y \rangle\right| \;\leq\; C\, e^{-\alpha c|x-y|}$$

with some constants $c > 0$, $C < \infty$ which are independent of ω. Since the lengths of the contours on the first and third segments are bounded and we may choose $\alpha = (c\xi)^{-1}$, this establishes the claimed exponential decay for those terms.

In order to bound the first term, we use the localization assumption

$$\begin{aligned} \mathbb{E}\left[\left|\langle \delta_x \,, Q_1(H - E_F)\delta_y \rangle\right|\right] \;&\leq\; \frac{1}{2\pi}\int_{-\alpha}^{\alpha} \mathbb{E}\left[\left|\langle \delta_x \,, (H - E_F - i\eta)^{-1} \delta_y \rangle\right|\right] d\eta \\ &\leq\; \frac{1}{2\pi}\int_{-\alpha}^{\alpha} \mathbb{E}\left[\left|\langle \delta_x \,, (H - E_F - i\eta)^{-1} \delta_y \rangle\right|^s\right] \frac{d\eta}{|\eta|^{1-s}} \\ &\leq\; \frac{C\alpha^s}{\pi s} e^{-|x-y|/\xi}\,. \end{aligned} \tag{13.42}$$

Here the second inequality results from the bound $\left|\langle \delta_x \,, (H - z)^{-1} \delta_y \rangle\right| \leq |\operatorname{Im} z|^{-1}$. □

In the setting of Theorem 13.6 one may also show that

$$\mathbb{E}\left[\left|\langle \delta_x \,,\; \theta^{\beta}_{E_F}(H)\delta_y \rangle\right|\right] \;\leq\; A\, e^{-|x-y|/\xi} \tag{13.43}$$

where the constants on the right side are independent of the inverse temperature $\beta \in [0, \infty]$. Moreover, similarly to Proposition 13.5, the theorem can be extended to cover unbounded random operators [**10**].

Notes

In more general models with homogeneous disorder one may have anomalous transport behavior, rather than just diffusion. The link between fine properties of the spectral measure was already discussed in Chapter 2. The connection between transport properties and properties of the velocity correlation measure entering the Kubo formula is highlighted in [**341**].

A justification of the linear response ansatz and the Kubo formula based on the relaxation time approximation, in which one adds a special dissipative term to the Hamiltonian, can be found in [**49, 342**]. Within the adiabatic approach the linear response ansatz has been investigated in [**310**] for the lattice models studied here, and in [**136**] for the Landau operator on $\mathbb{R}^2$. We refer the reader to [**65, 123**] for a more complete discussion of linear response theory and the mathematical setup involved in the derivation of the Green-Kubo-Streda formula, which was only sketched in Section 13.3.

Basic properties of the conductivity and the velocity correlation measure such as its self-averaging and regularity properties can be found in [**319, 341, 342, 194, 50, 248, 93, 250**].

The description of alternating currents (ac) requires a more general setup. The linear response theory of ac-conductivity is contained in [**248**], which justifies Mott's formula; see also [**323, 231, 250, 251**].

Conductivity may also be studied through considerations of heat production under the application of a weak electric field. Within this approach, the justification of linear response theory and a derivation of the Kubo formula for the electric conductivity was carried out for non-interacting fermions on the lattice in [**74, 75**]. The analysis was recently extended to disordered systems of also interacting fermions at positive temperature [**76, 77**].

Yet a different way to characterize transport is mesoscopically in the Landauer-Büttiker formalism. References can be found in the Notes of Chapter 2.

Exercises

13.1. Let H be a self-adjoint operator on $\ell^2(\mathbb{Z}^d)$ with the property that

$$S_\alpha := \sup_x \sum_y |H(x,y)| \left(e^{\alpha|x-y|} - 1\right) < \infty$$

for some $\alpha > 0$.

a. Show that for any $E \in \mathbb{R}$ and $\eta > 0$,

$$\begin{aligned}&\left|\langle \delta_x \,, \operatorname{Im}(H - E - i\eta)^{-1}\delta_y\rangle\right| \\ &\quad \leq \; 12\, e^{-\mu|x-y|} \sqrt{\langle \delta_x \,, \operatorname{Im}(H - E - i\eta)^{-1}\delta_x\rangle \, \langle \delta_y \,, \operatorname{Im}(H - E - i\eta)^{-1}\delta_y\rangle}\end{aligned}$$

for any $\mu \leq \alpha$ with the property that $S_\alpha \leq \eta/2$.

[Hint: Use the ideas in the proof of the Combes-Thomas estimate (Theorem 10.5).]

b. Conclude that for any function F which is bounded by $\|F\|_\infty$ and analytic in the strip $|\operatorname{Im} z| \leq \eta$ one has

$$|\langle \delta_x \,, F(H)\, \delta_y \rangle| \;\leq\; 18\sqrt{2}\,\|F\|_\infty\, e^{-\mu|x-y|}$$

for any $\mu \leq \alpha$ with the property that $S_\alpha \leq \eta/2$.

13.2. Give a proof of (13.43).

Chapter 14

Integer Quantum Hall Effect

The Hall effect has already played an interesting role in classical electrodynamics since it provides a unique tool for determining the sign of the current carriers. This task is challenging since the electric current produced by particles of charge q moving at velocity $\mathbf{v}$ is the same as that produced by particles of the opposite charge $(-q)$ moving in the opposite direction (at the same speed and density of the moving charges). The two cases can, however, be distinguished by the sign of the Hall resistance which is measured in the presence of the combination of crossed electric ($\mathbf{E}$) and magnetic ($\mathbf{B}$) fields.

The Lorentz force on moving charges $\mathbf{F} = q\mathbf{v} \times \mathbf{B}$ deflects the current in a transverse direction, e.g., towards a lateral edge of the conductor in which the current flows. The charge accumulation there results in a transversal voltage difference between the lateral edges of the conductor, which is named V_{Hall} after the discoverer of this effect. In the steady state the corresponding electric field counters the transversal drift; that is,

$$q\mathbf{v} \times \mathbf{B} + q\mathbf{E}_{\text{Hall}} = \mathbf{0}\,. \tag{14.1}$$

Since from macroscopic-scale measurements one can determine the direction of $q\mathbf{v}$ as well as $\mathbf{E}_{\text{Hall}}$, the above relation allows us to deduce the sign of the current carriers' charge q.

The Hall, i.e., transversal, resistance is

$$\rho_H = \frac{V_{\text{Hall}}}{I_{\text{channel}}}. \tag{14.2}$$

According to the classical explanation of electrical conduction, at a given density of the current carriers the Hall resistance would be proportional to the applied magnetic field (as is implied by (14.2) and the Drude picture of electrical conductance).

In their 1975 paper on a quantum calculation of the Hall conductance, T. Ando, Y. Matsumoto, and Y. Uemura [**29**] noted the following:

> [*The Green function G_N or X_N is determined by the self-consistency equation (2.8) of II.*] *From the above equation, one can conclude that $\Delta\sigma^{(1)}_{XY}$ vanishes and σ_{XY} becomes $-nec/H = -e^2(N+1)/2\pi\hbar$, when the Fermi level lies in energy gaps between adjacent N-th and $N+1$-th Landau levels at zero temperature....*

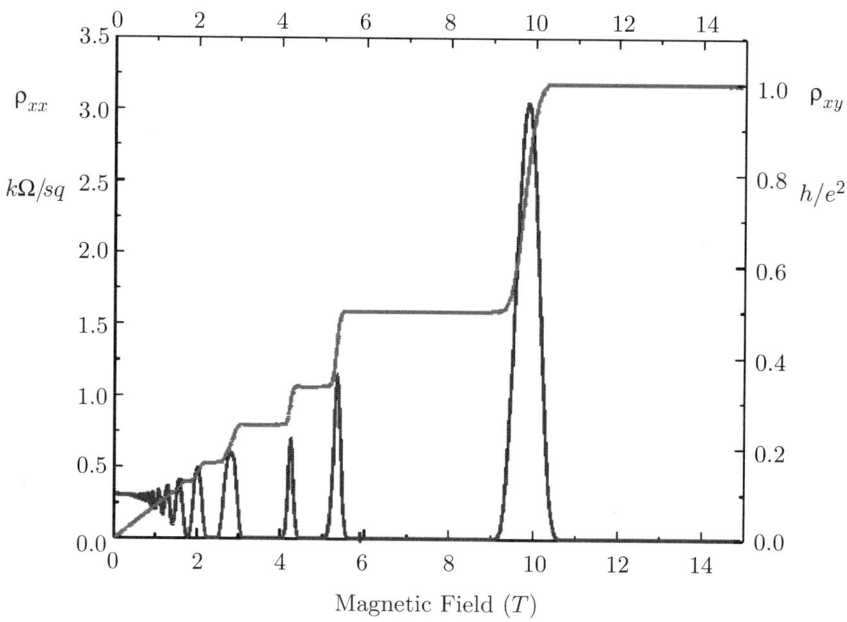

Figure 14.1. A sketch of the Hall resistance ($\rho_{x,y}$), plotted along with the direct resistance ($\rho_{x,x}$), for two-dimensional samples at low temperature. The Hall conductance exhibits plateaux of integer multiples of e^2/h as a function of the perpendicular magnetic field. The direct conductivity ($\sigma_{xx} = \rho_{x,x}/(\rho^2_{x,x} + \rho^2_{x,y})$), which is shown here in unrelated units, vanishes over the plateaux.

The significance of the above point was realized in the 1980 experimental work of K. von Klitzing who, working with samples provided by G. Dorda and M. Pepper [**253**], turned the quantization of the Hall conductance (within its plateaux) into a tool for a very precise determination of e^2/h. This constant's value was reportedly determined by such means up to the relative precision of 10^{-9}. Within five years the work was awarded the

Nobel prize. More surprises were in store, in particular the discovery of the fractional quantum Hall effect [**387**], i.e., observation of plateaux with Hall conductance at fractional multiples of e^2/h. We shall, however, not get to this subject here.

The successful observation of the integer quantum Hall effect (IQHE) required working at low temperatures and high magnetic fields but not necessarily with "clean" samples.[1] In fact, it was noted that the effect is improved by disorder (in the form of impurities in the sample).

Soon after the experiments focused attention on the IQHE phenomenon, theorists started to grasp the beautiful mathematical structure behind it. In fact, the quantization has a topological aspect. Its robust explanation combines functional analysis, topology (in the space of operators), and probability. In particular, it was understood that in the absence of direct conductance the quantum Hall conductance of two-dimensional systems can be expressed both as a Chern number and as a Fredholm index. The precision of integers is protected by their topological nature! The effect also provides an example of the interesting phenomenon that adiabatic transport tends to be quantized in situations which are enabled by the vanishing of the direct conductivity [**40, 42**].

14.1. Laughlin's charge pump

By now there have been a number of different derivations of the IQHE. Let us present here a slightly formal one, which fits neatly into the general setup of random operators discussed in this book. The argument starts from Laughlin's observation that Hall conductance can be probed in a charge pump Gedankenexperiment [**282**], which is outlined in Figure 14.2.

Deforming this picture, the cylinder can be stretched and flattened to form a "Corbino disk" whose boundaries consist of a pair of concentric rings. Thus, one considers a system in which the charges are confined to a plane and the magnetic field is changed through an adiabatic process which results in an increase of the flux through the smaller ring. As described by the Faraday law, responding to changes in the magnetic flux is a time-dependent electric field E whose integral along a loop $\mathcal{C}$ is

$$\oint_{\mathcal{C}} \mathbf{E} \cdot d\mathbf{l} \;=\; -\frac{d\,\Phi_{\mathcal{C}}}{d\,t}, \tag{14.3}$$

where $\Delta\Phi_{\mathcal{C}}$ is the change in the flux enclosed by $\mathcal{C}$. By Lenz's law, the current which would result from this field would tend to diminish the rate of change of the enclosed flux. However, we are now interested in the transversal

[1]The temperature range has since been brought up, in particular working with graphene [**316**]—the subject of another Nobel prize.

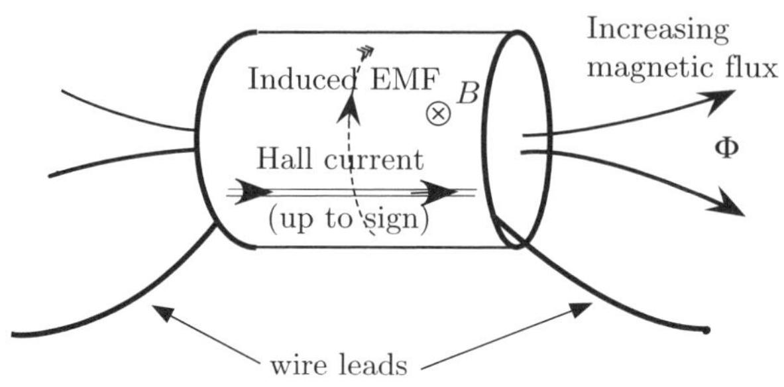

Figure 14.2. Laughlin's charge pump, based on the Hall effect (drawing based on [**42**]): A conducting sheet is wrapped into a cylinder bounded by a pair of rings connected to lead wires. In the presence of the non-vanishing transversal magnetic field B, increase in the enclosed magnetic flux Φ results in charge transport between the rings which form the cylinder boundary.

current. The latter may be present if in addition to the above electromotive force there is also a transversal magnetic field throughout the lattice ($\mathbf{B} \neq \mathbf{0}$).

The induced current density is

$$\mathbf{j} = \begin{pmatrix} \sigma_D & \sigma_H \\ -\sigma_H & \sigma_D \end{pmatrix} \mathbf{E}\,, \tag{14.4}$$

where σ_D and σ_H (the direct and the Hall conductance) are elements of the bulk homogenized conductivity tensor within the plane. Correspondingly, the rate of the induced charge transport across the contour $\mathcal{C}$ is

$$\begin{aligned} \frac{d\,Q}{d\,t} &= \oint_{\mathcal{C}} \mathbf{j} \cdot \mathbf{n}\, d\ell \\ &= \sigma_D \oint \mathbf{E} \cdot \mathbf{n}\, d\ell + \sigma_H \oint \mathbf{E} \cdot d\mathbf{l}\,. \end{aligned} \tag{14.5}$$

The last integral on the right-hand side of (14.5) is the induced electromotive force and is tied by Lenz's law (14.3) to the flux change rate. The first term vanishes in situations in which the direct conductivity σ_D is zero. To keep this condition, the process will be carried out adiabatically. Under this condition, integrating (14.5) and (14.3) over time one obtains an expression for the Hall conductance as the ratio of the transported charge to the flux change:

$$\boxed{\sigma_H = -\frac{\Delta Q}{\Delta \Phi_B}}\,. \tag{14.6}$$

14.2. Charge transport as an index

Let us recall from (3.12) that a natural way to incorporate the magnetic field into the Hamiltonian is to replace the regular Laplacian by its magnetic version; i.e., consider

$$H = -\Delta^{(A)} + V(\omega)$$

with

$$(\Delta^{(A)}\psi)(x) = \sum_{|x-y|=1} \left(e^{-iqA(x,y)}\psi(y) - \psi(x)\right)$$

where $A(x,y) = -A(y,x)\delta_{|x-y|,1}$ corresponds to the vector potential integrated over the edge from x to y. The magnetic flux through the lattice plaquettes (P, which carry orientation) is given by the discrete version (3.11) of the relation $\mathbf{B} = \text{Curl } \mathbf{A}$. An explicit choice of the vector potential for the present case of a constant perpendicular magnetic field can be found in (3.12).

Of relevance to the discussion which follows is the observation that the insertion of a unit flux (measured in units h/e) at a point $a \in \mathbb{R}^2 \backslash \mathbb{Z}^2$ in the plane corresponds to a gauge transformation which is given by the unitary mapping:

$$(U_a\psi)(x) := e^{-i\theta_a(x)}\psi(x), \tag{14.7}$$

where $\theta_a(x) := \text{Arg}(x-a) \in (-\pi, \pi]$ is the argument of the vector $x-a \in \mathbb{R}^2$, the latter being identified with $\mathbb{C}$.

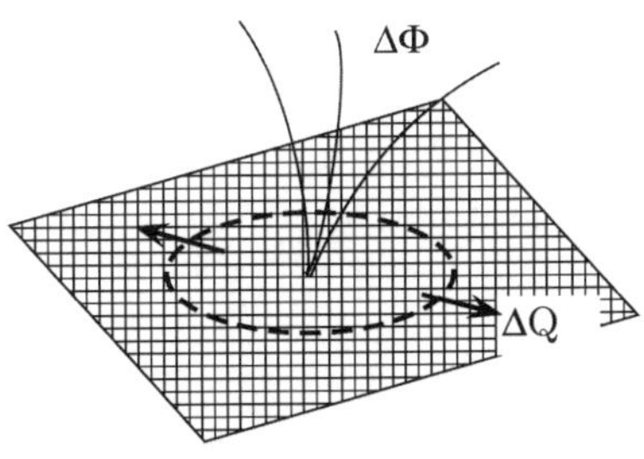

Figure 14.3. The Avron-Seiler-Simon "Hilbert hotel" based on the Hall effect. Charge is "added" to the infinite system through the insertion of a unit flux.

Y. Avron, R. Seiler, and B. Simon [**41**] stretched the Corbino disk picture even further and considered the setup in which the flux Φ is inserted through one of the plaquettes of a regular lattice and the outer ring is taken out to infinity (at least on the lattice scale). Furthermore, they in effect propose that the lattice itself can serve as a charge reservoir and instead of looking at the charge flow *through* the system one could try to make sense of the change in the total charge within the infinite lattice due to a finite increase in the magnetic flux. We now turn to a more explicit review of this approach to the IQHE, assuming now that the system's valence electrons (which have charge q) can be treated as a Fermi gas at zero temperature.

Under the assumption that the Fermi energy falls within a spectral gap, or, as emphasized by J. Bellissard [**49**], within a "mobility gap", i.e., a regime of Anderson localization, one may expect that under an adiabatic flux insertion, the system's state would evolve from the Fermi projection

$$P_{E_F} \;=\; 1[H \leq E_F] \tag{14.8}$$

to the corresponding projection for the modified Hamiltonian. The final state is particularly easy to write explicitly if the flux increment is h/e (or an integer multiple of it) since in that case the final state is given by $U_a P_{E_F} U_a^\dagger$ with U_a the gauge transformation (14.7). Thus, it is tempting to argue that

$$\Delta Q = q \ \mathrm{tr}\left(U_a P_{E_F} U_a^\dagger - P_{E_F}\right) , \tag{14.9}$$

to the extent that the difference makes sense.

The latter question is not quite trivial: (i) if the difference is calculated by evaluating the traces separately, it becomes $(\infty - \infty)$, which is ill-defined and (ii) had the difference of the two projection operators been trace class the difference trace would vanish since locally the densities are equal:

$$\langle \delta_x \, , \, P_{E_F} \, \delta_x \rangle \;=\; \langle \delta_x \, , \, U_a P_{E_F} U_a^\dagger \delta_x \rangle \, . \tag{14.10}$$

As it turns out, the difference $\left(U_a P_{E_F} U_a^\dagger - P_{E_F}\right)$ is not trace class, so the above trivial answer does not apply. However, if E_F is in the localization regime the difference is a compact operator, and thus instead of (14.9) one may take the index which is defined as follows.

Definition 14.1. The index of two orthogonal projections P, Q on a Hilbert space $\mathcal{H}$ whose difference $P - Q$ is a compact operator is

$$\boxed{\mathrm{Index}(P, Q) \;:=\; \dim \ker(P - Q - 1) - \dim \ker(Q - P - 1)} \, . \tag{14.11}$$

It may be added that the dimensions are well-defined since the assumed compactness of $P - Q$ ensures that $\dim \ker(P - Q \pm 1) < \infty$.

In the special case where P and Q are trace class the index is given by (14.9); that is, it equals the difference of the dimensions of $P\mathcal{H}$ and $Q\mathcal{H}$. A more generally useful formula is (14.14), which is explained within the proof of Proposition 14.2.

In this terminology, the tentative proposal (14.9) is revised to

$$\boxed{\Delta Q = q\ \text{Index}\left(U_a P_{E_F} U_a^\dagger,\ P_{E_F}\right)}\,. \tag{14.12}$$

When defined, this index takes integer values, and through (14.6) it yields $\sigma_H \in (e^2/h)\,\mathbb{Z}$. Furthermore, the value which it yields for the Hall conductance σ_H coincides with that provided by the Kubo formula, as was shown in [**41, 49**], and is explained below following [**10**].

14.3. A calculable expression for the index

It may be worth stressing that the notion of index comes with certain subtleties which are actually of relevance in the present context:

1. When defined, the index takes only integer values, and it is robust in its independence of the flux insertion point and by implication also independent of the disorder (in the "almost-sure" sense).
2. However, the relative index of two projections Index(P, Q) is defined only under some strict conditions. In the case of interest here, a necessary condition for Index $\left(U_a P_{E_F} U_a^\dagger,\ P_{E_F}\right)$ to make sense is that the Fermi projection P_{E_F} has rapidly decaying matrix elements. For that, localization plays an essential enabling role. This caveat well reflects also the observed physics, with the Hall conductance changing over regimes where it is not quantized.
3. Formal manipulations, in which questions of convergence are ignored, could be misleading. Splitting traces or invoking cyclicity of the trace without paying attention to trace-class restrictions (as is allowed in finite dimensions) may easily read to contradictions or to the wrong impression that the index is always zero.

From the index definition one may conclude the following elementary properties:

$$\begin{aligned} \text{Index}(P, Q) &= -\,\text{Index}(Q, P) = -\,\text{Index}(P^\perp, Q^\perp) \\ &= \text{Index}\left(UPU^{-1}, UQU^{-1}\right), \end{aligned} \tag{14.13}$$

where $P^\perp := 1 - P$ and $Q^\perp := 1 - Q$ stand for the projections onto the orthogonal complement and $U : \mathcal{H} \to \mathcal{H}$ is any invertible linear map.

In computing the index we shall rely on

Proposition 14.2. *Let P, Q be orthogonal projections on a Hilbert space $\mathcal{H}$. Then for any n for which $(P - Q)^{2n+1}$ is trace class, one has*

$$\boxed{\text{Index}(P, Q) = \text{tr}\,(P - Q)^{2n+1}}\,. \tag{14.14}$$

Proof. The strange dependence on n of the value in (14.14) is explained by the observation that the spectrum of $P - Q$ is antisymmetric, counting multiplicity, except possibly for the values 1 and (-1). This is established in [**41**] by considering the pair of anticommuting operators $C := P - Q$ and $S := P - Q^{\perp}$, which are easily seen to satisfy

$$S^2 + C^2 = 1 \qquad \text{and} \qquad SC + CS = 0\,. \tag{14.15}$$

For any eigenvalue $\lambda \in \sigma(C)\backslash\{0\}$ and a corresponding (normalized) eigenfunction $\varphi \in \mathcal{H}$ one has

$$CS\varphi = -SC\varphi = -\lambda S\varphi\,; \tag{14.16}$$

i.e., unless $S\varphi = 0$, the function $S\varphi$ is an eigenfunction corresponding to the reflected eigenvalue. In the exceptional case, $S\varphi = 0$ implies $\lambda^2\varphi = C^2\varphi = \varphi - S^2\varphi = \varphi$ and hence $\lambda \in \{\pm 1\}$.

Since with the exception of the eigenvalues ± 1 the spectrum of C (counted with its multiplicity) is invariant under reflections, at any odd power at which the eigenvalues are summable one gets cancellations, and the contribution left standing refers to just the terms which are counted in the definition of the index. Thus,

$$\begin{aligned} \operatorname{tr} C^{2n+1} &= \sum_{\lambda\in\sigma(C)} \lambda^{2n+1} \dim\ker(C-\lambda) \\ &= \dim\ker(C-1) - \dim\ker(C+1) \\ &= \operatorname{Index}(P,Q)\,. \end{aligned} \tag{14.17}$$

□

Guided by the criterion provided by Proposition 14.2, one would want to know under what conditions $T = \left(U_a P_{E_F} U_a^{\dagger} - P_{E_F}\right)$ is a compact operator and more precisely under what conditions and for what powers $p > 0$ is

$$\|T\|_p := (\operatorname{tr}|T|^p)^{1/p} < \infty\,. \tag{14.18}$$

In this case T is said to be in the Schatten-p class.

The following lemma gives a sufficient condition for that.

Lemma 14.3. *Let $T : \ell^2(\mathbb{Z}^d) \to \ell^2(\mathbb{Z}^d)$ be a bounded linear operator with kernel $T(x,y) = \langle \delta_x, T\delta_y\rangle$. Then for any $p \in [1,\infty)$,*

$$\|T\|_p \leq \sum_{y\in\mathbb{Z}^d} \Big(\sum_{x\in\mathbb{Z}^d} |T(x,y)|^p \Big)^{\frac{1}{p}}\,. \tag{14.19}$$

Proof. We rewrite $T = \sum_{\alpha\in\mathbb{Z}^d} T_\alpha$ in terms of the operators defined by $T_\alpha(x,y) := T(x,y)\,\delta_{x-\alpha,y}$. The triangle inequality yields

$$\|T\|_p \leq \sum_{\alpha\in\mathbb{Z}^d} \|T_\alpha\|_p = \sum_{\alpha\in\mathbb{Z}^d} \|T_\alpha^\dagger T_\alpha\|_{p/2}^{1/2} \tag{14.20}$$
$$= \sum_{\alpha\in\mathbb{Z}^d}\Big(\sum_{x\in\mathbb{Z}^d} |T(x+\alpha,x)|^p\Big)^{\frac{1}{p}}. \qquad \square$$

To answer the question posed in (14.18) let us note that

$$\langle \delta_x\,,\, U_a P_{E_F} U_a^\dagger \delta_y\rangle - P_{E_F}(x,y) \;=\; \langle \delta_x\,,\, P_{E_F}\,\delta_y\rangle\,\Big(e^{i[\theta_a(y)-\theta_a(x)]} - 1\Big). \tag{14.21}$$

Assuming the off-diagonal matrix elements $\langle \delta_x\,,\, P_{E_F}\,\delta_y\rangle$ decay rapidly, but terms with bounded distance $|x-y|$ do not, the factor involving

$$|e^{i[\theta_a(y)-\theta_a(x)]} - 1| = 2\Big|\sin\Big(\frac{1}{2}\angle(x,a,y)\Big)\Big|, \quad \angle(x,a,y) := \theta_a(y)-\theta_a(x),$$

yields decay at the rate of $O(1/|x|)$. Thus, one is led to expect that in the regime of exponential localization the condition (14.18) would be met in the two-dimensional system considered here for all $p>2$. In particular, the index should be computable through $\operatorname{tr}\Big(U_a P_{E_F} U_a^\dagger - P_{E_F}\Big)^3$.

In addition to the above estimates we shall also find the following property of the unitary gauge transformations U_a useful.

Lemma 14.4. *The unitary transformations $U_a : \ell^2(\mathbb{Z}^2) \to \ell^2(\mathbb{Z}^2)$, defined by the gauge transformation $(U_a\psi)(x) = e^{-i\theta_a(x)}\psi(x)$, with $a\in\mathbb{R}^2\backslash\mathbb{Z}^2$, satisfy the following:*

1. *The operator differences $U_b - U_a$ are in the Schatten-3 class, and furthermore for all $a,b\in\mathbb{R}^2\backslash\mathbb{Z}^2$ with $|a-b|\leq 1$,*

$$\|U_b - U_a\|_3 := (\operatorname{tr}|U_b-U_a|^3)^{1/3} \;\leq\; C\,|a-b|^{1/3}, \tag{14.22}$$

   with a uniform constant $C<\infty$.

2. *The unitary operators U_a transform covariantly,*

$$(U_x^{(A)})^\dagger U_a U_x^{(A)} \;=\; U_{a-x},$$

   under the magnetic translations $U_x^{(A)}$, $x\in\mathbb{Z}^2$, that are given by $U_x^{(A)}\psi(y) = e^{-iBq(y_1x_2-x_1y_2)/2}\psi(y-x)$.

Proof. For a proof of the first assertion we note that $U_b - U_a$ is diagonal in the basis of position eigenfunctions δ_x, and thus

$$\|U_b-U_a\|_3^3 \;=\; \sum_{x\in\mathbb{Z}^2}\Big|e^{-i\theta_a(x)} - e^{-i\theta_b(x)}\Big|^3. \tag{14.23}$$

By elementary geometric considerations,

$$|e^{i\theta_a(x)} - e^{i\theta_b(x)}| \leq \min\left\{2, \frac{|b-a|}{\sqrt{|x-a||x-b|}}\right\} \tag{14.24}$$

for all $a, b \in \mathbb{R}^2$ and $x \in \mathbb{Z}^2$. Therefore, for $\min\{|x-a|, |x-b|\} \geq |b-a|/2$ the summand in (14.23) is of the order $O(1/|x|^3)$, which is summable in $d = 2$ dimensions. The claimed bound (14.22) follows by summation over unit increments in $|b-a|$.

The second assertion is a direct consequence of the way the magnetic translations are set up. □

14.4. Evaluating the charge transport index in a mobility gap

The above preparatory statements enable the following general result which allows us to conclude the quantization of the quantum Hall conductance at $T = 0$ and Fermi energy E_F within the regime of Anderson localization.

Theorem 14.5 (Existence of charge transport index). *Let $P \in \mathcal{K}_{\mathrm{mc}}$ be a covariant family of orthogonal projections on $\ell^2(\mathbb{Z}^2)$ and assume that its kernel $P(x,y) = \langle \delta_x, P\delta_y \rangle$ satisfies*

$$\sum_{x \in \mathbb{Z}^2} |x| \left(\mathbb{E}\left[|P(0,x)|^3\right]\right)^{1/3} < \infty\,. \tag{14.25}$$

Then there is a set Ω_0 of full measure $\mathbb{P}(\Omega_0) = 1$ such that for all $\omega \in \Omega_0$ the following hold:

1. *$P(\omega) - U_a P(\omega) U_a^\dagger$ is Schatten-3 class for all $a \in \mathbb{R}^2 \backslash \mathbb{Z}^2$, so that the charge transport index*

$$C_a(P)(\omega) = \mathrm{Index}(U_a P(\omega) U_a^\dagger, P(\omega)) \tag{14.26}$$

   is well-defined and takes integer values.
2. *The value of $C_a(P)(\omega)$ does not change with $a \in \mathbb{R}^2 \backslash \mathbb{Z}^2$.*
3. *It is almost surely equal to its mean value.*

Proof. 1. For $n \in \mathbb{Z}^2$ let $B_n := \{a \in \mathbb{R}^2 \,|\, \|a-n\|_\infty < 1\}$. Lemma 14.3 allows us to deduce the uniform estimate:

(14.27)
$$\mathbb{E}\left[\sup_{a\in B_0\backslash\mathbb{Z}^2} \|P - U_a P U_a^\dagger\|_3\right]$$
$$\leq \sum_{y\in\mathbb{Z}^2} \mathbb{E}\Big[\sup_{a\in B_0\backslash\mathbb{Z}^2}\Big(\sum_{x\in\mathbb{Z}^2}\Big|P(x+y,x)\left(1-e^{i(\theta_a(x)-\theta_a(x+y))}\right)\Big|^3\Big)^{\frac{1}{3}}\Big]$$
$$\leq \sum_{y\in\mathbb{Z}^2}\Big(\sum_{x\in\mathbb{Z}^2}\mathbb{E}\left[|P(x,x+y)|^3\right]\sup_{a\in B_0\backslash\mathbb{Z}^2}\left|e^{-i\theta_a(x+y)}-e^{-i\theta_a(x)}\right|^3\Big)^{\frac{1}{3}}$$
$$= \sum_{y\in\mathbb{Z}^2}\mathbb{E}\left[|P(0,y)|^3\right]^{\frac{1}{3}}\Big(\sum_{x\in\mathbb{Z}^2}\sup_{a\in B_0\backslash\mathbb{Z}^2}\left|e^{-i\theta_a(x+y)}-e^{-i\theta_a(x)}\right|^3\Big)^{\frac{1}{3}} < \infty$$

where the second step is by Jensen's inequality. By an estimate similar to (14.24) the last term on the right side is shown to be bounded by $C(1+|y|)$. This proves the finiteness of the right side.

As a consequence, there is a full measure set of ω's for which $P(\omega) - U_a P(\omega) U_a^\dagger$ is Schatten-3 class for all $a \in B_0\backslash\mathbb{Z}^2$. Let Ω_0 be the countable intersection of lattice shifts of this set. Then Ω_0 is still of full measure and for all $\omega \in \Omega_0$, the operator difference $P(\omega) - U_a P(\omega) U_a^\dagger$ is Schatten-3 class for all $a \in \mathbb{R}^2\backslash\mathbb{Z}^2$. On this set, $\mathrm{Index}(U_a P(\omega) U_a^\dagger, P(\omega))$ is therefore well-defined and integer-valued for all $a \in \mathbb{R}^2\backslash\mathbb{Z}^2$.

2. The independence of $C_a(P)(\omega)$ from a is implied by its quantization and continuity in a over the connected domain $\mathbb{R}^2\backslash\mathbb{Z}^2$. The continuity relies on the continuity of the Schatten-3 norm $\|\cdot\|_3 = (\mathrm{tr}\,|\cdot|^3)^{1/3}$ of the gauge transformations, which yields

$$\begin{aligned}(14.28)\qquad d(a,b;\omega) &:= \left\|U_a P(\omega) U_a^\dagger - U_b P(\omega) U_b^\dagger\right\|_3 \\ &\leq \left\|(U_a - U_b) P(\omega) U_a^\dagger\right\|_3 + \left\|U_b P(\omega)(U_a^\dagger - U_b^\dagger)\right\|_3 \\ &\leq 2\,\|U_a - U_b\|_3 \;\leq\; C\,|a-b|^{1/3}\,.\end{aligned}$$

Here the last inequality is by Lemma 14.4 and holds for $|a-b| \leq 1$. The proof of the continuity of the index is completed by the estimate

$$\begin{aligned}(14.29)\qquad &\left|\mathrm{tr}(P(\omega) - U_a P(\omega) U_a^\dagger)^3 - \mathrm{tr}(P(\omega) - U_b P(\omega) U_b^\dagger)^3\right| \\ &\leq\; d(a,b;\omega)^3 + 3\,\|P - U_a P(\omega) U_a^\dagger\|_3\, d(a,b;\omega)^2 \\ &\quad + 3\,\|P(\omega) - U_a P(\omega) U_a^\dagger\|_3^2\, d(a,b;\omega)\,,\end{aligned}$$

which is a consequence of an elementary algebraic identity for the difference of $(P - U_a P U_a^\dagger)^3 - (P - U_b P U_b^\dagger)^3$ and Hölder's inequality for Schatten norms.

3. The invariance of $C_a(P)$ under lattice shifts of a and the covariance of the Schrödinger operators under magnetic shifts imply that $C_a(P)$, as a function of ω, is invariant under the ergodic action of this group. It is therefore almost surely constant and hence also equal to its mean value. □

14.5. Quantization of the Kubo-Streda-Hall conductance

The next result will show that under the localization assumption (14.25) on the Fermi projection P_{E_F}, the explicit formula for the **charge-transport index**,

$$\boxed{C(P_{E_F}) \;=\; \mathrm{Index}(U_a P_{E_F} U_a^\dagger, P_{E_F})} \tag{14.30}$$

with $a \in (\mathbb{Z}^2)^*$ an arbitrary point on the dual lattice, also agrees with the Kubo-Streda formula (13.21) for Hall conductance (at zero temperature). That is somewhat reassuring since the justification of the above approach did include a certain creative ansatz. The relation can also be used to explain the Hall conductance plateaux.

The following statement, in a slightly different form, dates back to [**49**]. The present version is based on [**10**]. Use is also made here of the continuity of the integrated density of states, which is known for all ergodic random operators of the subsequent form [**114**].

Theorem 14.6 (Quantization and plateaux of Hall conductance). *Let $H(\omega) = -\Delta^{(A)} + V(\omega)$ be a random operator on $\ell^2(\mathbb{Z}^2)$ with a constant perpendicular magnetic field $B \neq 0$ and a bounded ergodic random potential V. Then over any value of the Fermi energy E_F at which*

$$\xi(E_F) \;:=\; \sum_{x\in\mathbb{Z}^2} |x|\, \mathbb{E}\left[|\langle \delta_0\,, P_{E_F}(H)\, \delta_x\rangle|^3 \right]^{1/3} \;<\; \infty \tag{14.31}$$

the zero-temperature Kubo-Streda-Hall conductance $\sigma_{1,2}(E_F)$ defined by (13.21) *coincides with the charge transport index described by Theorem* 14.5 *times $q^2/2\pi \equiv q^2/h$; i.e., almost surely,*

$$\boxed{\sigma_{1,2}(E_F) \;=\; \frac{q^2}{h} C(P_{E_F})}\,. \tag{14.32}$$

Furthermore, the value of $\sigma_{1,2}(E_F)$, which in this case is an integer multiple of e^2/h, stays constant as E_F ranges over any interval I in which (14.31) *holds with $\sup_{E\in I} \xi(E) < \infty$.*

Proof. 1. In the setting of Theorem 14.5 with $P = P_{E_F}$ for the charge transport index one gets

(14.33)

$$\begin{aligned}
C(P) &:= \mathbb{E}\left[\operatorname{tr}\left(U_a P U_a^\dagger - P\right)^3\right] \\
&= \sum_{x,u,v\in\mathbb{Z}^2} \mathbb{E}\left[P(x,u)\,P(u,v)\,P(v,x)\right] \\
&\quad \times \left(e^{i(\theta_a(u)-\theta_a(x))}-1\right)\left(e^{i(\theta_a(v)-\theta_a(u))}-1\right)\left(e^{i(\theta_a(x)-\theta_a(v))}-1\right) \\
&= -2i \sum_{x,u,v\in\mathbb{Z}^2} \mathbb{E}\left[P(x,u)\,P(u,v)\,P(v,x)\right] \\
&\quad \times \left(\sin\angle(u,a,x)+\sin\angle(v,a,u)+\sin\angle(x,a,v)\right) \\
&= -2i \sum_{\substack{u,v\in\mathbb{Z}^2\\ a\in(\mathbb{Z}^2)^*}} \mathbb{E}\left[P(0,u)\,P(u,v)\,P(v,0)\right] \\
&\quad \times \left(\sin\angle(u,a,0)+\sin\angle(v,a,u)+\sin\angle(0,a,v)\right).
\end{aligned}$$

Here $\angle(u,a,x) = \theta_a(x) - \theta_a(u)$ is the angle of the segment enclosed by the half-lines anchoring in a and passing through u and x, respectively. In the last line, we used homogeneity, which allows us to shift x to the origin and sum over $a \in (\mathbb{Z}^2)^*$ instead.

In the last expression the sum over $a \in \mathbb{Z}^2$ can be carried out with the help of a striking formula of A. Connes [**99**] which is presented in Proposition 14.7 below. Applying it (with $g(\alpha) = \sin\alpha$) one has

$$\begin{aligned}
\sum_{a\in(\mathbb{Z}^2)^*} &\left(\sin\angle(u,a,0)+\sin\angle(v,a,u)+\sin\angle(0,a,v)\right) \\
&= \pi\, u\wedge v = \pi(u_2v_1 - u_1v_2)\,.
\end{aligned} \tag{14.34}$$

Therefore the charge transport index is given by

$$\begin{aligned}
C(P) &= -2\pi i \sum_{u,v\in\mathbb{Z}^2} \mathbb{E}\left[P(0,u)\,P(u,v)\,P(v,0)\right]\, u\wedge v \\
&= -2\pi i\; \mathbb{E}\left[\langle\delta_0, PX_2PX_1P\delta_0\rangle - \langle\delta_0, PX_1PX_2P\delta_0\rangle\right] \\
&= -2\pi i\; \mathbb{E}\left[\langle\delta_0, P\left[[X_2,P],[X_1,P]\right]P\delta_0\rangle\right] = \frac{2\pi}{q^2}\,\sigma_{1,2}\,,
\end{aligned} \tag{14.35}$$

where the last expression relates directly to the transversal conductance as given by the Kubo-Streda formula (13.21).

2. The constancy of the Hall conductance over I under the uniformity assumption follows from its quantization and continuity. To prove the latter, let $E \geq E'$ in I. Then applying the representation (14.35), an

elementary telescopic decomposition, translation invariance, and Hölder's inequality with $2q^{-1} + r^{-1} = 1$, one gets

(14.36)

$$\begin{aligned}
&\left|\sigma_{1,2}(E) - \sigma_{1,2}(E')\right| \\
&\leq 6 \sum_{u,v\in\mathbb{Z}^2} \mathbb{E}\left[\left|P^{\#}(0,u)\right|^q\right]^{1/q} \mathbb{E}\left[\left|P^{\#}(u,v)\right|^q\right]^{1/q} \mathbb{E}\left[\left|\Delta P(v,0)\right|\right]^{1/r} |u|\,|u-v| \\
&\leq 6 \left(\sum_{u\in Z^2} |u|\, \mathbb{E}\left[\left|P^{\#}(0,u)\right|^q\right]^{1/q}\right)^2 \mathbb{E}\left[\left|\Delta P(0,0)\right|\right]^{1/r}
\end{aligned}$$

where $P^{\#}$ is either $P_{(-\infty,E]}$ or $P_{(-\infty,E']}$ and $\Delta P := P_{(-\infty,E]} - P_{(-\infty,E']} = P_{(E',E]}$. For the last inequality use was also made of the Cauchy-Schwarz inequality in Hilbert space, $|\Delta P(v,0)|^2 \leq |\Delta P(0,0)|\,|\Delta P(v,v)|$, followed by the Cauchy-Schwarz inequality for the expected value and translation invariance.

By the continuity of the integrated density of states [**114**], established in Exercise 3.8, the expected value $\mathbb{E}\left[|\Delta P(0,0)|\right]$ tends to zero as $E \downarrow E'$. Thus, under the assumption that $\xi(E)$ is uniformly bounded over I the quantized Hall conductance is also continuous there. □

14.6. The Connes area formula

In relating the charge-transport index to Hall conductivity as given by the Kubo formula, a vital role is played by the Connes area formula. Following is its formulation and derivation, using the streamlined argument of Y. Colin de Verdière ([**40**]).

Proposition 14.7. *For a fixed triplet $x_1, x_2, x_3 \in \mathbb{Z}^2$, let $\alpha_j(a) \in (-\pi,\pi)$ be the angle of view from $a \in (\mathbb{Z}^2)^*$ of x_{j+1} relative to x_j in the positive angular orientation, with the index interpreted cyclicly and $\alpha_j(a) = 0$ if a lies between them. Then the following equality holds:*

$$\sum_{a\in(\mathbb{Z}^2)^*} \sum_{j=1}^{3} g(\alpha_j(a)) \;=\; 2\pi\ \mathrm{Area}(x_1,x_2,x_3) \tag{14.37}$$

for any antisymmetric bounded function $g : (-\pi,\pi) \to \mathbb{R}$ which for small $\alpha = 0$ behaves as

$$g(\alpha) = \alpha + O(\alpha^3) \tag{14.38}$$

with Area$(\cdot)$ *the triangle's oriented area.*

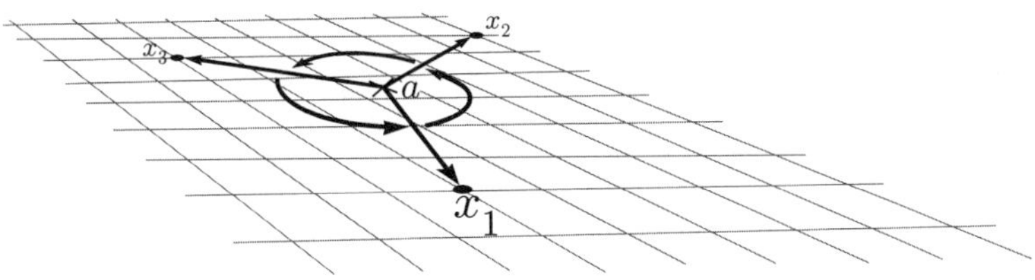

Figure 14.4. The three angles in the sum (14.37).

Proof. We may assume the triangle to be positively oriented. The statement (14.37) is true for $g(\alpha) = \alpha$. Indeed, for each $\alpha \in (\mathbb{Z}^2)^*$,

$$\sum_{j=1}^{3} \alpha_j(a) \;=\; 2\pi \left\{ \begin{array}{c} 1, \\ \frac{1}{2}, \\ 0 \end{array} \right\} \quad \text{for } a \left\{ \begin{array}{c} \text{inside,} \\ \text{on the boundary,} \\ \text{outside} \end{array} \right\} \text{the triangle.}$$

Thus, for $g(\alpha) = \alpha$ the left side of (14.37) is 2π times the number of dual lattice sites inside the triangle (counting a boundary site with weight $\frac{1}{2}$). This number is the same for triangles obtained by the lattice translation and reflection symmetry operations. Since this set of triangles tiles the plane, the number of enclosed dual sites must equal the triangle's area.

The above observation reduces (14.37) to the statement that for $f(\alpha) = g(\alpha) - \alpha$

$$\sum_{a \in (\mathbb{Z}^2)^*} \sum_{j=1}^{3} f(\alpha_j(a)) \;=\; 0\,. \tag{14.39}$$

A significant difference between f and g is that the individual terms $f(\alpha_j(a))$ are summable in $a \in (\mathbb{Z}^2)^*$ since by assumption $f(\alpha_j(a)) = O(|a|^{-3})$ for $|a| \to \infty$. This allows us to split the sum into three terms: $\sum_{a\in(\mathbb{Z}^2)^*} f(\alpha_j(a))$, $j = 1, 2, 3$. Each term is antisymmetric with respect to a reflection which is a symmetry of the lattice $\mathbb{Z}^2$. More explicitly, the jth term is antisymmetric under the reflection of $(\mathbb{Z}^2)^*$ with respect to the midpoint $(x_j + x_{j+1})/2$. Thus the individual sums (at given j) vanish, and (14.39) holds. □

Notes

We used the notation $\sigma_H = \sigma_{1,2} = \sigma_{xy}$ for the Hall conductance and $\rho_H = \rho_{1,2} = \rho_{x,y}$ for the Hall resistance. The relation $\sigma_{x,y} = -\rho_{x,y}^{-1}$ is valid only over the regime where the direct conductance vanishes, i.e., $\sigma_D = \sigma_{xx} = \sigma_{yy} = 0$, which coincides with the Hall plateaux. Transport quantization is also present in other situations of quantum "charge pumps", which are driven adiabatically by a cyclic increase in some of a quantum system's parameters ([**385**]). In this more general context it is the analog of the Hall resistance, rather than conductance, which is expected to be quantized.

Exercise 14.1 concerns the relation, which was introduced in [**41**], of the charge-transport index with the Fredholm-Noether index of the operator PU_aP on the Hilbert space $P\ell^2(\mathbb{Z}^2)$:

$$C_a(P) = \text{Index}_{FN}(PU_aP) .$$

The independence of $C_a(P)$ from a can be viewed as homotopy invariance within a norm-continuous family of operators of finite index; cf. [**284**]. This can also be deduced from the invariance of the Fredholm-Noether index under compact perturbations (since by Lemma 14.4 $U_a - U_b$ are compact operators).

Early on, it was noted that the Laughlin argument [**282**] for the quantization of the Hall conductance is due to its geometric content and therefore also applies more generally even to interacting systems [**39**]. The topic is the subject of an ongoing work [**191**].

For completeness it should be pointed out that there are now quite a number of explanations of the IQHE. Many of the arguments, including those of Laughlin [**282**] and Halperin [**185**], start from the Landau picture of the eigenvalues, slightly perturbed by disorder. Another perspective, emphasizing the role of disorder using a semiclassical picture of the dynamics under a magnetic field and a random potential, has been provided by Chalker and Coddington [**86**]. Proofs of localization for this model can be found in [**31, 32**].

The Hall conductance can also be interpreted as a Chern number. In the situation without disorder, this was noticed early on by Thouless, Kohmoto, Nightingale, and den Nijs [**386**] and by Avron, Seiler, and Simon [**38**]. Bellissard and coauthors [**48, 49**] have extended the use of non-commutative geometry to identify the Hall conductance as a Chern number in the situation with disorder.

The index approach to charge transport was presented by Bellissard [**48, 314**] and Avron, Seiler, and Simon [**41**]. The role which the spectral gap plays in [**275**] and [**41**] was replaced by the mobility gap in [**49**], whose analysis was completed in Aizenman and Graf [**10**]. See also [**179**] for a survey.

In the continuum situation $\mathbb{R}^2$ with a perpendicular magnetic field, i.e., for the Landau Hamiltonian with random potential, similar results can be established. As was already mentioned, one even knows the existence of transport due to energies near the Landau levels [**171, 172, 173**].

More recently, the descriptions of disordered topological insulators have brought other applications for these geometric ideas; cf. [**327**] and the references therein.

In finite samples, the presence of a Hall conductance in the bulk is tied to the simultaneous occurrence of an edge conductance [**185**]. The latter originates from delocalized edge states in the disordered sample. Since these currents are intrinsically tied to each other, the bulk conductance implies the quantization of the edge conductance and vice versa, yielding a yet equivalent description of the quantum Hall effect in finite samples. Their equality has been established (under a gap assumption) in [**343**] using K-theory and in [**134**] by more elementary methods.

Exercises

14.1. Let P be an orthogonal projection, let U be a unitary operator on some Hilbert space $\mathcal{H}$, and set $Q := UPU^\dagger$. Assume that for some $n \in \mathbb{N}_0$, $\operatorname{tr}|P-Q|^{2n+1} < \infty$, i.e., that $(P-Q)$ is Schatten-$(2n+1)$ class.

a. Show that $(P-PQP)^{n+1} = (P-Q)^{2n+2}P$ (and similarly for $P \leftrightarrow Q$), and prove that

$$\operatorname{Index}(P,Q) \;=\; \operatorname{tr}(P-PQP)^{n+1} - \operatorname{tr}(Q-QPQ)^{n+1}\,.$$

b. For $F := PUP$ on the Hilbert space $P\mathcal{H}$ prove that $1-F^\dagger F$ and $1-FF^\dagger$ are Schatten-$(n+1)$ class on $P\mathcal{H}$ and that $\operatorname{Index}(Q,P) = \operatorname{tr}(1-F^\dagger F)^{n+1} - \operatorname{tr}(1-FF^\dagger)^{n+1}$, where the trace extends over $P\mathcal{H}$.

c. Use the isospectrality of $1-F^\dagger F$ on $(\ker F)^\perp$ with $1-FF^\dagger$ on $(\ker F^\dagger)^\perp$ to prove

$$\operatorname{Index}(Q,P) \;= \dim\ker F - \dim\ker F^\dagger := \operatorname{Index}_{FN}(F)\,.$$

[The right side defines the **Fredholm-Noether index** of F.]

Chapter 15

Resonant Delocalization

The mathematical analysis of delocalization has so far turned out to be more elusive than the proofs of localization, perhaps contrary to the initial reactions to Anderson's arguments.

The Bloch-Floquet mechanism for the formation of bands of ac spectrum for periodic operators (which commute with a lattice group of translations and hence can be diagonalized simultaneously with it) is unstable with respect to even weak homogeneous disorder. That is drastically demonstrated by the one-dimensional example, where at arbitrarily weak disorder the entire spectrum changes to pure point (Chapter 12). Some continuity arguments, by which ac spectrum can be shown to persist into weak disorder, have been developed, but so far that can be done only for random operators on tree graphs [**246, 15, 161**], and even there only in a very limited range of energies, which in both the quantitative sense and qualitative sense is short of the full range of the continuous spectrum [**18, 21**].

There is therefore the need to develop insight into other possible mechanisms by which extended states may form in the presence of extensive, though possibly weak, frozen disorder. One such mechanism (and the list is woefully short) is **resonant delocalization**. The underlying phenomenon is quantum tunneling. It will be discussed here in the context of the following:

1. the **hybridization** of resonating **local quasi-modes**, through which may emerge eigenstates of wide geometric spread,
2. **resonant delocalization** (RD), through which may emerge delocalized states and continuous spectrum.

The argument for resonant delocalization which is discussed next would be disabled by the presence of ac spectrum. Thus RD can be viewed as an instability mode of other than ac spectra. Yet for a number of random operators, including the operators on tree graphs which are discussed in the next section, the general criterion presented below was found to correctly indicate the extent of continuous spectrum, up to the mobility edge.

15.1. Quasi-modes and pairwise tunneling amplitude

In its simplest form, eigenfunction hybridization is seen in the context of the two-level system with a Hamiltonian of the form

$$H = \begin{pmatrix} E(1) & -\tau \\ -\tau^* & E(2) \end{pmatrix}. \tag{15.1}$$

In case $\tau = 0$ the operator's two eigenvectors are $\psi_1 = \begin{pmatrix} 1 \\ 0 \end{pmatrix}$ and $\psi_2 = \begin{pmatrix} 0 \\ 1 \end{pmatrix}$, and these persist as approximate eigenvectors as long as

$$|\tau| \ll |E(2) - E(1)| =: 2\,\Delta\mathcal{E}\,. \tag{15.2}$$

In such a situation we refer to ψ_j as the operator's quasi-modes, with $E(j)$ the approximate energies and τ the **tunneling amplitude** between the two states. A simple calculation reveals that the picture is different for the case when the relation between the energy gap and the tunneling amplitude is reversed. When $|\tau| \gg \Delta E$ the operator's eigenfunctions are approximately given by the pair of vectors

$$\phi_1 \approx \frac{1}{\sqrt{2}}(\psi_1 + \psi_2) \quad \text{and} \quad \phi_2 \approx \frac{1}{\sqrt{2}}(\psi_1 - \psi_2)\,. \tag{15.3}$$

That is, the eigenfunctions are no longer localized in the basis provided by $\{\psi_1, \psi_2\}$ and instead are given by hybridized functions which are coherent combinations of the two quasi-modes.

More explicitly, the eigenvalues of the 2×2 matrix, $\mathcal{E}_1 \leq \mathcal{E}_2$, are

$$\mathcal{E}_j \;=\; \overline{\mathcal{E}} + (-1)^j \sqrt{(\Delta\mathcal{E})^2 + |\tau|^2} \tag{15.4}$$

with $\overline{\mathcal{E}} = (E(2)+E(1))/2$ and $\Delta\mathcal{E} = |E(2)-E(1)|/2$, and the corresponding eigenvectors ϕ_1 and ϕ_2 satisfy, for each value of $j = 1, 2$,

(15.5)

$$\frac{\min\{|(\phi_j,\psi_1)|, |(\phi_j,\psi_2)|\}}{\max\{|(\phi_j,\psi_1)|, |(\phi_j,\psi_2)|\}} = \left(\sqrt{1 + \left|\frac{\Delta\mathcal{E}}{\tau}\right|^2} + \frac{\Delta\mathcal{E}}{|\tau|}\right)^{-1} \approx \begin{cases} 1, & |\tau| \gg \Delta\mathcal{E}, \\ 0, & |\tau| \ll \Delta\mathcal{E}. \end{cases}$$

Eigenfunction hybridization also occurs in larger systems. In referring to it within the context of operators with random potential we employ the following terminology.

Definition 15.1. For a self-adjoint operator H on $\mathcal{H}$, a **quasi-mode** with discrepancy d is a pair (E, ψ) such that

$$\|(H - E)\psi\| \leq d\,\|\psi\|\,. \tag{15.6}$$

For instance, quasi-modes may be provided by eigenfunctions of an operator with Dirichlet or other decoupling boundary conditions along a chosen surface, provided the eigenfunction's ℓ^2-norm is mainly supported away from the surface along which the operator is modified.

A question of interest is then whether two, or more, quasi-modes (E_j, ψ_j) of small energy gaps, which for convenience we take to be orthogonal to each other, hybridize in the sense that they are equally represented in the actual eigenfunctions. For a relevant criterion, we write the resolvent restricted to the subspace $P\mathcal{H} = \operatorname{span}\{\psi_j, \psi_k\}$ spanned by two such orthonormal functions using the Krein-Feshbach-Schur complement representation (5.55):

$$\begin{aligned} &\begin{pmatrix} \langle\psi_j, (H-z)^{-1}\psi_j\rangle & \langle\psi_j, (H-z)^{-1}\psi_k\rangle \\ \langle\psi_k, (H-z)^{-1}\psi_j\rangle & \langle\psi_k, (H-z)^{-1}\psi_k\rangle \end{pmatrix} \\ &\qquad = \begin{bmatrix} E(j) - \sigma(j;z) & -\tau(j,k;z) \\ -\tau(k,j;z) & E(k) - \sigma(k;z) \end{bmatrix}^{-1} \end{aligned} \tag{15.7}$$

where $E(n) = \langle\psi_n, H\psi_n\rangle$ for $n = j, k$ and

$$\tau(j,k;z) \;=\; \langle\psi_j, T\,(H_Q - z)^{-1}\,T^\dagger\psi_k\rangle - \langle\psi_j, H\psi_k\rangle\,, \tag{15.8}$$

$$\sigma(n;z) \;=\; z + \langle\psi_n, T\,(H_Q - z)^{-1}\,T^\dagger\psi_n\rangle\,, \tag{15.9}$$

$Q = 1 - P$, $H_Q = QHQ$, and $T = PHQ$. We refer to $\tau(j,k;z)$ as the (pairwise) **tunneling amplitude**, between the two modes, at energy z.

The condition for E to be an eigenvalue of a finite system (to be precise, within the cyclic subspace generated by the action of H on $\{\psi_1, \psi_2\}$) is the vanishing of the determinant of the 2×2 matrix in (15.7). Denoting $U(n;E) := E(n) - \sigma(n;E)$, this condition reads

$$U(j;E)\,U(k;E) - |\tau(j,k;E)|^2 \;=\; 0\,. \tag{15.10}$$

If at that point also

$$|U(j;E) - U(k;E)| \lessapprox |\tau(j,k;E)|\,, \tag{15.11}$$

then the corresponding eigenstates have comparable projections on the two quasi-modes. When these correspond to functions which are well-separated in the real space, the **hybridized eigenstates** are **delocalized** in the sense of the spread of their support.

15.2. Delocalization through resonant tunneling

The above observations, and in particular equation (15.5) for the ratio of the overlaps of eigenfunctions ϕ with a pair of orthonormal functions (ψ_1, ψ_2), are not limited to quasi-modes (which form approximate eigenfunctions). To present an argument for the occurrence of continuous spectra (whose eigenfunctions are then extended) we next apply these considerations to the collection of localized δ-function states ($\{\delta_x\}_{x\in\mathbb{G}}$) on a graph which represents the quantum system's configuration space.

Our analysis proceeds by focusing on the Green function at specified energies, taking into account the fact that potential values are sufficiently uncorrelated. For simplicity of presentation, we focus on iid potentials with a continuous distribution (and an additional assumption which is given in Theorem 15.3).

15.2.1. The condition to prove. The basic tool for proving the existence of a continuous spectrum is the boosted Simon-Wolff criterion. The criterion requires the almost-sure divergence of the ℓ^2-sums:

$$\gamma_x(E,\omega) \;=\; \lim_{\eta\downarrow 0} \sum_{y\in\mathbb{G}} |G(x,y;E+i\eta,\omega)|^2 \,. \tag{15.12}$$

For the criterion's boosted version it suffices to prove that at the specified values of E the above sum diverges with positive probability.

Let us recall that by the zero-one law of Theorem 5.13, for any iid random potential with non-atomic distribution, the range of energies can be decomposed into two sets, $\mathbb{R} = Q \cup Q^{\mathrm{c}}$, such that, up to an additional exclusion of a zero measure subset $I_0 \subset \mathbb{R}$,

$$\begin{aligned} &\text{for each } E\in Q, && \mathbb{P}\{\gamma_x(E)=\infty\}=1 && \text{for some } x\in\mathbb{G}\,,\\ &\text{for each } E\in Q^{\mathrm{c}}\backslash I_o, && \mathbb{P}\{\gamma_x(E)<\infty\}=1 && \text{for all } x\in\mathbb{G}\,, \end{aligned} \tag{15.13}$$

where I_0 is a set of $E\in\mathbb{R}$ for which the limiting values of the Green function in $\mathbb{G}$ fail to exist at $E+i0$ (a possibility which is limited to a zero measure set of energies by the de la Vallée-Poussin theorem, Proposition B.3).

The Simon-Wolff criteria (Theorem 5.7) then readily imply that the set Q forms the (deterministic) support of the continuous spectrum of H, in the following sense:

i. For almost every realization there is no pure-point spectrum in Q; i.e.,

$$\mu_{\delta_x}^{\mathrm{pp}}(Q) \;=\; 0$$

for all $x \in \mathbb{G}$.

ii. For any $\varepsilon > 0$ and almost any $E \in Q$,

$$\mu_{\delta_x}^{\mathrm{c}}([E-\varepsilon, E+\varepsilon]) \;\neq\; 0$$

for some $x \in \mathbb{G}$.

The above holds true for any operator of the form $H(\omega) = T + V(\omega)$, but more can be said in the homogeneous case, on a transitive graph, where (15.13) holds simultaneously for all $x \in \mathbb{G}$.

Thus, to prove the presence of continuous spectrum within an interval I it suffices to prove that for a positive measure set of energies $E \in I$ and for some $0 \in \mathbb{G}$,

$$\boxed{\mathbb{P}\left\{\lim_{\eta \downarrow 0} \sum_{x \in \mathbb{G}} |G(0,x;E+i\eta,\omega)|^2 = \infty\right\} \;>\; 0}\;. \tag{15.14}$$

The representation provided in (5.57) for the above ℓ^2-sum makes it clear that, with the possible exception of a zero measure set of energies, the condition $\gamma_0(E,\omega) = \infty$ is satisfied for each E at which

$$\operatorname{Im} G(0,0;E+i0) \;\neq\; 0\,. \tag{15.15}$$

Our discussion will therefore focus on conditions implying $\gamma_0(E,\omega) = \infty$ assuming $G(0,0;E+i0)$ is real (an event whose probability also satisfies a zero-one law).

For brevity of notation, from this point on we shall omit the explicit reference to the limit, denoting

$$G(x,y;E) \;:=\; \lim_{\eta \downarrow 0} G(x,y;E+i\eta)\,. \tag{15.16}$$

The limit is known to exist for almost all (E,ω), simultaneously for all $x, y \in \mathbb{G}$, as well as all subgraphs obtained by excluding a finite subset from $\mathbb{G}$.

15.2.2. Rare but destabilizing resonances. For any $R \in \mathbb{N}$ one has the estimate

$$\lim_{\eta \downarrow 0} \sum_{x \in \mathbb{G}} |G(0,x;E+i\eta)|^2 \geq \sum_{x:d(0,x)=R} |G(0,x;E)|^2 \tag{15.17}$$
$$= |G(0,0;E)|^2 \sum_{x:d(0,x)=R} |r(x;E)|^2 ,$$

in terms of the ratios

$$r(x;E) := \frac{G(0,x;E)}{G(0,0;E)} = \frac{\tau(0,x;E)}{V(x)-\sigma(x;E)}, \tag{15.18}$$

with $\tau(\cdot)$ and $\sigma(\cdot)$ defined by (15.7).

Our goal now is to present a scenario under which a given site 0 is resonant at energy E with many other sites $x \in \mathbb{G}$, in the sense that $|r(x;E)| \approx 1$. For this purpose it will be convenient to express $r(x;E)$ through the so-called "self-energy" term $\Sigma(x;E)$ which is defined through the first of the following pair of relations:

(15.19)
$$G(x,x;z) =: \frac{1}{V(x)-\Sigma(x;z)}, \quad \Sigma(x;z) = \sigma(x;z) + \frac{\tau(x,0;z)\,\tau(0,x;z)}{V(0)-\sigma(0;z)},$$

where the second equality is by (15.7).

The resonances on which we shall focus are expressed in the joint occurrences of the following three events, at a large but random collection of sites $x \in \mathbb{G}$:

$$\boxed{\begin{aligned} \mathcal{T}_x &:= \{\, |\tau(0,x;E)| \geq t(0,x;E) \,\}, \\ \mathcal{E}_x &:= \{\, |V(x)-\Sigma(x;E)| \leq t(0,x;E) \,\}, \\ \mathcal{N}_x &:= \{\, |V(0)-\sigma(0;E)| \geq |\tau(x,0;E)| \,\} \end{aligned}} . \tag{15.20}$$

with $t(x,y;E) > 0$ the "75-percentile value of $\tau(x,y;E)$", for which for all $x,y \in \mathbb{G}$,

$$\mathbb{P}\left(|\tau(x,y;E)| \geq t(x,y;E)\right) \geq 3/4 . \tag{15.21}$$

A related quantifier of the tunneling amplitude's distribution which will play a role is its truncated average

$$T(x,y;E) := \mathbb{E}\left[\min\{|\tau(x,y;E)|,1\}\right] \tag{15.22}$$
$$\geq \frac{3}{4}\, t(x,y;E) .$$

The events $\mathcal{T}_x$ and $\mathcal{N}_x$ do not depend on $V(x)$ and neither one would be a rare event. In contrast, $\mathcal{E}_x$ depends on the value of $V(x)$, and under the conditions described below, it is required to fall within an extremely narrow range of values (near $\Sigma(x; E)$, which depends on $V_{\neq x}$).

Lemma 15.2. *At each specified realization of the potential and any $x \in \mathbb{G}$ for which the three condition are met, i.e., under the event $\mathcal{T}_x \cap \mathcal{E}_x \cap \mathcal{N}_x$, one has*

$$\boxed{|r(x, E)| \geq \frac{1}{2}}\,. \tag{15.23}$$

Proof. By (15.19), under the event $\mathcal{N}_x$ one has

$$|\Sigma(x; E) - \sigma(x; E)| \;\leq\; |\tau(0, x; E)|\,. \tag{15.24}$$

Claim (15.23) can be deduced by combining this bound with (15.18) and the condition defining $\mathcal{N}_x$. □

Now let $\mathcal{A}(E)$ denote the following pair of conditions:

A1. The partial sums, over the R-spheres, diverge:

$$\boxed{\lim_{R\to\infty} \sum_{x:d(0,x)=R} t(0, x; E) \;=\; \infty}\,. \tag{15.25}$$

A2. The tunneling amplitude's truncated averages decay with the distance, i.e.,

$$\lim_{R\to\infty} \max_{x:d(x,0)=R} T(x, 0; E) \;=\; 0\,, \tag{15.26}$$

and also satisfy, for all $R < \infty$ and x with $d(0, x) = R$,

$$\boxed{\sum_{y:d(0,y)=R} T(x, y; E) \;\leq\; C \sum_{x:d(0,x)=R} t(0, x; E)} \tag{15.27}$$

with some $C < \infty$.

For insight into these conditions it may be of relevance to consider the case of $\mathbb{G}$ a regular tree graph of branching ratio K. On tree graphs the tunneling between any two sites is naturally presentable through a sequence of steps. That yields a natural factorization of $\tau(x, y)$ into a product of "single step" random variables. As will be explained in the next chapter, both $t(x, y; E)$ and $T(x, y; E)$ decay exponentially:

$$\begin{aligned} t(x, y; E) &= A_0\, e^{-\mathcal{L}_0(E)\ \mathrm{dist}(x,y)}\,, \\ T(x, y; E) &= A_1\, e^{-\mathcal{L}_1(E)\,\mathrm{dist}(x,y)} \end{aligned} \tag{15.28}$$

with $\mathcal{L}_0(E)$ which can be identified as the Lyapunov exponent of a transfer-matrix-driven dynamics and $\mathcal{L}_1(E) \;\leq\; \mathcal{L}_0(E)$ by (15.22).

In that situation condition **A1** is valid when

$$\boxed{\mathcal{L}_0(E) \ < \ \log K}\,, \tag{15.29}$$

that is, provided the (geometric) surface growth rate exceeds the typical tunneling decay rate. Assumption **A2** is valid provided

$$\boxed{\mathcal{L}_0(E)\,/\,2 \ < \ \mathcal{L}_1(E)}\,. \tag{15.30}$$

This expresses the fact that in the hyperbolic geometry of the tree, for each x with $\mathrm{dist}(x,0) = R$ most of the R sphere is asymptotically further from x than from the center 0 by a factor which asymptotically can be chosen arbitrarily close to 2. Due to this, the surface average (over y) of $T(x,y)$ decays at asymptotically twice the decay rate of the surface average of $T(0,y)$ (the exact calculation is elementary).

As explained in greater detail in the next chapter

$$\text{for } \lambda = 0 \text{ and all } E \in \sigma(T), \qquad \mathcal{L}_0(E) \ = \ \mathcal{L}_1(E) \ = \ \frac{1}{2}\log K\,, \tag{15.31}$$

which expresses the facts that (i) there are no fluctuations and (ii) at zero disorder the ℓ^2-sum $\sum_{x:\,\mathrm{dist}(x,0)=R} |G(0,x;E)|^2$ tends to a finite constant as $R \to \infty$. While at zero disorder there are no potential fluctuations (which plays into the argument presented next), in this case sufficient Lyaponov exponent bounds are in place, and continuity arguments for the exponents can be developed making the next argument applicable at weak disorder throughout the spectrum of T and even beyond it.

The general result for which we aim here is the following criterion for the occurrence of continuous spectrum in specified energy ranges.

Theorem 15.3. *For an infinite transitive graph $\mathbb{G}$, let $H(\omega) = T + V(\omega)$ be a random operator on $\ell^2(\mathbb{G})$ with the following:*

i. *T an operator which is shift-invariant modulo gauge transformations—in the sense of* (3.4),

ii. *V an iid potential whose single-site distribution is absolutely continuous with bounded density ϱ, satisfying $\inf_{|v|\le S} \varrho(v) > 0$ for all $S > 0$.*

Then for any Borel set $I \subset \mathbb{R}$ over which assumptions **A1** *and* **A2** *hold with respect to $0 \in \mathbb{G}$,*

$$\mu^{(c)}_{\delta_0}(I;\omega) \ > \ 0 \tag{15.32}$$

for almost every realization of the potential (i.e., almost every ω).

For a proof we proceed along the lines sketched above and consider the number of sites resonant with 0:

$$\boxed{N_R(E) := \sum_{x:d(x,0)=R} 1_{\mathcal{T}_x \cap \mathcal{E}_x \cap \mathcal{N}_x}} . \tag{15.33}$$

As stated above (15.15), it suffices for us to focus on energies for which almost surely, for all $x \in \mathbb{G}$,

$$\boxed{\operatorname{Im} \Sigma(x; E) = 0} . \tag{15.34}$$

This condition is important for the argument which follows since $\mathcal{E}_x$ would fail to occur if the alternative condition (15.15) holds almost surely. On the other hand, over the set where the alternative condition (15.15) holds, we already know that the conclusion (15.32) is satisfied. The key remaining step is

Lemma 15.4. *Under the assumptions of Theorem* 15.3 *for each* $E \in \mathbb{R}$ *at which* (15.34), *as well as* **A1** *and* **A2**, *holds for almost every realization of the potential, for any* $M < \infty$ *and all* R *large enough,*

$$\mathbb{P}\left(N_R(E) \geq M\right) \geq p_0 \tag{15.35}$$

with some $p_0 > 0$ *which does not depend on* M *and* R.

Once this is established, Theorem 15.3 readily follows through the zero-one law of Theorem 5.13 and the Simon-Wolff criterion of Theorem 5.7.

15.2.3. The second-moment method. Lemma 15.4 is proven by a two-step argument: The first is to show that the mean (i.e., the first moment of N_R) diverges:

$$\lim_{R \to \infty} \mathbb{E}\left[N_R\right] = \infty . \tag{15.36}$$

Then, the second-moment test will be used to establish a uniformly positive lower bound on the probability that the random variables N_R assume values comparable with their mean. The alternative which needs to be ruled out here is that the mean diverges only due to some very large contributions of very rare events, while the typical range of values (e.g., the median) remains finite. A convenient tool for a such purpose is the following:

Lemma 15.5 (Paley-Zygmund). *For a random variable* N *and any* $\theta \in (0,1)$,

$$\mathbb{P}\left(N \geq \theta\, \mathbb{E}\left[N\right]\right) \geq (1-\theta)^2 \frac{\mathbb{E}\left[N\right]^2}{\mathbb{E}[N^2]} . \tag{15.37}$$

The proof is left as Exercise 15.1.

In summary, to employ the above strategy one needs to derive a **lower bound** on $\mathbb{E}[N_R]$ and an **upper bound** on $\mathbb{E}[N_R^2]$.

The lower bound. In order to estimate the joint probability $\mathbb{P}(\mathcal{T}_x \cap \mathcal{E}_x \cap \mathcal{N}_x)$ from below, under the assumptions $\mathcal{A}(E)$, we gather the following facts:

1. By our selection of the function $t(0,x;E)$, for all $x \in \mathbb{G}$,

$$\mathbb{P}(\mathcal{T}_x) \geq 3/4 . \tag{15.38}$$

2. Since the random variable $V(0)$ is independent of $\sigma(0;E)$ and its distribution is absolutely continuous with density $\varrho \in L^\infty(\mathbb{R})$, we also have—for R large enough

$$\mathbb{P}(\mathcal{N}_x) \geq 1 - \mathbb{E}[\min\{2\|\varrho\|_\infty |\tau(x,0;E)|, 1\}] \geq 3/4 \tag{15.39}$$

   for all x with $\operatorname{dist}(0,x) \geq R$. The first inequality is generally true and the second holds for sufficiently large R by assumption **A2**.

3. In contrast to the above events, $\mathcal{E}_x$ is extremely rare and has a non-zero probability only if (15.34) holds. It is at this stage of the argument that use is made of the *no-ac* hypothesis, which we aim to contradict.

Let $S < \infty$ be the minimal value for which

$$\mathbb{P}(\{|\Sigma(x;E)| \leq S\}) \geq 3/4 , \tag{15.40}$$

and let $\mathcal{S}_x$ be the event $\{|\Sigma(x;E)| \leq S\}$. Assuming **A1**, the average over $V(x)$ yields for the conditional probability

$$\begin{aligned} \mathbb{P}(\mathcal{E}_x | V_{\neq x})(\omega) &\geq \mathbb{P}(\mathcal{E}_x \cap \mathcal{S}_x | V_{\neq x})(\omega) \\ &\geq t(0,x)\, \varrho_s\, 1_{\mathcal{S}_x}(\omega) \end{aligned} \tag{15.41}$$

with $\varrho_S := \min_{|v| \leq S+1} \varrho(v)$, where the second inequality is derived through the conditional expectation of $\mathcal{E}_x \cap \mathcal{S}_x$ conditioned on the values of $V_{\neq x}$.

We now choose $R \in \mathbb{N}$ such that both (15.38) and (15.39) hold. Combining (15.41) with the observation that the event $\mathcal{T}_x \cap \mathcal{N}_x$ is measurable with respect to $V_{\neq x}$ we conclude that

$$\begin{aligned} \mathbb{E}[N_R] &= \sum_{x:d(x,0)=R} \mathbb{P}(\mathcal{T}_x \cap \mathcal{E}_x \cap \mathcal{N}_x) \\ &= \sum_{x:d(x,0)=R} \mathbb{E}[1_{\mathcal{T}_x \cap \mathcal{N}_x} \mathbb{P}(\mathcal{E}_x \mid V_{\neq x})] \\ &\geq \varrho_s \sum_{x:d(x,0)=R} t(0,x)\, \mathbb{P}(\mathcal{T}_x \cap \mathcal{N}_x \cap \mathcal{S}_x) \geq \frac{\varrho_s}{4} \sum_{x:d(x,0)=R} t(0,x) , \end{aligned} \tag{15.42}$$

where in the last step $\frac{1}{4} = 1 - 3(1 - \frac{3}{4})$ is the sum of the probabilities of the complements of three events. The claimed divergence (15.36) for $R \to \infty$ is then implied by assumption **A1**.

The upper bound. For the second-moment (upper) bound we start from

$$\begin{aligned} \mathbb{E}\left[N_R(N_R-1)\right] &= \sum_{x\neq y}^{(R)} \mathbb{P}\left(\mathcal{T}_x \cap \mathcal{E}_x \cap \mathcal{N}_x \cap \mathcal{T}_y \cap \mathcal{E}_y \cap \mathcal{N}_y\right) \\ &\leq \sum_{x\neq y}^{(R)} \mathbb{P}\left(\mathcal{E}_x \cap \mathcal{E}_y\right), \end{aligned} \tag{15.43}$$

where the sums in $\sum^{(R)}$ are over sites of $\mathbb{G}$ at distance R from 0.

By (15.19) the event $\mathcal{E}_x$ corresponds to $\{|G(x,x;E+i0)| \geq 1/t(0,x;E)\}$, and likewise for y. Thus the challenge is therefore to bound the effects of the correlations between such rare events, which exist but decay with the distance.

For insight into the joint probability distribution of $\mathcal{E}_x$ and $\mathcal{E}_y$ it is natural to consider the restriction of the resolvent kernel to the two-dimensional space spanned by δ_y and δ_y. Lemma 15.6 yields

$$\begin{aligned} \mathbb{P}\left(\mathcal{E}_x \cap \mathcal{E}_y\right) \leq C \Big(& t(0,x)t(0,y) \\ & + \sqrt{t(0,x)t(0,y)}\, \mathbb{E}\left[\min\left\{\sqrt{|\tau(x,y;E)\tau(y,x;E)|}, 1\right\}\right]\Big). \end{aligned} \tag{15.44}$$

Under assumption (15.34) one may conclude that $\operatorname{Im}\langle\psi, (H-E-i0)^{-1}\psi\rangle = 0$ for all $\psi \in \ell^2(\mathbb{G})$ and thus $|\tau(x,y;E)| = |\tau(y,x;E)|$. The above average may therefore be identified with $T(x,y)$, of (15.22), and hence, applying the Cauchy-Schwarz inequality,

$$\mathbb{E}\left[N_R(N_R-1)\right] \leq C\, \mathbb{E}\left[N_R\right]^2 + C \sum_{x:d(x,0)=R} t(0,x) \sum_{y:d(y,0)=R} T(x,y)\,. \tag{15.45}$$

The assumption (15.26) of **A2** then allows us to conclude the second-moment bound:

$$\limsup_{R\to\infty} \frac{\mathbb{E}\left[N_R(N_R-1)\right]}{\mathbb{E}\left[N_R\right]^2} \leq C\,. \tag{15.46}$$

Through the Paley-Zygmund criterion, of Lemma 15.5, the pair of moment bounds (15.36) and (15.46) yields Lemma 15.4.

15.2.4. Correlations among local resonances.

Essential use was made above of the following bound on correlations [**21**].

Lemma 15.6 (Resonance correlation bound). *Let H_0 be a self-adjoint operator in $\ell^2(\mathbb{G})$ and let $\rho(du\,dv) = \varrho_1(u)\,\varrho_2(v)du\,dv$ be an absolutely continuous probability measure on $\mathbb{R}^2$ with bounded densities $\varrho_j \in L^\infty(\mathbb{R})$ $(j=1,2)$. Then there is some $C<\infty$ such that the Green function of*

$$H_{u,v} := H_0 + u\,1_{\{y\}} + v\,1_{\{x\}}$$

in $\ell^2(\mathbb{G})$ satisfies for any $z \in \mathbb{C}\backslash\mathbb{R}$ and any $a, b > 0$

$$\rho_2\left(\{(u,v) \in \mathbb{R}^2 \mid |G_{u,v}(x,x;z)| > a^{-1} \text{ and } |G_{u,v}(y,y;z)| > b^{-1}\}\right) \leq C\sqrt{ab}\,\min\left\{2\sqrt{ab} + 2\sqrt{|\tau(x,y;z)||\tau(y,x;z)|}, 1\right\} \tag{15.47}$$

with $\tau(x,y;z)$ the tunneling amplitude associated with (δ_x, δ_y) at z.

Proof. The rank-2 Schur complement formula (15.7) reveals the dependence of the diagonal Green functions on (u,v). Abbreviating

$$U := \sqrt{\frac{b}{a}}\,(u - \sigma(x;z))\,, \quad V := \sqrt{\frac{a}{b}}\,(v - \sigma(y;z))\,, \tag{15.48}$$

and $\gamma := \tau(x,y;z)\tau(y,x;z)$, the lower bounds on $|G_{u,v}(x,x;z)|$ and $|G_{u,v}(y,y;z)|$ translate to

$$\left|U - \frac{\gamma}{V}\right| \leq \sqrt{ab}\,, \qquad \left|V - \frac{\gamma}{U}\right| \leq \sqrt{ab}\,. \tag{15.49}$$

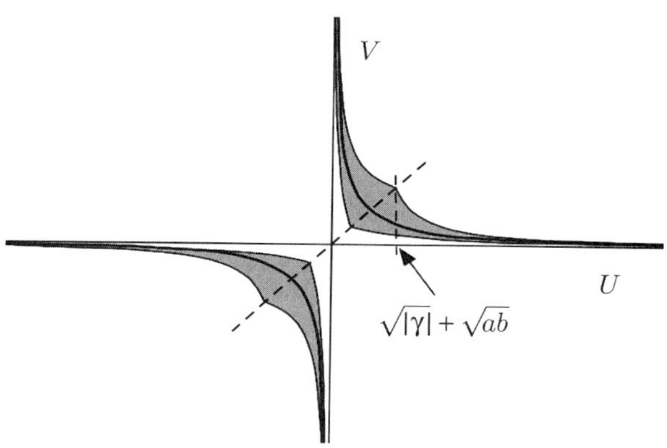

Figure 15.1. The solution set of (15.47) in the (u,v)-plane (for real γ and σ).

The claim can be proven through the following two observations about the set in the (u,v)-plane over which the conditions (15.49) are met (the set's shape is indicated in Figure 15.1).

i. For any solution,

$$\min\{|U|,|V|\} \leq \sqrt{|\gamma|} + \sqrt{ab}\,. \tag{15.50}$$

ii. For specified v, the set of u for which (15.49) holds is an interval of length at most $2\sqrt{ab}$, and a similar statement holds for v and u interchanged.

The area bound which these conditions yield upon integration translates directly into (15.47). □

15.3.* Exploring the argument's limits

There are two directions in which the above analysis can be extended, with interesting outcomes:

1. The restriction to unbounded potentials can be **relaxed**—at the cost of some added technicalities.
2. The **range of energies** for which a continuous spectrum can be established by the resonant delocalization mechanism may be **enlarged** by considering the rare fluctuations due to which the tunneling amplitude $\tau(0,x)$ to certain sites is significantly larger than its fixed percentile value $t(0,x)$.

The restriction on the potential's probability distribution which features in Theorem 15.3 was used only in the lower bound (15.41). An extension to bounded potentials is possible, with the ρ_S factor in (15.3) replaced by a more technical estimate in which appears the local density of states at energy E (defined in (4.4)) which under the *no-ac* hypothesis (15.34) is

$$\nu_x(E) \;=\; \mathbb{E}\left[\rho(\Sigma(x;E))\right] . \tag{15.51}$$

An interesting outcome derived along these lines was the surprising result in [**19**] that for the random operators on trees, for bounded potentials there is a threshold of minimal disorder which is required for the existence of a mobility edge beyond which the spectrum is pure point only. At the intuitive level that is so since on trees the Lyapunov exponent conditions (15.29) and (15.30) hold at zero disorder not just in $\sigma(T)$ but throughout a larger set, which for small disorder includes the entire spectrum of $H(\omega)$.

The other extension is based on the following "optimistic picture" of the possible range of reach of the resonant delocalization argument.

For an infinite transitive metric graph, let

$$\chi(R) \;:=\; \log\left[\operatorname{card}\{x \in \mathbb{G} |\ \operatorname{dist}(0,x) \in [R, R+1]\}\right] \tag{15.52}$$

so that the number of sites at distance R from $0 \in \mathbb{G}$ grows as $e^{\chi(R)}$. E.g., on a regular tree of degree $K+1$, $\chi(R)$ is approximately linear in the graph distance,

$$\chi(R) \approx R \log K\,, \tag{15.53}$$

while for the (finite) hypercube the analogous function grows much faster.

Assume now that the tunneling amplitude is exponentially small in $\chi(\operatorname{dist}(0,x))$, at a rate which exhibits large deviation behavior. Roughly speaking, this means that for $\operatorname{dist}(0,x) \to \infty$ the probability that $\tau(0,x) \approx$

$\exp\left(-[L+o(1)]\chi(\mathrm{dist}(0,x))\right)$ is itself of an exponential order, with (for any fixed $\delta > 0$)

$$\mathbb{P}\left\{\left|\frac{\log \tau(0,x)}{\chi(\mathrm{dist}(0,x))} + L\right| < \delta\right\} \approx e^{-[\gamma(L)+o(1)]\,\chi(\mathrm{dist}(0,x))} \tag{15.54}$$

with a good rate function $\gamma(L)$. (The vocabulary of large deviation theory and its relevant basic results may be found in, e.g., [**116**].)

In order to explore the contribution of resonances with sites at which an unusually low value of L is realized, let us consider the three events described by (15.20) with the cutoff function modified to

$$t(0,x) \;=\; e^{-L\,\chi(\mathrm{dist}(0,x))}\,. \tag{15.55}$$

For L at which $\gamma(L) > 0$, $\mathcal{T}_x$ is now made into a rare event. However, among the sites where it is realized, the probability of $\mathcal{E}_x$ is now larger than before since the very small set to which $V(x)$ is restricted is now enlarged. Consequently, instead of (15.42) we get the first-moment lower bound:

$$\mathbb{E}\left[N_R\right] \;\geq\; \text{Const.}\; e^{\chi(R)}\, e^{-[\gamma(L)+L]\,\chi(R)} \tag{15.56}$$

where the two exponential factors are the surface area and the fraction of sites at which the events $\mathcal{T}_x \cap \mathcal{E}_x$ occur.

Thus, assuming the validity of the above-sketched large deviation structure, under the modified criteria a sufficient condition for passing the first-moment test (15.36) is that

$$-\varphi(1) := \inf_{L\geq 0}\left[\gamma(L) + L\right] \;<\; 1 \tag{15.57}$$

where $\varphi(\cdot)$ is the Legendre transform of the function γ (extended to $\gamma(L) = +\infty$ for $L < 0$):

$$-\varphi(s) \;:=\; \inf_{L\geq 0}\left[\gamma(L) + sL\right]. \tag{15.58}$$

The (regrettable) negative sign is for keeping notational consistency with [**21**].

As in the standard large deviation theory, one can learn about the Legendre transform from the decay rate of the moments, for which the natural considerations yield

$$\varphi(s) \;=\; \lim_{\mathrm{dist}(0,x)\to\infty} \frac{1}{\chi(\mathrm{dist}(0,x))} \log \mathbb{E}\left(|\tau(0,x;E)|^s\right), \tag{15.59}$$

within a range of s dictated by the extent of the good large deviations behavior. In view of the divergence of moments $s \geq 1$ due to purely local fluctuations (except in the presence of ac spectrum) it will be natural to approach the case $s \in (0,1)$, through the limit $s \nearrow 1$ or, alternatively, through the moments of the truncated quantity

$$(15.60) \qquad |\tau(0,x;E)| \wedge 1 \ := \ \min\{\, |\tau(0,x;E)|\, , \, 1\}\,.$$

This may be summarized by noting that in the above scenario a condition which indicates the validity of the first step in the moment test for **delocalization** is

$$(15.61) \qquad \boxed{\text{for some } \varepsilon > 0, \quad \sum_x \mathbb{E}\left(|\tau(0,x;E) \wedge 1|^{1+\varepsilon} \right) \ = \ \infty}\,.$$

It is of interest to note that condition (15.61) is practically **complementary** to our established fractional moment **localization criterion**:

$$(15.62) \qquad \boxed{\text{for some } \varepsilon > 0, \quad \sum_x \mathbb{E}\left(|\tau(0,x;E)|^{1-\varepsilon} \right) \ < \ \infty}\,.$$

(Missing is a proof that the transition between convergence and divergence occurs along a reasonably regular boundary in the energy-disorder plane.)

However, to turn the above discussion into a proof one also needs to establish the second-moment upper bound (15.37). That is a bit more involved since it now requires bounding correlations between not just pairs of essentially local events, at (x,y) as above, but also between large deviations of the corresponding path-related tunneling amplitudes. In the next chapter we discuss the tree graph case for which that could be done.

Notes

The considerations of tunneling between **pairs of states** may not yet yield the full picture of the operator's eigenfunctions. Among the other issues of relevance are the following:

1. the possibility of simultaneous resonances among many quasi-modes,
2. the fact that in a large finite system the entries of the 2×2 matrix of (15.7) are singular at a collection of energies E (corresponding to the spectrum of the full operator) which becomes dense in the infinite-volume limit.

Hence a more complete analysis is required for establishing the full picture of the eigenfunctions in large finite systems. Nevertheless, in certain model examples, of, e.g., the random Schödinger operator on the complete graph [**12**], the criterion (15.10) does yield the correct phase diagram in which in addition to regimes of localization there are bands of energies whose eigenstates are delocalized in terms of the spacial spread of the eigenfunctions (though not necessarily in terms of the inverse participation ratio $\|\phi\|_4^4/\|\phi\|_2^2$ or, equivalently, the vanishing of $\|\phi\|_\infty/\|\phi\|_2$). The full proof, which can be found in [**12**], proceeds through other means. However, the results are in accordance with the criterion expressed by (15.11).

It may be of interest to see assumptions **A1** and **A2** of Theorem 15.3, and hence spectral delocalization, established for graphs other than regular trees discussed in the next chapter. Natural directions to consider are expander/hyperbolic graphs.

Exercises

15.1. Prove Lemma 15.5 (the Paley-Zygmund criterion).

Chapter 16

Phase Diagrams for Regular Tree Graphs

Often problems which are hard in finite dimensions, $d \in (1, \infty)$, are more approachable on regular tree graphs, and there is much that can be learned from analyzing the topic in that simplified setup. That also holds true for the Anderson transition—albeit with some surprises which were missed in the foundational papers on the subject [**2, 3**]. The latter relate to aspects which were not seen in the theory of one-particle localization in finite dimensions, though perhaps are of relevance for multi-particle systems.

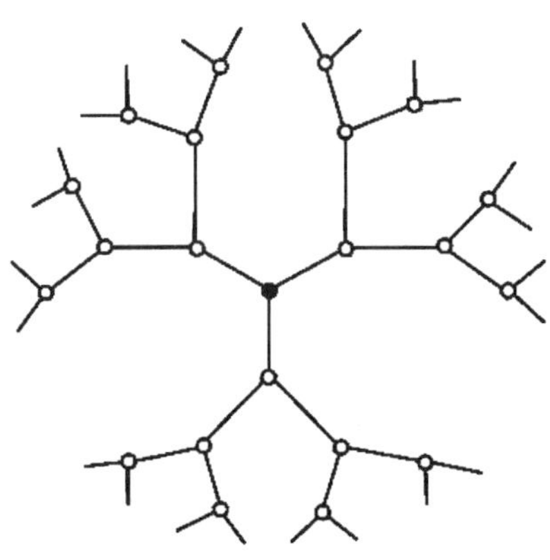

Figure 16.1. The Bethe lattice $\mathbb{B}$ with degree $3 = K + 1$.

A tree is a graph with no loops. Its vertex set will be generically denoted by $\mathbb{T}$. Singling out a special vertex 0 from that set renders any tree a rooted tree. We will be mainly concerned with the following two examples: the Bethe lattice $\mathbb{B}$ which is the **regular tree graph** whose vertices have a common degree $K+1$ and the corresponding **rooted regular tree graphs** $\mathbb{B}_0$ for which all vertices including the root 0 have K forward neighbors.

The analysis of various topics on homogeneous tree graphs often proceeds through the study of the given issue on the rooted tree $\mathbb{B}_0$. In that case, upon the removal of the root the graph breaks into K disconnected graphs which are isomorphic to the original one. This invariably leads to useful recursion relations. Once the case of the rooted graphs $\mathbb{B}_0$ is understood, it is typically easy to extend the results to the fully homogeneous graphs $\mathbb{B}$. This strategy also serves well the spectral analysis of the Schrödinger operator on homogeneous tree graphs.

16.1. Summary of the main results

As a guide to this section let us start with a list of the main results presented here for the random operator

$$(H_\lambda(\omega)\,\psi)\,(x) = -\sum_{y:d(x,y)=1} \psi(y) + \lambda V(x;\omega)\,\psi(x)\,, \tag{16.1}$$

acting in $\ell^2(\mathbb{B})$—omitting the exact conditions on the iid random potential under which each statement is proven. The first term coincides with the negative adjacency operator on the graph, which is the Laplacian with omitted diagonal terms.

i. Quantifying the Green function's decay rate

For the random Schrödinger operator on the homogeneous tree with $K>1$, the Green function $G_\lambda(x,y;z) = \langle \delta_x, (H_\lambda - z)^{-1}\delta_y\rangle$ exhibits exponential decay. The typical decay rate is quantified through the observation that for any preselected simple path γ to infinity, the following limit exists almost surely:

$$-\lim_{x\in\gamma\,:\,d(0,x)\to\infty} \frac{1}{d(0,x)} \ln|G_\lambda(0,x;E+i0)| \;=\; \mathcal{L}_\lambda(E) \tag{16.2}$$

with $\mathcal{L}_\lambda(E)$ referred to as the Lyapunov exponent.

Since the number of paths reaching distance R grows exponentially as K^R, for typical realizations of the random operator other decay rates will also be observed. The corresponding rate can be studied, using the standard methods of large deviation theory, through properties of the cumulant moment-generating function:

$$\varphi_\lambda(s;E) \;:=\; \lim_{d(0,x)\to\infty} \frac{\ln \mathbb{E}\left[|G_\lambda B(0,x;E+i0)|^s\right]}{d(0,x)}\,. \tag{16.3}$$

This function is convex and non-increasing in s over a range which includes $[0,1)$, where it satisfies for Lebsgue-almost all $E \in \mathbb{R}$:

$$\frac{\partial \varphi_\lambda}{\partial s}(0;E) = -\mathcal{L}_\lambda(E)\,, \qquad -s\,\mathcal{L}_\lambda(E) \le \varphi_\lambda(s;E) \le -s\,\ln\sqrt{K}\,. \tag{16.4}$$

ii. Spectrum of the free operator

The spectrum of the adjacency operator on $\ell^2(\mathbb{B})$ is purely absolutely continuous and given by

$$\sigma\left(H_0\right) \;=\; \left[-2\sqrt{K}, 2\sqrt{K}\right]\,. \tag{16.5}$$

The corresponding Lyapunov exponent is constant along the spectrum:

$$\mathcal{L}_0(E) \;=\; -\,\ln\sqrt{K} \quad \text{for all } E \in \left[-2\sqrt{K}, 2\sqrt{K}\right] \tag{16.6}$$

and for $|E| > 2\sqrt{K}$ it is strictly monotone increasing in $|E|$ with $\mathcal{L}_0(K+1) = -\ln K$. The energy value $E = K+1$ is significant: The spectral radius of the adjacency operator on finite, K-regular graphs is $K+1$, and for such finite bipartite graphs both $\pm(K+1)$ form the ℓ^2-spectral extrema. Because of the regular trees' expander property or hyperbolicity, this value does not show up in $\sigma\left(H_0\right)$ and there is the significant spectral gap to $2\sqrt{K}$. However, as can be inferred from the Lyapunov exponent, the operator's ℓ^1-spectrum is $[-(K+1), K+1]$.

iii. Regime of absolutely continuous spectrum

In contrast to one dimension where disorder produces total localization, on trees the absolutely continuous spectrum is rather stable. Its occurrence for operators of the form (16.1) has been established through a number of methods. These can be lumped into the following:

a) *Continuity arguments.* A number of distinct methods, starting with [**244**] and also including those of [**14, 161**], yield proofs that absolutely continuous spectrum persists in the vicinity of the unperturbed spectrum, $\sigma\left(H_0\right)$, at weak enough disorder.

b) *Resonant delocalization.* The mechanism which is discussed in detail in Chapter 15 is very relevant to tree graphs. It yields proofs of the existence of absolutely continuous spectrum even far from $\sigma\left(H_0\right)$, where the minute size of the density of states suggests localization.

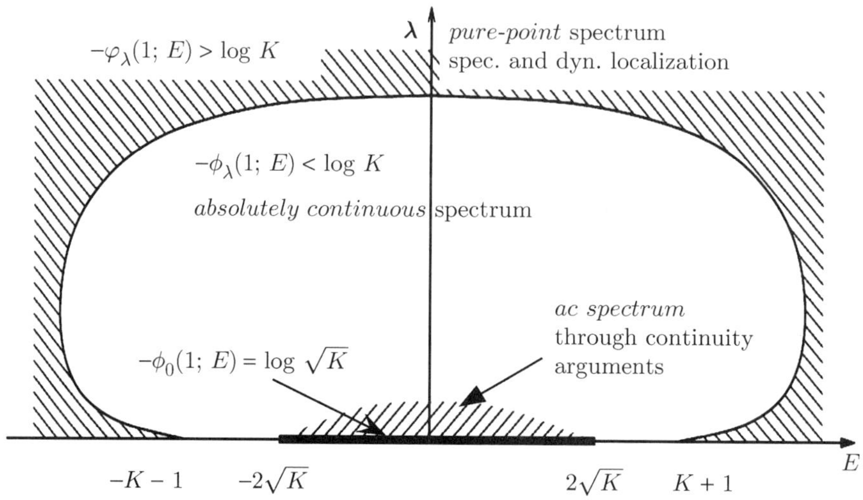

Figure 16.2. The spectral phase diagram (sketched) for unbounded random potentials, assuming $\phi_\lambda(s;E)$ is continuous and its level set at the value $\ln K$ is a simple curve. For $\lambda \downarrow 0$ this set is proven to connect to $|E| = K+1$. Also depicted is the difference between the larger regime where the *ac* spectrum is proven to exist due to the singular spectrum's instability to *resonant delocalization* and the much smaller regime in which the *ac* spectrum was earlier proven to persist through continuity arguments.

More explicitly, the resonant delocalization argument (which amounts to proof by contradiction) yields that, under some auxiliary assumptions on the distribution of the random potential, the random Schrödinger operator has absolutely continuous spectrum over the set of energies at which there is spectrum, and

$$-\varphi_\lambda(1;E) \le \log K\,. \tag{16.7}$$

In particular, the above applies throughout the regime where

$$\mathcal{L}_\lambda(E) \le \log K\,. \tag{16.8}$$

One may note that, by (16.6), for $\lambda = 0$ this condition both covers and extends significantly beyond the unperturbed spectrum. This suggest that not only will the absolutely continuous spectrum in $|E| < 2\sqrt{K}$ be stable under weak disorder, but for $2\sqrt{K} \le |E| < K+1$ absolutely continuous spectrum will appear as soon as some spectrum is produced there, even if that is through rare fluctuations of the random potential. To be able to make such a conclusion it will be of value to know that $\mathcal{L}_\lambda(E)$ is continuous in λ, and some such **weak continuity** results are presented below.

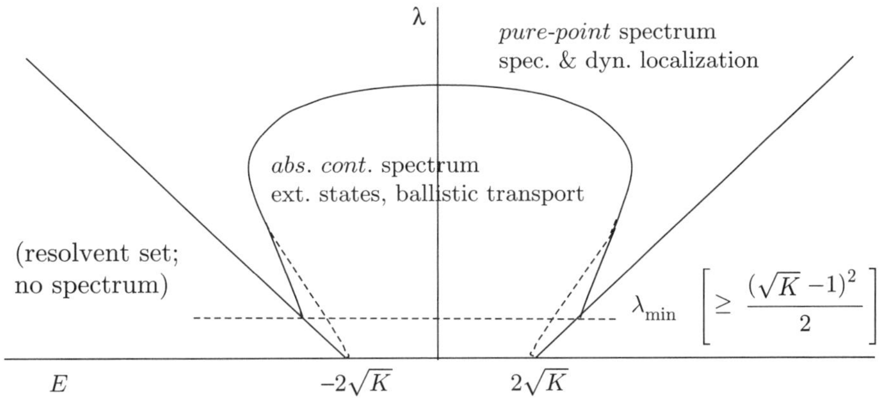

Figure 16.3. Sketch of the phase diagram for the Anderson model on the Bethe lattice for bounded potential. The analysis suggests that at weak disorder there is no localization and the spectrum is purely absolutely continuous. While the proof of that is incomplete, it is proven that for $\lambda \leq (\sqrt{K}-1)^2/2$ near the spectral edges the spectrum is purely absolutely continuous. The slanted dashed lines represent the previously expected behavior of the mobility edge at weak disorder.

iv. Regime of localization

Applying the fractional moment localization proof one may conclude that spectral and dynamical localization occur over the regime of energies where for some $\varepsilon > 0$

$$-\varphi_\lambda(1-\varepsilon; E) \; > \; \log K\,. \tag{16.9}$$

Up to questions of continuity and regularity of $\varphi_\lambda(s;E)$ as a function of (λ, E), the two conditions seem gratifyingly complementary to each other. This indicates the relevance of the two criteria, for localization and for delocalization, at a non-perturbative level.

v. Surprises in the phase diagram

The results summarized above also yield the curious implication that in the case of bounded random potential a mobility edge beyond which the operator H_λ has only pure-point spectrum does not occur until the disorder strength exceeds a certain positive threshold [**21**]; cf. Figure 16.3. This appeared as an unexpected result of the rigorous analysis of the Anderson transition on trees.

16.2. Recursion and factorization of the Green function

While the main results presented here for the random Schrödinger operators apply only to the homogeneous case, some properties of the resolvent are also valid for general tree graphs $\mathbb{T}$. To explain this, we highlight the dependence of the operator on the domain in our notation and at our convenience write (16.10)

$$(H_{\mathbb{T}}\psi)(x) = -(A_{\mathbb{T}}\psi)(x) + V(x)\,\psi(x)\,, \qquad (A_{\mathbb{T}}\psi)(x) := \sum_{y:d(x,y)=1} \psi(y)\,,$$

acting in $\ell^2(\mathbb{T})$. The negative first term coincides with the **adjacency operator** on the graph $\mathbb{T}$. The spectral analysis of $H_{\mathbb{T}}$, which amounts to the study of the Green function

$$G_{\mathbb{T}}(x,y;z) = \langle \delta_x, (H_{\mathbb{T}} - z)^{-1}\delta_y \rangle\,, \qquad x,y \in \mathbb{T}\,,\ z \in \mathbb{C}\backslash\mathbb{R}, \tag{16.11}$$

is aided by the observation that upon the removal of any vertex $v \in \mathbb{T}$ the tree graph splits into a collection of disconnected components. More generally we have the following:

1. For a collection of vertices $v_1, \ldots, v_n \in \mathbb{T}$ we denote by $\mathbb{T}_{v_1,\ldots,v_n}$ the disconnected subgraph obtained by deleting this collection from $\mathbb{T}$.
2. for $\mathbb{T}' \subset \mathbb{T}$ we denote by $H_{\mathbb{T}'}$ the restriction of $H_{\mathbb{T}}$ to $\ell^2(\mathbb{T}')$. E.g., $H_{\mathbb{T}_{v_1,\ldots,v_n}}$ is the operator obtained by eliminating all the matrix elements of H involving any of the removed sites. The corresponding Green function $G_{\mathbb{T}'}(x,y;z)$ vanishes if x and y belong to different connected components of $\mathbb{T}'$, and otherwise it stands for the Green function corresponding to the component which contains the two.

Given an oriented simple path in a tree $\mathbb{T}$ which passes through u, we abbreviate as follows:

$$\begin{aligned} \Gamma(y;z) := \Gamma_-(u;z) &:= G_{\mathbb{T}_{u_-}}(u,u;z)\,, \\ \Gamma_+(u;z) &:= G_{\mathbb{T}_{u_+}}(u,u;z)\,, \end{aligned} \tag{16.12}$$

where u_- and u_+ are the neighboring sites of u on that path. Rooted trees are naturally partially ordered by the relation $x \prec y$ and $x \preceq y$, respectively, which means that x lies on the unique path from the root to y possibly coinciding with y.

The following recursion and factorization relation is the key to the spectral analysis of operators of the form (16.10) on trees.

Proposition 16.1. *The Green function of any self-adjoint operator $H_{\mathbb{T}} = A_{\mathbb{T}} + V$ on $\ell^2(\mathbb{T})$ on a rooted tree $\mathbb{T}$ with a uniformly bounded degree satisfies the following for all $z \in \mathbb{C}\backslash\mathbb{R}$:*

1. *For any $x \in \mathbb{T}$,*

$$\boxed{G_{\mathbb{T}}(x,x;z) = \Big(V(x) - z - \sum_{y:d(x,y)=1} G_{\mathbb{T}_x}(y,y;z)\Big)^{-1}}\,. \tag{16.13}$$

2. *For any pair of partially ordered sites,* $0 \prec x \prec y$,

(16.14)
$$\boxed{G_{\mathbb{T}}(x,y;z) = G_{\mathbb{T}}(x,x;z) \prod_{x \prec u \preceq y} \Gamma_-(u;z) = G_{\mathbb{T}}(y,y;z) \prod_{x \preceq u \prec y} \Gamma_+(u;z)},$$

where the $\pm$ *subscripts on* Γ *are defined relative to the root.*

Since both properties follow by straightforward application of the resolvent equation or, alternatively, with the help of the Feenberg expansion, we will leave the proof of this proposition as an exercise. Let us comment on two special cases, which are of particular interest:

1. Upon the removal of any vertex x of a regular tree graph $\mathbb{B}$, the graph $\mathbb{B}_x$ decomposes into $K+1$ disjoint rooted trees $\mathbb{B}_y$ with $d(x,y) = 1$. The diagonal part of the Green function on the regular tree can hence be computed from the diagonal part of the rooted components; i.e., (16.13) holds with $\mathbb{T} = \mathbb{B}$ and $G_{\mathbb{T}_x}(y,y;z) = G_{\mathbb{B}_y}(y,y;z)$. In particular, this entails that the spectral measure of $H_{\mathbb{B}}$ and δ_x can be obtained from $V(x)$ and the spectral measures of $H_{\mathbb{B}_y}$ and all neighbors y of x.
2. In the special case of a regular rooted tree $\mathbb{B}_0$, the relation (16.13) yields the recursion relation for the forward Green functions:

$$\Gamma(0;z) := G_{\mathbb{B}_0}(0,0;z) = \Big(V(0) - z - \sum_{y:d(0,y)=1} \Gamma(y;z)\Big)^{-1}. \tag{16.15}$$

As a special case of (16.14), the Green function $G_{\mathbb{B}_0}(0,x;z)$ factorizes into the following product:

$$G_{\mathbb{B}_0}(0,x;z) := \prod_{0 \preceq u \preceq x} \Gamma(u;z)\,. \tag{16.16}$$

16.3. Spectrum and DOS of the adjacency operator

The above observations allow us to determine the spectrum of the adjacency operator on the regular and regular rooted trees.

Corollary 16.2. *Let* $\mathbb{B}$ *stand for the regular tree with degree* $K+1$ *and let* $\mathbb{B}_0$ *be the regular rooted tree with* K *forward neighbors.*

1. *The spectral measure of the adjacency operator* $A_{\mathbb{B}_0}$ *on* $\ell^2(\mathbb{B}_0)$ *associated with the root is purely absolutely continuous and is given by the semicircle measure*

$$\boxed{\mu_{\delta_0}(dE) = \sqrt{(4K - E^2)_+}\,\frac{dE}{2\pi K}} \tag{16.17}$$

with $(\cdot)_+$ *denoting the positive part.*

2. *The spectral measure of the adjacency operator $A_{\mathbb{B}}$ on $\ell^2(\mathbb{B})$ associated with any vertex x is purely absolutely continuous and is given by the Kesten-McKay law*

$$\boxed{\mu_{\delta_x}(dE) = \sqrt{(4K-E^2)_+}\,\frac{(K+1)\,dE}{2\pi\left((K+1)^2-E^2\right)}}\,. \tag{16.18}$$

3. *Over both the homogeneous $(K+1)$-regular tree $\mathbb{B}$ and the corresponding rooted tree $\mathbb{B}_0$, the spectrum of the adjacency operator A is purely absolutely continuous and is given by*

$$\sigma(A) = \left[-2\sqrt{K}, 2\sqrt{K}\right]. \tag{16.19}$$

Proof. 1. By (16.15) the Green function $\Gamma(0;z) = G_{\mathbb{B}_0}(0,0;z)$ of the adjacency operator on the rooted tree is for any $z \in \mathbb{C}^+$ given by the unique value $\Gamma \in \mathbb{C}^+$ which satisfies the quadratic equation

$$K\Gamma^2 + z\Gamma + 1 = 0\,. \tag{16.20}$$

From its explicit solution for real energies

$$\sqrt{K}\,\Gamma(0;E+i0) = -\frac{E}{2\sqrt{K}} + \begin{cases} i\sqrt{1-\frac{E^2}{4K}} & \text{if } |E| \le 2\sqrt{K}\,, \\ \operatorname{sgn}(E)\sqrt{\frac{E^2}{4K}-1} & \text{if } |E| > 2\sqrt{K}\,, \end{cases}$$

one can directly determine the spectral measure

$$\mu_{\delta_0}(dE) = \operatorname{Im}\Gamma(0;E+i0)\,\frac{dE}{\pi}$$

which turns out to be absolutely continuous with a semicircle density, supported in $[-2\sqrt{K}, \sqrt{K}]$.

2. The Green function of the adjacency operator on $\mathbb{B}$ then derives from the above solution $\Gamma(0;z) = \Gamma$ using a special case of (16.13):

$$G_{\mathbb{B}}(0,0;z) = \left(-z-(K+1)\,\Gamma\right)^{-1}. \tag{16.21}$$

Taking imaginary parts yields the assertion about the spectral measure $\mu_{\delta_0}(dE) = \operatorname{Im} G_{\mathbb{B}}(0,0;E+i0)\,\frac{dE}{\pi}$. By the graph's homogeneity, the above extends to all other sites $x \in \mathbb{B}$.

3. In the case of the fully homogeneous tree the third assertion can be concluded from the fact that by symmetry (16.18) is valid for the spectral measures associated with the functions $\{\delta_x\}$ for all sites $x \in \mathbb{B}$, whose collection spans $\ell^2(\mathbb{B})$.

The last argument does not apply to a rooted tree which lacks homogeneity. Noting that the essential spectrum is not affected by a finite-rank perturbation, one may conclude that the only difference in the spectrum between $\mathbb{B}_0$ and $\mathbb{B}$ is limited to possibly a finite number of proper eigenvalues. The exclusion of that possibility is left as Exercise 16.4, where a path towards a proof is hinted at. □

Thanks to the factorization property (16.16) of the off-diagonal Green function, its decay is determined by the **Lyapunov exponent**

$$\mathcal{L}_0(z) := -\ln|\Gamma(0;z)|\,. \tag{16.22}$$

It dictates the decay of the Green function of the adjacency operator on the regular tree $\mathbb{B}$ as well as on the rooted regular tree $\mathbb{B}_0$:

$$\begin{aligned} |G_{\mathbb{B}}(x,y;z)| &= |G_{\mathbb{B}}(x,x,;z)|\, e^{-\mathcal{L}_0(z)\, d(x,y)}\,, \\ |G_{\mathbb{B}_0}(0,x;z)| &= |\Gamma(0;z)|\, e^{-\mathcal{L}_0(z)\, d(0,x)}\,. \end{aligned} \tag{16.23}$$

As can be read from the above explicit solution, the boundary value, $\mathcal{L}_0(E) := \lim_{\eta\downarrow 0}\mathcal{L}_0(E+i\eta)$, is a monotone decreasing function of $|E|$ with the property

$$\begin{aligned} \mathcal{L}_0(E) &= \sqrt{K} \quad \text{for } |E| \leq 2\sqrt{K}, \\ \mathcal{L}_0(E) &> K \quad \text{for } |E| > K+1. \end{aligned} \tag{16.24}$$

This entails that the Green function on the (rooted) regular tree is ℓ^1-summable if and only if $|E| > K+1$ and not for any energy outside the ℓ^2-spectrum of the adjacency operator—a fact which is of relevance for the onset of localization at weak disorder.

16.4. Decay of the Green function

Let us now focus the discussion on the random operator (16.1) on $\ell^2(\mathbb{B})$. Unless stated otherwise, we will assume here that $\lambda \in \mathbb{R}$ and:

> The random variables $(V(x))_{x\in\mathbb{B}}$ are iid with absolutely continuous single-site distribution with bounded density $\varrho \in L^\infty(\mathbb{R})$ for which
>
> i. the local Lipschitz condition (8.39) holds with $\tau = 1$ and
> ii. $\int |v|^\varsigma \varrho(v)dv < \infty$ for some $\varsigma \in (0,1)$.

The **Lyapunov exponent** for the random operator $H_\lambda(\omega)$ is defined in complete analogy to the unperturbed case for all $z \in \mathbb{C}^+$ in terms of the operator's Green function restricted to the rooted tree $\mathbb{B}_0$:

$$\boxed{\mathcal{L}_\lambda(z) := -\mathbb{E}\left[\ln|\Gamma_\lambda(0;z)|\right]}\,. \tag{16.25}$$

As we will see in Corollary 16.5 below, the Lyapunov exponent dictates the typical decay of the Green function on the regular tree graph $\mathbb{B}$. Among its other notable basic properties are the following:

1. As a function of $z \in \mathbb{C}^+$, the Lyapunov exponent is the negative real part of the Herglotz-Pick function $z \mapsto \mathbb{E}\left[\ln \Gamma_\lambda(0; z)\right]$. Its boundary value

$$\mathcal{L}_\lambda(E) := \lim_{\eta \downarrow 0} \mathcal{L}_\lambda(E + i\eta) \tag{16.26}$$

hence exists for Lebesgue-almost all $E \in \mathbb{R}$.

2. As a special case of (16.30) below, we have the bound $\mathcal{L}_\lambda(E) \geq \ln \sqrt{K}$.
3. In the special case of the Cauchy distribution, $\varrho(v) = \pi^{-1}(1 + v^2)^{-1}$, one may calculate the Lyapunov exponent explicitly,

$$\mathcal{L}_\lambda(E) = \mathcal{L}_0(E + i|\lambda|)\,, \tag{16.27}$$

using a contour integration argument; cf. Exercise 4.4.

Generally, it is an open problem whether $\mathcal{L}_\lambda(E)$ is continuous in the disorder parameter λ. Energy averages are, however, easily seen to be **continuous**.

Lemma 16.3. *For any bounded Borel set $I \subset \mathbb{R}$ the function $\mathbb{R} \ni \lambda \mapsto \int_I \mathcal{L}_\lambda(E)\, dE$ is continuous, and, in particular,*

$$\lim_{\lambda \to 0} \int_I \mathcal{L}_\lambda(E)\, dE \;=\; \int_I \mathcal{L}_0(E)\, dE\,. \tag{16.28}$$

Proof. Since the harmonic measure $\sigma_\lambda(I) := \int_I \mathcal{L}_\lambda(E)\, dE$ associated with $\mathcal{L}_\lambda(z) = \pi^{-1} \int \operatorname{Im}(E - z)^{-1} \sigma_\lambda(dE)$ is absolutely continuous, the asserted continuity follows from the vague continuity of σ_λ, which in turn follows from the (weak) resolvent convergence of the operators as $\lambda \to \lambda_0$. □

In order to describe the full phase diagram of the random operator $H_\lambda(\omega)$ more quantitative information on the decay rate is needed. This is contained in the **moment-generating function**:

$$\boxed{\varphi_\lambda(s; z) \;:=\; \lim_{d(0,x) \to \infty} \frac{\ln \mathbb{E}\left[|G_\lambda(0, x; z)|^s\right]}{d(0, x)}}\,. \tag{16.29}$$

The existence of the limit is guaranteed in the follow two cases:

Complex energies: $z \in \mathbb{C}^+$ and $s \in [-\varsigma, \infty)$.

Real energies: $z = E + i0$, where the boundary value of the Green function exists for all $x \in \mathbb{B}$ and almost all $E \in \mathbb{R}$. In this case, we restrict ourselves to $s \in [-\varsigma, 1)$.

Further properties are summarized in the following theorem.

Theorem 16.4. *For the random operator* (16.1) *with a random potential satisfying the above assumptions:*

1. *The function* $[-\varsigma,\infty) \ni s \mapsto \varphi_\lambda(s;\zeta)$ *is well-defined as the limit* (16.29) *for all* $z \in \mathbb{C}^+$ *and has the following properties:*
   (i) $\varphi_\lambda(s;z)$ *is convex and non-increasing in* $s \in [-\varsigma,\infty)$.
   (ii) *For* $s \in [0,2]$,

$$\boxed{-s\,\mathcal{L}_\lambda(z) \;\le\; \varphi_\lambda(s;z) \;\le\; -s\,\ln\sqrt{K}} \tag{16.30}$$

   and one has

$$\frac{\partial\varphi_\lambda}{\partial s}(0;z) = -\mathcal{L}_\lambda(z)\,. \tag{16.31}$$

2. *The function* $[-\varsigma,1) \ni s \mapsto \varphi_\lambda(s;E)$ *is well-defined as the limit* (16.29) *for* $z = E{+}i0$ *at Lebesgue-almost all* $E \in \mathbb{R}$. *It coincides with the limiting value*

$$\lim_{\eta\downarrow 0} \varphi_\lambda(s;E+i\eta) = \varphi_\lambda(s;E)\,, \tag{16.32}$$

   and $\varphi_\lambda(s;E)$ *shares properties* (i) *and* (ii) *above.*

3. *For any* $s \in [-\varsigma,\infty)$ *there are* $c_s(z), C_s(z) \in (0,\infty)$ *such that for all* $x,y \in \mathbb{B}$,

$$\boxed{c_s(z)^{-1}\, e^{\varphi_\lambda(s;z)\,d(x,y)} \;\le\; \mathbb{E}\left[|G_\lambda(x,y;z)|^s\right] \;\le\; C_s(z)\, e^{\varphi_\lambda(s;z)\,d(x,y)}}\,. \tag{16.33}$$

   Moreover:
   (i) *The constants* $c_s(z), C_s(z)$ *are bounded uniformly in* $z \in K + i(0,1]$ *and* $s \in S$ *for any compact* $K \subset \mathbb{R}$ *and* $S \subset [-\varsigma,1)$.
   (ii) *The same bounds also apply to the Green function* $\widehat{G}_\lambda(x_-,y_-;z)$ *of the restriction of the operator to the truncated tree* $\mathbb{B}_{x,y}$.

The proof of this theorem parallels the corresponding results of Theorem 12.8 in the one-dimensional situation. The only new feature on trees is the upper bound in (16.30). The proofs will be left as Exercise 16.5.

Let us note an immediate consequence concerning the typical decay of the Green function.

Corollary 16.5. *In case* $z \in \mathbb{C}^+$ *or* $z = E + i0$ *and almost all* $E \in \mathbb{R}$ *one has for any* $\varepsilon > 0$,

$$\boxed{\lim_{d(x,y)\to\infty} \mathbb{P}\left(|G_\lambda(x,y;z)| > e^{-(\mathcal{L}_\lambda(z)+\varepsilon)\,d(x,y)}\right) = 1}\,. \tag{16.34}$$

The same applies with $G_\lambda(x,y;z)$ *replaced by* $\widehat{G}_\lambda(x_-,y_-;z)$.

Proof. Chebyshev's inequality yields

$$\mathbb{P}\left(|G_\lambda(x,y;z)| \leq e^{-(\mathcal{L}_\lambda(z)+\varepsilon)\, d(x,y)}\right) \leq e^{-s\,(\mathcal{L}_\lambda(z)+\varepsilon)\, d(x,y)}\, \mathbb{E}\left[|G_\lambda(x,y;z)|^{-s}\right] . \tag{16.35}$$

The right side is bounded for any $s \in (0,\varsigma)$ using (16.33). Optimizing

$$\kappa(\varepsilon;z) := \sup_{s\in(0,\varsigma)} \left[(\mathcal{L}_\lambda(z)+\varepsilon)\, s - \varphi_\lambda(-s;z)\right] , \tag{16.36}$$

we hence conclude that the right side in (16.35) is bounded by $e^{-\kappa(\varepsilon;z)\, d(x,y)}$. Thanks to the convexity of the moment-generating function and the fact that $-\mathcal{L}_\lambda(z)$ is the derivative of $\varphi_\lambda(s;z)$ at $s = 0$, we have $\kappa(\varepsilon;z) > 0$ for any $\varepsilon > 0$. This concludes the proof of (16.34).

In the case of $\widehat{G}_\lambda(x_-,y_-;z)$ we proceed similiary. We start with the Chebyshev estimate and use the upper bound corresponding to (16.33). □

16.5. Resonant delocalization and localization

Let us now discuss how the general result concerning **resonant delocalization** applies in the tree setup. The application of Theorem 15.3 first requires us to derive an explicit expression for the tunneling amplitude on $\mathbb{B}$.

Lemma 16.6. *For any self-adjoint operator $H_{\mathbb{B}}$ on $\ell^2(\mathbb{B})$ the tunneling amplitude associated with $x,y \in \mathbb{B}$ and $z \in \mathbb{C}^+$ is given by*

$$\tau_{\mathbb{B}}(x,y;z) = G_{\mathbb{B}_{x,y}}(x_-,y_-;z) , \tag{16.37}$$

where x_- and y_- are the neighbors of x and y on the unique path connecting these vertices in $\mathbb{B}$.

Proof. We rewrite using the definition (15.7) of the tunneling amplitude and Cramér's rule:

$$\tau_{\mathbb{B}}(x,y;z) = \frac{G_{\mathbb{B}}(x,y;z)}{G_{\mathbb{B}}(x,x;z)G_{\mathbb{B}}(y,y;z) - G_{\mathbb{B}}(x,y;z)G_{\mathbb{B}}(y,x;z)} . \tag{16.38}$$

Since the left side does not depend on $\langle\delta_x, H\delta_x\rangle$ and $\langle\delta_y, H\delta_y\rangle$, we may take them to infinity. In this limit the ratio

$$G_{\mathbb{B}}(x,y;z)/[G_{\mathbb{B}}(x,x;z)\, G_{\mathbb{B}}(y,y;z)] \tag{16.39}$$

tends to the right side in (16.37) while the last factor in the denominator vanishes. □

Since the decay of the tunneling amplitude is, by implication, the same as that of the Green function, Corollary 16.5 ensures that the maximal function $t(x,y;E)$ as defined in (15.21) satisfies the following for all $d(x,y)$ large enough:

$$t(x,y;E) \geq e^{-L\,d(x,y)} \quad \text{with any } L > \mathcal{L}_\lambda(E). \tag{16.40}$$

Moreover, the function $T(x,y;E)$ defined in (15.22) is also bounded exponentially; i.e.,

$$T(x,y;E) \leq \mathbb{E}\left[\left|\widehat{G}_\lambda(x_-,y_-;E+i0)\right|^s\right] \leq C_s(E)\, e^{-s\,d(x,y)\,\ln\sqrt{K}}, \tag{16.41}$$

for any $s \in (0,1)$ due to the upper bounds in (16.33) and (16.30). In order to check the assumptions in Theorem 15.3 we note the following:

A1. The partial sums over the spheres,

$$\sum_{x:d(0,x)=R} t(x,y;E) \;\geq\; K^R e^{-LR}, \tag{16.42}$$

diverge in the limit $R \to \infty$ under the assumption $\mathcal{L}_\lambda(E) < L < \ln K$.

A2. The decay (15.26) is ensured by the bound (16.41):

$$\begin{aligned}\sum_{y:d(0,y)=R} T(x,y;E) &\leq C_s(E) \sum_{n=0}^{2R} \sum_{y:d(x,y)=n} e^{-sn\ln\sqrt{K}} \\ &\leq C_s(E) \sum_{j=0}^{R} K^{j(1-s)} \leq C_s(E) \frac{K^{1-s}}{K^{1-s}-1} K^{R(1-s)}.\end{aligned} \tag{16.43}$$

For the choice $s = L/\ln K < 1$, the right side is bounded for any $R \geq 0$ by the left side in (16.42) up to a multiplicative constant, which is independent of R.

We have thus established resonant delocalization under the condition (16.44).

Theorem 16.7. *Consider the random operator* (16.1) *with a random potential satisfying the above assumptions with a single-site density for which* $\inf_{|v|\leq S} \varrho(v) > 0$ *for all* $S > 0$. *If for a Borel set* $I \subset \mathbb{R}$ *one has for Lebesgue-almost all* $E \in I$

$$\boxed{\mathcal{L}_\lambda(E) < \ln K}, \tag{16.44}$$

then $\mu^{\mathrm{c}}_{\delta_x}(I;\omega) > 0$ *for almost every* ω *and all* $x \in \mathbb{B}$.

There exist three notable extensions of this theorem:

1. Theorem 16.7 remains valid in case the single-site distribution ϱ has a bounded support (under some technical condition on the single-site density).

2. One may further strengthen Theorem 16.7 by replacing (16.44) with

$$\boxed{\varphi_\lambda(1;E) \;>\; -\ln K}\,. \tag{16.45}$$

This is of interest since it complements the localization condition (16.48). Moreover, in the tree setup, the analysis which underlies the main result on resonant delocalization (Theorem 15.3) may be refined in order to establish the presence of ac spectrum; i.e., almost surely

$$\operatorname{Im} G_\lambda(x,x;E+i0) > 0\,, \tag{16.46}$$

at Lebesgue-almost all energies E at which (16.44) and, respectively, (16.45) hold.

The details can be found in [**21**].

The above results on delocalization provide a fitting supplement to the fractional moment criteria for **localization**. On the regular tree graphs this criterion may be characterized in terms of the left limit

$$\varphi_\lambda(1;E) := \lim_{s\uparrow 1} \varphi_\lambda(s;E) \tag{16.47}$$

which exists due to convexity.

Theorem 16.8. *For the random operator* (16.1) *with a random potential satisfying the above assumptions, if for $I \subset \mathbb{R}$ and Lebesgue-almost all $E \in I$*

$$\boxed{\varphi_\lambda(1;E) \;<\; -\ln K}\,, \tag{16.48}$$

then the continuous component of the operator's spectrum in I is empty; i.e., $\mu^{c}_{\delta_x}(I;\omega) = 0$ for all $x \in \mathbb{B}$ and almost all ω.

Proof. The Simon-Wolff criterion (Theorem 5.7) is applicable here. For almost every $E \in I$ there is $s \in (0,1)$ and $\eta_0 \in (0,1)$ such that for all $\eta \in (0,\eta_0)$, we have $\varphi_\lambda(s;E) := \sup_{\eta\in(0,\eta_0)} \varphi_\lambda(s;E+i\eta) < -\ln K$ and hence, using Fatou's lemma, Jensen's inequality, and (16.33),

$$\begin{aligned}
&\mathbb{E}\left[\left(\lim_{\eta\downarrow 0} \sum_{y\in\mathbb{B}} |G_\lambda(x,y;E+i\eta)|^2\right)^{s/2}\right] \\
&\quad\le \liminf_{\eta\downarrow 0} \sum_{y\in\mathbb{B}} \mathbb{E}\left[|G_\lambda(x,y;E+i\eta)|^s\right] \\
&\quad\le \liminf_{\eta\downarrow 0} C_s(E+i\eta) \sum_{y\in\mathbb{B}} e^{\varphi_\lambda(s;E+i\eta)\, d(x,y)} \\
&\quad\le C(s;E)\left(1 + (K+1)\sum_{n=0}^{\infty} K^n e^{\varphi_\lambda(s;E)\,(n+1)}\right).
\end{aligned} \tag{16.49}$$

Here we have abbreviated that $C(s;E) := \sup_{\eta\in(0,\eta_0)} C_s(E+\eta) < \infty$. The right side of (16.49) is finite. Hence the Green function at any $x \in \mathbb{B}$ is almost surely square summable. □

In case $\limsup_{s\uparrow 1} \sup_{E\in I} \varphi_\lambda(s;E) < -\ln K$ one can even conclude strong exponential dynamical localization in I.

In the regime of **weak disorder**, we already established a more explicit criterion for the absence of continuous spectrum. Part of Theorem 10.4 can be restated as follows:

Corollary 16.9. *Consider the random operator* (16.1) *with a random potential satisfying the above assumptions with a density which is* 1*-moment regular in the sense of Definition* 8.5. *Then for any* $|E| > K+1$ *there is some* $\widehat{\lambda}(E) > 0$ *such that the continuous component of the operator's spectrum in* I *is empty if for Lebesgue-almost all* $E \in I$

$$0 < \lambda < \widehat{\lambda}(E) \quad \textit{and} \quad |E| > K+1\,. \tag{16.50}$$

The proof is left as an exercise (Exercise 16.6). The absence of continuous spectrum can mean either the presence of pure-point spectrum or the absence of spectrum. Since the spectrum of the adjacency operator is characterized by $|E| \leq 2\sqrt{K}$, the above corollary will be void unless the support of the potential's distribution ϱ is unbounded (Figure 16.2). Examples of distributions to which the above corollary applies include the Gaussian and Cauchy cases.

Let us now summarize the findings by sketching the spectral phase diagram of the operator (16.1). Since the free group, which generates the regular tree graph $\mathbb{B}$, acts ergodically on the respective probability space, the results of Chapter 3 in particular imply that almost surely

$$\sigma\left(H_\lambda\right) \;=\; \left[-2\sqrt{K}, 2\sqrt{K}\right] + \lambda \operatorname{supp} \varrho\,. \tag{16.51}$$

The spectral characteristics were the subject of Theorem 16.8 and (16.45). These can summarized as follows:

$$\begin{aligned} &\textbf{Regime of continuous spectrum:} && \varphi_\lambda(1;E) \;>\; -\ln K,\\ &\textbf{Absence of continuous spectrum:} && \varphi_\lambda(1;E) \;<\; -\ln K. \end{aligned}$$

Again, the absence of continuous spectrum can mean either the presence of pure-point spectrum or the absence of spectrum. The condition $\varphi_\lambda(1;E) < -\ln K$ has to be intersected with the almost-sure spectrum (16.51).

It is too much to ask to be able to compute the moment-generating function explicitly. However, in order to determine qualitative information on the shape of the phase boundary

$$\varphi_\lambda(1;E) = \ln K\,, \tag{16.52}$$

in the energy vs disorder (E,λ)-diagram, it would be of value to be able to answer the following questions:

1. Does equation (16.52) describe a curve in the (E,λ)-diagram?
2. Is the boundary value of the free-energy function $\varphi_\lambda(1;E)$ continuous in λ?

Both are open problems.

In the regime of small disorder λ, the moment-generating function should ensemble the unperturbed one, for which $\varphi_0(s;z) = -s\mathcal{L}_0(z)$. In fact, in this regime not much should be lost by characterizing the regime of continuous spectrum through the sufficiency criterion of Theorem 16.7,

$$\mathcal{L}_\lambda(E) < \ln K \quad \Rightarrow \quad \varphi_\lambda(1;E) \ > \ -\ln K\,. \tag{16.53}$$

Working with this Lyapunov exponent criterion has the advantage that the weak continuity of Lemma 16.3 can be employed to obtain information on the shape of the mobility edge at weak disorder. This leads to the following result for unbounded random potentials.

Corollary 16.10. *For unbounded random potentials with* $\operatorname{supp}\varrho = \mathbb{R}$, *on every closed interval* $I \subset (-K-1\,,K+1)$ *the operator* (16.1) *almost surely has continuous spectrum in* I *at sufficiently low disorder.*

Proof. Since $\mathcal{L}_\lambda(E) \geq \ln\sqrt{K}$, Chebyshev inequality can be used to control the Lebesgue measure of that subset of I on which (16.44) is violated:

$$|\{E\in I\,|\,\mathcal{L}_\lambda(E)\geq \ln K\}| \ \leq \int_I \frac{\mathcal{L}_\lambda(E)-\ln\sqrt{K}}{\ln\sqrt{K}}\,dE = |I|\,\frac{M_\lambda(I)-\ln\sqrt{K}}{\ln\sqrt{K}}\,, \tag{16.54}$$

where we abbreviated the mean value of the Lyapunov exponent by

$$M_\lambda(I) \ := \ \frac{1}{|I|}\int_I \mathcal{L}_\lambda(E)\,dE\,. \tag{16.55}$$

The latter is continuous in $\lambda\in\mathbb{R}$ by Lemma 16.3. The proof is concluded by noting that $\ln\sqrt{K} \leq M_0(I) < \ln K$ for all closed intervals $I \subset (-K-1, K+1)$ by (16.24). □

The above result should be contrasted with Corollary 16.9 which identifies the points $|E| = K+1$ as the phase boundary at $\lambda = 0$ towards the localization regime. With some leap of faith, one might expect that the Lyapunov exponent is actually continuous, which would lead to the phase diagram depicted in Figure 16.2. In fact, for the Cauchy distribution this is evident from the explicit expression (16.27).

The proven emergence of extended states in the regime $2\sqrt{K} < |E| < K+1$ and small $|\lambda|$ is certainly surprising. After all, this region in the phase diagram belongs to the regime of Lifshitz tails in which the states only occur due to large deviations of the potential values.

A similar surprise is encountered at the band edges of the spectrum (16.51) in the case of random potentials with bounded support. Based on a more elaborate continuity analysis of the Lyapunov exponent, one may show that if $\operatorname{supp} \varrho = [-1, 1]$ (and some further technical condition on ϱ) in the regime

$$\lambda < [\sqrt{K} - 1]^2/2 , \tag{16.56}$$

the operator $H_{\mathbb{B}}(\omega)$ almost surely has purely absolutely continuous spectrum at the spectral edges, i.e., within a range of energies of the form

$$|E_\lambda| - \delta(\lambda) \leq |E| \leq |E_\lambda| \tag{16.57}$$

at some $\delta(\lambda) > 0$, with $E_\lambda = -2\sqrt{K} - \lambda$ the lower spectral edge. The situation is depicted in Figure 16.3; we refer to [**21**] for further details.

Notes

Random Schrödinger operators on trees have attracted attention since the very early works on the localization/delocalization transition. First were the works by R. Abou-Chacra, P. W. Anderson, and D. J. Thouless [**2, 3**], who recognized in (16.13) a self-consistency relation which is exact on trees but could also be of guidance for the general case.

A different point stressing the potential relevance of the study of tree graphs was made by B. Altshuler, Y. Gefen, A. Kamenev, and L. S. Levitov [**25**], who argued that loop effects may not play a significant role in the dynamics of multi-particle systems. Thus tree graphs of very high degree may be of relevance for a theory of localization/delocalization in the multi-particle systems. Further proposals for a theory of the latter were advances in [**25, 44**]. The topic continues to be the subject of current research and is one of the frontier areas in this field.

The Kesten-McKay law (16.18) first appeared in Kesten's work on random walks on groups [**222**]. Given the set of d-regular graphs on n vertices, the ensemble of random d-regular graphs corresponds to the equidistribution on that set. The average eigenvalue distribution of the adjacency operator on a typical d-regular graph was identified in the limit $n \to \infty$ to coincide with (16.18) by McKay [**294**]. An alternative derivation is the subject of Exercise 17.3.

Concerning the natural question of transport rates, it can be shown that for any initial state $\psi = f(H_{\mathbb{B}})\delta_0$, with $f \in L^2(\mathbb{R})$, which is energetically supported in the regime of ac states, $\{E \in \mathbb{R} \mid \operatorname{Im} \langle \delta_0, (H_{\mathbb{B}} - E - i0)^{-1}\delta_0\rangle > 0 \text{ a.s.}\}$, the transport is ballistic, in the sense that for all $b > 0$,

$$\mathbb{E}\left[\eta \int_0^\infty e^{-\eta t} \sum_{|x|<b/\eta} \left|\langle \delta_x\,, e^{-itH_{\mathbb{B}}}\,\psi\rangle\right|^2 dt\right] \;\leq\; C(f)\, b \;+\; o(\eta)\,,$$

with some $C(f) < \infty$ and $o(\eta)$ a quantity which vanishes for $\eta \to 0$. Further explanation and proof of this statement for iid random potentials whose single-site distribution has a sufficiently high moment can be found in [**20**]. In a perturbative regime, ballistic transport was first established by A. Klein [**245**].

These results do not contradict the generally expected picture that transport in random media is at most diffusive since on the Bethe lattice the diffusion is also ballistic; i.e.,

$$\int_0^\infty \langle \delta_x, e^{t\Delta}\delta_0\rangle\, dt = \langle \delta_x\,, (-\Delta)^{-1}\delta_0\rangle \;=\; \frac{C}{K^{|x|}}$$

where $(\Delta\psi)(x) := A_{\mathbb{B}}\psi(y) - (K+1)\psi(x)$ stands for the diffusion generator.

The moment-generating (aka the free-energy) function informs us of the rate of large-deviations in the decay rate of the Green function along different paths to infinity. More is said on this in [**21**], where large deviation considerations were used in narrowing the defining conditions for the mobility edge. Bounds on the moment-generating function which are particularly useful in the large connectivity limit, $K \to \infty$, are derived in [**45**].

Further interesting results on the rich spectral theory of random operators on a variety of tree graphs can be found in [**160, 162, 159, 183, 252, 218, 219, 220, 221, 345, 163**].

Exercises

16.1. Prove the Green function relations listed in Proposition 16.1.

16.2. To identify possible proper eigenfunctions of the adjacency operator $A_{\mathbb{B}_0}$ on the infinite rooted tree $\mathbb{B}_0$, let R be a graph isomorphism which leaves the root invariant and permutes in a cyclic order the K subtrees emerging from it.

a. Noting that the unitary map U_R induced by R on $\ell^2(\mathbb{B}_0)$ commutes with $A_{\mathbb{B}_0}$, show how it may be deduced that if $A_{\mathbb{B}_0}$ has proper eigenfunctions, a complete orthonormal set of such can be found with the following property:

$$\text{either } \psi(0) = 0 \quad \text{or} \quad \psi(Rx) = \psi(x) \text{ for all } x \in \mathbb{B}_0.$$

b. Prove that there are no proper eigenfunction of $A_{\mathbb{B}_0}$ in $\ell^2(\mathbb{B}_0)$ satisfying the above conditions.

16.3. Prove that the Green function at any $x \in \mathbb{B}_0$ of $H_{\mathbb{B}_0} = -A_{\mathbb{B}_0} + V$ is related to that of $H_{\mathbb{B}} = -A_{\mathbb{B}} + V$ through

$$G_{\mathbb{B}_0}(x, x; z) = G_{\mathbb{B}}(x, x; z) + G_{\mathbb{B}_0}(x, 0; z)\, G_{\mathbb{B}}(0_-, x; z)$$

where $0_- \in \mathbb{B}$ is the unique backward neighbor of the root in $\mathbb{B}_0 \subset \mathbb{B}$. Show how this relation can be used to conclude that in case $V \equiv 0$ the limit $G_{\mathbb{B}_0}(x, x; E) = \lim_{\eta \downarrow 0} G_{\mathbb{B}_0}(x, x; E + i\eta)$ exists for all $E \in \mathbb{R}$ at any $x \in \mathbb{B}_0$.

16.4. Show that the spectral measure of $A_{\mathbb{B}_0}$ and $x \in \mathbb{B}_0$ is purely absolutely continuous and its support is independent of x.
[Hint: The previous exercise could be of use here.]

16.5. Provide a proof of Theorem 16.4, e.g., proceeding along the lines of reasoning in the proof of Theorem 12.8.

Hint: For a proof of the upper bound (16.30) one may use the bound for $z \in \mathbb{C}^+$:

$$\sum_{x:\, \mathrm{dist}(x,0)=n} |G_\lambda(0, x; z)|^2 \leq \sum_{x \in \mathbb{B}} |G_\lambda(0, x; z)|^2 = \frac{\operatorname{Im} G_\lambda(0, 0; z)}{\operatorname{Im} z} \leq \frac{1}{(\operatorname{Im} z)^2}\,.$$

16.6. Give the proof of Corollary 16.9.

Chapter 17

The Eigenvalue Point Process and a Conjectured Dichotomy

We now turn to the relation of the spectral properties of the infinite-volume random operator and the local statistics of the eigenvalues of its finite-volume counterparts. As the main result we present a proof of the Poisson statistics throughout the localization regime for the basic Anderson-type model $H(\omega) = -\Delta + V(\omega)$ in $\ell^2(\mathbb{Z}^d)$. We also comment on the corresponding question and available results in the setting of tree graphs. The latter is of interest since (i) some of the arguments which are natural for finite dimensions do not work on graphs of exponential growth such as trees and (ii) the tree example allows us to highlight an important point concerning how the general conjecture needs to be interpreted.

17.1. Poisson statistics versus level repulsion

Let $H(\omega)$ be an Anderson-type operator in $\ell^2(\mathbb{Z}^d)$ with a homogeneous random potential and $(E_{n,L}(\omega))$ the eigenvalues of the related finite-volume operator

$$H_L(\omega) := 1_{\Lambda_L} H(\omega) 1_{\Lambda_L} \quad \text{in } \ell^2(\Lambda_L) \tag{17.1}$$

with $\Lambda_L := [-L, L]^d \cap \mathbb{Z}^d$. By the Wegner estimate (Theorem 4.1) these eigenvalues accumulate at a density proportional to the number of sites $|\Lambda_L|$.

At a finer level, one may enquire about the joint distribution of the eigenvalues at it appears on the scale of the mean eigenvalue spacing $(\nu(E)|\Lambda_L|)^{-1}$ with $\nu(E)$ the density of states of Theorem 4.1 (which due to the stochastic homogeneity does not depend on $x \in \mathbb{Z}^d$).

For simplicity of notation, we will scale by volume, omitting the bounded factor $\nu(E)$. However, we restrict our attention to energies at which the integrated density of states is differentiable or, equivalently, the set of Lebesgue points of $\nu(E)$, which is the set of energies

$$\mathcal{L}_{\text{DOS}} := \left\{ E \in \mathbb{R} \,\middle|\, \nu(E) = \lim_{\varepsilon \to 0} \frac{1}{2\varepsilon} \int_{-\varepsilon}^{\varepsilon} \nu(E+a)\, da \right\}. \tag{17.2}$$

By the Lebesgue differentiation theorem this set is of full measure (i.e., its complement is of zero Lebesgue measure).

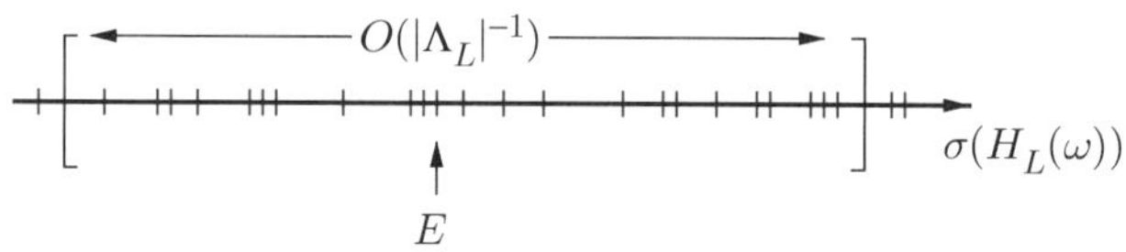

Figure 17.1. Sketch of the random process of eigenvalues of the finite-volume operator $H_L(\omega)$ in a window about an energy E on the scale of the mean eigenvalue spacing.

Thus, we shall discuss here the point process $\{|\Lambda_L|\,(E_{n,L}(\omega) - E)\} \subset \mathbb{R}$ of rescaled eigenvalues of the finite-volume operators, where E is a preselected energy in $\mathcal{L}_{\text{DOS}}$ about which we magnify. The point process can also be described by the corresponding **random counting measure**:

$$\boxed{\mu_L^E(\cdot\,;\omega) := \sum_n \delta_{|\Lambda_L|(E_{n,L}(\omega)-E)}(\cdot)}\,. \tag{17.3}$$

The sum is over the eigenvalues of (17.1), which in principle are to be counted with their multiplicity, though in our setup they are known to be almost surely simple (Theorem 5.8). The place holder $(\cdot)$ refers to the argument of the measure function, which ranges over Borel subsets of $\mathbb{R}$. Hence, $\mu_L^E(I;\omega)$ is just the number of eigenvalues of $H_L(\omega)$ in the interval $E + |\Lambda_L|^{-1} I$.

Intuitive arguments based on perturbation theory carried to the second order, as in Dyson's derivation of the eigenvalue dynamics [**129**], suggest that for random operators the eigenvalue process will or will not exhibit level repulsion, depending on the typical order of magnitude of the overlap sums which form the numerator of the relevant ratio:

$$R_{j,k} = \frac{\sum_{x\in\Lambda_L} |\psi_{j,L}(x)|^2 |\psi_{k,L}(x)|^2}{E_{j,L} - E_{k,L}}\,. \tag{17.4}$$

For energies in the localization regime, the overlap of eigenstates which are localized in different regions would typically be much smaller than the energy differences. Consequently, as the disorder is refreshed, levels may move independently, except for brief instances where the energies are about to cross, at which point the small denominator in (17.4) results in the phenomenon of *avoided crossings*. Under this scenario on the scale of the typical level spacing the energies may be expected to be distributed fairly independently of each other. On the other hand, in the regime of extended states, nearby levels may repel under the potential fluctuations in a manner similar to the picture familiar from the **random matrix theory**. This has motivated the following **spectral statistics conjectures**.

Conjecture 17.1. *The spectral characteristics of the infinite-volume operator $H(\omega)$ are reflected in the following behavior of the eigenvalue process of its finite-volume versions $H_L(\omega)$ in the limit $L \to \infty$, at energies in the vicinity of $E \in \mathcal{L}_{\mathrm{DOS}}$:*

1. *For energies within the regime of exponential localization (and thus pure-point spectrum) of $H(\omega)$ the rescaled eigenvalue process converges to a Poisson process of density $\nu(E)$.*
2. *For energies within the regime of absolutely continuous spectrum the rescaled eigenvalue process converges to the GOE eigenvalue process, and this changes to the GUE eigenvalue process under the addition in H of an external magnetic field.*

GOE and GUE refer here to the Gaussian random matrix ensembles with the orthogonal or, correspondingly, unitary symmetry of the measure. Their rescaled eigenvalues are known to converge to determinantal point processes which are uniquely characterized by their two-point functions (which in the case of GUE is the famous **sine kernel** $K(x,y) = \frac{\sin \pi(x-y)}{\pi(x-y)}$ [**28**]).

The above-stated conjecture resonates with its counterparts in quantum chaos theory: O. Bohigas, M.-J. Giannoni, and C. Schmit [**62**] assert and numerically support the conjecture that random matrix behavior (GOE or GUE depending on the presence of time reversal symmetry) is universally found in the level statistics of chaotic billiards. Conversely, for billiards which in the semiclassical limit $E/\hbar \to \infty$ yield integrable systems (in the sense of classical Hamiltonian flows), M. V. Berry and M. Tabor numerically found asymptotic absence of level repulsion, corresponding to a Poisson process [**55**].

The second part of Conjecture 17.1 is not yet proven and is also not beyond debate in terms of some fine-point adjustments (see the Notes for further information and recent progress). Its first part is on more solid ground and was established for energies at which somewhat more explicit localization conditions are satisfied, in particular for energies in the set

$$\mathcal{L}_{\mathrm{FM}} := \Big\{ E \in \mathbb{R} \,\Big|\, \text{at some } s \in (0,1) \text{ and } \mu_s, C_s \in (0,\infty), \text{ for all } x, y \in \mathbb{Z}^d: \\ \sup_{\Lambda \subseteq \mathbb{Z}^d} \sup_{\eta > 0} \mathbb{E}\left[|G_\Lambda(x,y; E + i\eta)|^s \right] \leq C_s \, e^{-\mu_s |x-y|} \Big\}. \tag{17.5}$$

By the finite-volume analysis (Theorem 11.3), this set coincides with the regime of exponential localization for the infinite-volume operator. Before turning to a precise formulation of the result, Theorem 17.6 below, let us provide a description of the limiting process in this case and a clarification of the notion of the convergence in the above conjecture.

17.2. Essential characteristics of the Poisson point processes

A Poisson process on $\mathbb{R}$, whose configurations are described by a random counting measure on the Borel sets over $\mathbb{R}$ which we denote by μ, is uniquely parametrized by its intensity. For any point process the latter is defined as the measure

$$\bar{\mu}(A) := \mathbb{E}\left(\mu(A)\right)$$

on Borel sets $A \subset \mathbb{R}$.

Definition 17.2. A Poisson process on $\mathbb{R}$ is a point process whose random counting measure μ satisfies the following two conditions:

1. For disjoint Borel sets $A_1, \ldots, A_n \subset \mathbb{R}$, the random variables $\mu(A_1), \ldots, \mu(A_n)$ are independent.
2. For each bounded Borel set $A \subset \mathbb{R}$ the random variable $\mu(A)$ has a Poisson distribution with mean $\bar{\mu}(A)$; i.e., for all $k \in \mathbb{N}_0$

$$\mathbb{P}\left(\mu(A) = k\right) = \frac{\bar{\mu}(A)^k}{k!} \, e^{-\bar{\mu}(A)}. \tag{17.6}$$

In case the intensity measure $\bar{\mu}$ has no atoms, i.e., $\bar{\mu}(\{a\}) = 0$ for all $a \in \mathbb{R}$, the Poisson process is simple in the sense that it has almost surely no double points; i.e., $\mu(\{a\}) \leq 1$ for all $a \in \mathbb{R}$ almost surely. Remarkably, for any simple point processes μ on $\mathbb{R}$, each of the above two properties already implies the other [**227, 106**]. The sufficiency of the second condition is due to A. Rényi [**333**] who also showed that (17.7) alone implies the rest of (17.6); see also [**227**]. Some of the equivalent characterizations of the Poisson process are summarized in the following proposition.

Proposition 17.3 ([**333, 227, 106**]). *For a simple point process μ on $\mathbb{R}$ whose intensity $\bar{\mu} = \mathbb{E}[\mu]$ is a Borel measure without atoms, the following conditions are all individually equivalent to μ being a Poisson process.*

1. *For any bounded Borel set $A \in \mathbb{R}$, the probability that A is void is*

$$\mathbb{P}(\mu(A) = 0) \;=\; e^{-\bar{\mu}(A)}\,. \tag{17.7}$$

2. *When the points of μ are labeled as $(u_n)_{n\in\mathbb{Z}\setminus\{0\}}$ according to their rank relative to an a priori chosen $u_0 \in \mathbb{R}$, the lengths $\bar{\mu}([u_{n+1} - u_n])$ of successive gaps as measured by $\bar{\mu}$ form independent random variables with the exponential distribution:*

$$\mathbb{P}(\bar{\mu}(\,[u_{n+1} - u_n]\,) \geq t) \;=\; e^{-t}\,. \tag{17.8}$$

3. *The process is infinitely divisible in the sense that for any $n \in \mathbb{N}$, μ can be presented as a sum $\mu = \sum_{j=1}^{n} \mu_j$ of n independent and identically distributed point processes μ_j.*

Any point process μ on $\mathbb{R}$ can also be uniquely characterized [**106, 216**] through the joint distribution of random variables of the form

$$\mu(g) := \int_{\mathbb{R}} g(u)\,\mu(du) \tag{17.9}$$

with non-negative, continuous functions of compact support, $g \in C_c^+(\mathbb{R})$. In fact, since $C_c^+(\mathbb{R})$ is closed under positive linear combinations, it suffices to know the distribution of $\mu(g)$ for a single, but arbitrary, $g \in C_c^+(\mathbb{R})$. This distribution is uniquely characterized by its **moment-generating function** which in the case of a Poisson process is given by

$$\mathbb{E}\left[e^{-\mu(g)}\right] \;=\; \exp\left(-\int_{\mathbb{R}} \left(1 - e^{-g(u)}\right) \bar{\mu}(du)\right). \tag{17.10}$$

The moment-generating function may also be used to define the notion of distributional convergence of processes.

Definition 17.4. A sequence of point processes (μ_n) on $\mathbb{R}$ converges as $n \to \infty$ in distribution to μ if $\lim_{n\to\infty} \mathbb{E}\left[e^{-\mu_n(g)}\right] = \mathbb{E}\left[e^{-\mu(g)}\right]$ for all $g \in C_c^+(\mathbb{R})$.

For further discussion of the implications of this convergence, we refer the reader to texts on random point processes, e.g., [**106, 216**].

In the proof of the main result of this chapter (Theorem 17.6) we shall apply the following known sufficiency criterion for the convergence to a Poisson process. For completeness we include its derivation here. The proof, which can be found in [**216**], demonstrates how one can work with the moment-generating function.

Proposition 17.5. *Let $\left\{\mu_{n,j} \,|\, j = 1, \ldots, N_n,\ n \in \mathbb{N}\right\}$ be a triangular array of random processes, which for each $n \in \mathbb{N}$ are independent among different*

$j \in \{1, \dots, N_n\}$. Assume the following:

1. *The array is null in the sense that for any bounded Borel set $I \subset \mathbb{R}$,*

$$\lim_{n\to\infty} \sup_{j\in\{1,\dots,N_n\}} \mathbb{P}\left(\mu_{n,j}(I) \geq 1\right) = 0\,. \tag{17.11}$$

2. *For some positive Borel measure $\overline{\mu}$ on $\mathbb{R}$ and all bounded Borel sets $I \subset \mathbb{R}$ with $\overline{\mu}(\partial I) = 0$,*

$$\lim_{n\to\infty} \sum_{j=1}^{N_n} \mathbb{P}\left(\mu_{n,j}(I) \geq 1\right) \;= \overline{\mu}(I)\,, \tag{17.12}$$

$$\lim_{n\to\infty} \sum_{j=1}^{N_n} \mathbb{P}\left(\mu_{n,j}(I) \geq 2\right) \;=\; 0\,. \tag{17.13}$$

Then $\sum_{j=1}^{N_n} \mu_{n,j}$ converges in distribution as $n \to \infty$ to a Poisson process μ with intensity measure $\overline{\mu}$.

Proof. It is sufficient to show that for any bounded Borel set $I \subset \mathbb{R}$ with $\overline{\mu}(\partial I) = 0$, the characteristic function converges:

$$\lim_{n\to\infty} \mathbb{E}\left[\exp\left(-t\sum_{j=1}^{N_n} \mu_{n,j}(I)\right)\right] \;=\; \exp\left(\left(e^{-t} - 1\right)\,\overline{\mu}(I)\right)\,. \tag{17.14}$$

Thanks to independence, the prelimit factorizes:

$$\mathbb{E}\left[\exp\left(-t\sum_{j=1}^{N_n} \mu_{n,j}(I)\right)\right] \;=\; \prod_{j=1}^{N_n} \mathbb{E}\left[e^{-t\mu_{n,j}(I)}\right]\,. \tag{17.15}$$

Since $\mu_{n,j}(I)$ takes only integer values, each term can be expressed in terms of a series:

$$\begin{aligned} X_{n,j} \;:=\; \mathbb{E}\left[e^{-t\mu_{n,j}(I)}\right] \;&=\; \sum_{m=0}^{\infty} e^{-tm}\,\mathbb{P}\left(\mu_{n,j}(I) = m\right) \\ &= 1 + \left(e^{-t} - 1\right)\mathbb{P}\left(\mu_{n,j}(I) \geq 1\right) - R_{n,j}(t)\,, \end{aligned} \tag{17.16}$$

with a remainder term:

$$R_{n,j} := \sum_{m=2}^{\infty} \left(e^{-t} - e^{-tm}\right)\mathbb{P}\left(\mu_{n,j}(I) = m\right) \;\leq\; \mathbb{P}\left(\mu_{n,j}(I) \geq 2\right)\,. \tag{17.17}$$

Since $R_{n,j} \geq 0$, we may bound the logarithm of each term in (17.15):

$$\ln X_{n,j} \leq X_{n,j} - 1 \leq \left(e^{-t} - 1\right) \mathbb{P}\left(\mu_{n,j}(I) \geq 1\right) . \tag{17.18}$$

Summing over j, the right side converges to $\left(e^{-t} - 1\right) \overline{\mu}(I)$ as $n \to \infty$ by assumption. The upper bound in (17.17) is used in the second step of the following estimate:

$$\ln X_{n,j} \geq \frac{X_{n,j} - 1}{X_{n,j}} \geq \frac{\left(e^{-t} - 1\right) \mathbb{P}\left(\mu_{n,j}(I) \geq 1\right) - \mathbb{P}\left(\mu_{n,j}(I) \geq 2\right)}{1 - \sup_j \mathbb{P}\left(\mu_{n,j}(I) \geq 1\right)} . \tag{17.19}$$

Applying the assumption to the various terms, the sum over j is guaranteed to converge to $\left(e^{-t} - 1\right) \overline{\mu}(I)$ as $n \to \infty$. □

17.3. Poisson statistics in finite dimensions in the localization regime

The localization part of the spectral statistics conjecture, i.e., the first part of Conjecture 17.1, has been established. The initial result, which pointed out the phenomenon, was presented by S. Molchanov in a one-dimensional setting [**302**]. The more general statement, which addressed the multi-dimensional lattice case, is due to N. Minami [**301**]. In presenting it we refer to the sets of energies defined in (17.2) and (17.5).

Theorem 17.6. *Let $H(\omega) = -\Delta + V(\omega)$ in $\ell^2(\mathbb{Z}^d)$ with the potential given by iid random variables whose single-site distribution $\mathbb{P}\left(V(x;\cdot) \in dv\right) = \varrho(v)\, dv$ is of a bounded density $\varrho \in L^\infty(\mathbb{R})$. Then for any $E \in \mathcal{L}_{\mathrm{FM}} \cap \mathcal{L}_{\mathrm{DOS}}$ the random process of rescaled eigenvalues μ_L^E converges in distribution, as $L \to \infty$, to a Poisson process with density $\nu(E)$.*

As is apparent from the proof, which is presented in the following subsections, the assumptions of the theorem can be relaxed. In particular, the fractional moment's exponential decay could also be replaced by a suitably fast power-law decay.

17.3.1. Construction of a null array.

It is natural to expect that for energies in the localization regime most of the eigenvalues of $H_L(\omega)$ in the localization regime ($\mathcal{L}_{\mathrm{FM}}$) would not be much affected by the nullification of the operator terms along walls which split Λ_L into M_L boxes $\Lambda_{L,j}$ of equal volume (with $j = 1, \ldots, M_L$), as depicted in Figure 17.2, provided their linear size is much larger than the localization length.

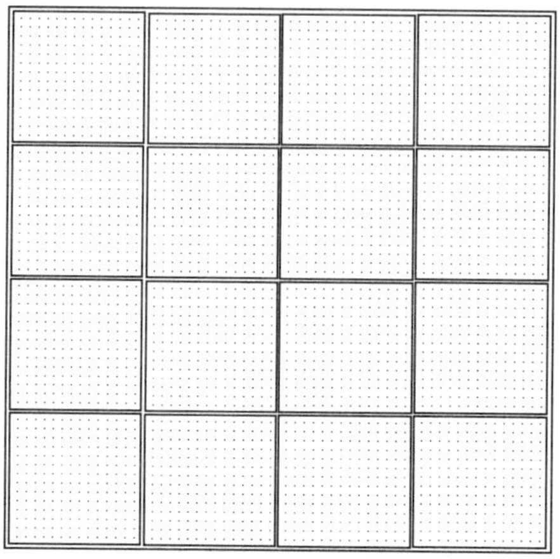

Figure 17.2. A box decomposition like the one used in the proof that at energies within the localization regime the limiting eigenvalue process is divisible.

If so, then the point measure μ_L^E should be well approximated by the sum $\sum_{j=1}^{M_L} \mu_{L,j}^E$ of random point measures

$$\mu_{L,j}^E(\omega) := \sum_n \delta_{|\Lambda_L|\left(E_{n,L,j}(\omega)-E\right)} \tag{17.20}$$

which represent the eigenvalue process of the operator's restriction to the corresponding boxes $H_{L,j}(\omega) := 1_{\Lambda_{L,j}}\, H(\omega)\, 1_{\Lambda_{L,j}}$ in $\ell^2(\Lambda_{L,j})$ (counted with their multiplicity). The local agreement of this pair of processes is established in the following lemma.

Lemma 17.7. *For the above construction, with the linear dimension of the boxes $\Lambda_{L,j}$ growing as L^α, at some $\alpha \in (0,1]$, for any $E \in \mathcal{L}_{\mathrm{FM}}$ and any $g \in L^1(\mathbb{R})$,*

$$\lim_{L\to\infty} \mathbb{E}\left[\left|\mu_L^E(g) - \sum_{j=1}^{M_L} \mu_{L,j}^E(g)\right|\right] = 0\,. \tag{17.21}$$

Proof. The convergence will be proven first for the special collection of functions of the form

$$\varphi_z(u) = \pi^{-1}\,\mathrm{Im}(u-z)^{-1} \;=\; \frac{\pi^{-1}\,\mathrm{Im}\, z}{(u-\mathrm{Re}\, z)^2 + \mathrm{Im}\, z^2} \tag{17.22}$$

with arbitrary $z \in \mathbb{C}^+$. Then an approximation argument will be applied to extend the statement to all $g \in L^1(\mathbb{R})$.

For functions in the above class, writing $z_L^E := E + |\Lambda_L|^{-1} z$, we have

(17.23)

$$\begin{aligned} &\mu_L^E(\varphi_z) - \sum_{j=1}^{M_L} \mu_{L,j}^E(\varphi_z) \\ &= \frac{1}{\pi |\Lambda_L|}\Big(\operatorname{tr} \operatorname{Im}(H_L - E - |\Lambda_L|^{-1}z)^{-1} - \sum_{j=1}^{M_L} \operatorname{tr} \operatorname{Im}(H_{L,j} - E - |\Lambda_L|^{-1}z)^{-1}\Big) \\ &= \frac{1}{\pi |\Lambda_L|} \sum_{j=1}^{M_L} \sum_{x \in \Lambda_{L,j}} \left(\operatorname{Im} G_{\Lambda_L}(x,x;z_L^E) - \operatorname{Im} G_{\Lambda_{L,j}}(x,x;z_L^E)\right) . \end{aligned}$$

We shall next see that under the assumed localization condition the above summand is very small except for x which lie within the distance of localization length to the walls of partition, i.e., the boundary of the set $\Lambda_{L,j}$ to which it belongs. Let us split each box into an inner part and a boundary layer, denoted $\Lambda_{L,j}^i$ and $\Lambda_{L,j}^b$, respectively.

The contribution to (17.23) from the boundary layers can be bounded using the Wegner estimate:

$$(17.24) \quad \begin{aligned} &\frac{1}{|\Lambda_L|} \sum_{j=1}^{M_L} \sum_{x \in \Lambda_{L,j}^b} \left(\mathbb{E}\left[\operatorname{Im} G_{\Lambda_L}(x,x;z_L^E)\right] + \mathbb{E}\left[\operatorname{Im} G_{\Lambda_{L,j}}(x,x;z_L^E)\right]\right) \\ &\qquad \le 2\|\varrho\|_\infty \frac{M_L\,|\Lambda_{L,j}^b|}{|\Lambda_L|} = 2\|\varrho\|_\infty \frac{|\Lambda_{L,j}^b|}{|\Lambda_{L,j}|} . \end{aligned}$$

It tends to zero in the limit $L \to \infty$ provided the boundary layer is defined so that its volume fraction vanishes in the limit.

The inner sum is estimated with the help of the localization estimate in Lemma 17.9 below:

$$(17.25) \quad \begin{aligned} &\frac{1}{|\Lambda_L|} \sum_{j=1}^{M_L} \sum_{x \in \Lambda_{L,j}^i} \mathbb{E}\left[\left|\operatorname{Im} G_{\Lambda_L}(x,x;z_L^E) - \operatorname{Im} G_{\Lambda_{L,j}}(x,x;z_L^E)\right|\right] \\ &\qquad \le C_s\, M_L \frac{|\Lambda_{L,j}^i|\,|\Lambda_L|^{1-2s}}{(\operatorname{Im} z)^{2(1-s)}} \exp\left(-\mu_s \operatorname{dist}\left(\Lambda_{L,j}^i, \partial\Lambda_{L,j}\right)\right) . \end{aligned}$$

Choosing the inner boxes $\Lambda_{L,j}^i$ such that its distance to $\partial\Lambda_{L,j}$ is proportional to $L^{a\alpha}$ with some $a \in (0,1)$, the right side converges to zero as $L \to \infty$. This concludes the proof of (17.21) in case $g = \varphi_z$.

The above can be extended to other functions using the standard approximation result which is included as Proposition 17.8 below, by which every $g \in L^1(\mathbb{R})$ is well approximated by its convolutions $g_\varepsilon = \varphi_{i\varepsilon} * g$, in the sense that $\lim_{\varepsilon\downarrow 0}\|g-g_\varepsilon\|_1 = 0$. Using such approximations, the claim follows from two observations concerning the difference measures $\nu_L := \mu_L^E - \sum_{j=1}^{M_L}\mu_{L,j}^E$:

i. For each $\varepsilon > 0$, by the above argument combined with the dominated convergence theorem,

$$\mathbb{E}\Big[\big|\nu_L(g_\varepsilon)\big|\Big] \le \int |g(x)|\,\mathbb{E}\Big[\big|\nu_L(\varphi_{x+i\varepsilon})\big|\Big]\,dx \underset{L\to\infty}{\longrightarrow} 0\,. \tag{17.26}$$

ii. For the difference $h = (g - g_\varepsilon) \in L^1(\mathbb{R})$ using the Wegner bound one has

$$\mathbb{E}\big[|\nu_L(h)|\big] \le \mathbb{E}\big[\mu_L^E(|h|)\big] + \sum_{j=1}^{M_L}\mathbb{E}\big[\mu_{L,j}^E(|h|)\big] \le 2\|\rho\|_\infty\|h\|_1\,. \tag{17.27}$$ □

In the above proof use was made of the following standard approximation argument [**87**, Thm. 5.1] for L^p-functions and their convolution $(\varphi_{i\eta} * g)(x) := \int \varphi_{i\eta}(x-y)g(y)dy$ with the Poisson kernel φ_z of (17.22). (It serves here as a **mollifier**, though not being of compact support φ_z does not meet all the requirements of the standard definition of mollifier.)

Proposition 17.8. *For any $p \in [1,\infty)$ and for any $g \in L^p(\mathbb{R})$ the convolution $\varphi_{i\eta} * g$ is in $L^p(\mathbb{R})$, and*

$$\lim_{\eta\downarrow 0}\|g - \varphi_{i\eta} * g\|_p = 0\,. \tag{17.28}$$

Lemma 17.7 allows us to reduce the proof of Theorem 17.6 to the claim that $\sum_{j=1}^{M_L}\mu_{L,j}^E$ converges to a Poisson process. This will be done by checking the different assumptions in Proposition 17.5.

17.3.2. Convergence of the density. Next it will be shown that the mean density of the eigenvalue counting measure μ_L^E converges vaguely to Lebesgue measure scaled by the density of states at E. Let us note that a scenario which would be in conflict with this statement is if the density of states were to show oscillations on some mesoscopic scale, i.e., one which is much larger that $1/|\Lambda_L|$ yet smaller than the scale $O(1)$ on which the integrated density of states $n(E)$ and its derivative $\nu(E)$ were determined. To rule this possibility out, and thus establish (17.12), we shall make use of the following estimate.

Lemma 17.9. *For any $\operatorname{Re} z \in \mathcal{L}_{\mathrm{FM}}$ there exist some $s \in (0,1/2)$ and $\mu_s, C_s \in (0,\infty)$ such that for all $\Lambda \subseteq \Lambda' \subseteq \mathbb{Z}^d$ and $x \in \Lambda$ and $\operatorname{Im} z > 0$,*

$$\mathbb{E}\left[|G_\Lambda(x,x;z) - G_{\Lambda'}(x,x;z)|\right] \le \frac{C_s}{(\operatorname{Im} z)^{2(1-s)}}\,e^{-\mu_s \operatorname{dist}(x,\partial\Lambda)}\,, \tag{17.29}$$

with the boundary set $\partial\Lambda := \big\{x \in \mathbb{Z}^d \,|\, \text{either } \operatorname{dist}(x,\Lambda) = 1 \text{ or } \operatorname{dist}(x,\Lambda^c) = 1\big\}$.

Proof. The resolvent equation yields

$$(17.30)\quad \begin{aligned} |G_\Lambda(x,x;z) - G_{\Lambda'}(x,x;z)| &\le \sum_{\substack{\mathrm{dist}(u,v)=1\\ u\in\Lambda, v\notin\Lambda}} |G_\Lambda(x,u;z)|\,|G_{\Lambda'}(v,x;z)| \\ &\le (\operatorname{Im} z)^{2(s-1)} \sum_{\substack{\mathrm{dist}(u,v)=1\\ u\in\Lambda, v\in\Lambda^c}} |G_\Lambda(x,u;z)|^s\,|G_{\Lambda'}(v,x;z)|^s, \end{aligned}$$

where the last line relied on the estimate $|G_\Lambda(x,u;z)| \le (\operatorname{Im} z)^{-1}$. Taking the expectation value and using the Cauchy-Schwarz inequality, the left side of (17.29) is seen to be bounded by

$$(17.31)\quad (\operatorname{Im} z)^{2(s-1)} \sum_{\substack{(u,v)\in\partial\Lambda\\ u\in\Lambda, v\in\Lambda^c}} \left(\mathbb{E}\left[|G_\Lambda(x,u;z)|^{2s}\right]\,\mathbb{E}\left[|G_{\Lambda'}(v,x;z)|^{2s}\right]\right)^{1/2}.$$

Each term in the above product can be estimated with the help of the localization assumption $\operatorname{Re} z \in \mathcal{L}_{\mathrm{FM}}$. Since one may always lower the fractional power with the help of Jensen's inequality, this guarantees the existence of $s \in (0,1/2)$ and $\mu_s, C_s \in (0,\infty)$ such that the sum over pairs (u,v) on the boundary $\partial\Lambda$ is estimated using

$$(17.32)\quad \sum_{u\in\Lambda\cap\partial\Lambda} e^{-\mu_s|u-x|} \le C_s \exp\left(-\frac{\mu_s}{2}\,\mathrm{dist}(x,\partial\Lambda)\right)$$

and similarly for the sum over $v \in \Lambda^c \cap \partial\Lambda$. □

The above estimate will now be used in the proof that the mean density

$$\overline{\mu}_L^E := \mathbb{E}\left[\mu_L^E\right]$$

converges vaguely to Lebesgue measure multiplied by the constant $\nu(E)$.

Lemma 17.10. *Let $E \in \mathcal{L}_{\mathrm{FM}} \cap \mathcal{L}_{\mathrm{DOS}}$. Then for any $g \in L^1(\mathbb{R})$,*

$$(17.33)\quad \lim_{L\to\infty} \overline{\mu}_L^E(g) = \nu(E)\int_{\mathbb{R}} g(u)\,du.$$

Proof. We shall again prove the statement first for the functions φ_z defined in (17.22), in which case it asserts that

$$(17.34)\quad \lim_{L\to\infty} \overline{\mu}_L^E(\varphi_z) = \nu(E)\int_{\mathbb{R}} \varphi_z(u)\,du = \nu(E)$$

for every $z \in \mathbb{C}^+$. Abbreviating $z_L^E := E + z\,|\Lambda_L|^{-1}$, we split the prelimit into two terms, corresponding to the boundary layer Λ_L^b and the inner part Λ_L^i of the (large) cube:

$$\pi\,\overline{\mu}_L^E(\varphi_z) = \mathbb{E}\Big[\operatorname{Im}\operatorname{tr}\frac{1}{|\Lambda_L|\,(H_L - E) - z}\Big] = \frac{1}{|\Lambda_L|}\sum_{x\in\Lambda_L}\mathbb{E}\left[\operatorname{Im} G_{\Lambda_L}\left(x,x;z_L^E\right)\right].$$

Choosing the boundary layer so that it occupies a negligible fraction of the volume, $\lim_{L\to\infty}|\Lambda_L^b|/|\Lambda_L| = 0$, the Wegner estimate yields for the corresponding part:

$$\frac{1}{|\Lambda_L|}\sum_{x\in\Lambda_L^b}\mathbb{E}\left[\operatorname{Im} G_{\Lambda_L}\left(x,x;z_L^E\right)\right] \le \|\varrho\|_\infty\frac{|\Lambda_L^b|}{|\Lambda_L|}\underset{L\to\infty}{\longrightarrow} 0\,. \tag{17.35}$$

Before turning to the inner contribution, let us note that since E is a Lebesgue point of the density of states,

$$\begin{aligned}\pi\,\nu(E) &= \lim_{L\to\infty}\operatorname{Im}\int_{\mathbb{R}}\frac{\nu(dx)}{x - z_L^E} = \lim_{L\to\infty}\mathbb{E}\left[\operatorname{Im} G(0,0;z_L^E)\right]\\ &= \lim_{L\to\infty}\frac{1}{|\Lambda_L|}\sum_{x\in\Lambda_L^i}\mathbb{E}\left[\operatorname{Im} G(x,x;z_L^E)\right],\end{aligned} \tag{17.36}$$

where we again used the fact that $\lim_{L\to\infty}|\Lambda_L^b|/|\Lambda_L| = 0$. To complete the proof of (17.34), it thus remains to show that the following difference stemming from the inner contributions tends to zero in the limit $L\to\infty$:

$$\begin{aligned}&\frac{1}{|\Lambda_L|}\sum_{x\in\Lambda_L^i}\left|\mathbb{E}\left[\operatorname{Im} G_{\Lambda_L}\left(x,x;z_L^E\right)\right] - \mathbb{E}\left[\operatorname{Im} G(x,x;z_L^E)\right]\right|\\ &\le \frac{1}{|\Lambda_L|}\sum_{x\in\Lambda_L^i}\mathbb{E}\left[\left|G_{\Lambda_L}\left(x,x;z_L^E\right) - G(x,x;z_L^E)\right|\right]\\ &\le C_s\,\frac{|\Lambda_L^i|\,|\Lambda_L|^{1-2s}}{(\operatorname{Im} z)^{2(1-s)}}\,\exp\left(-\mu_s\operatorname{dist}(\Lambda_L^i,\partial\Lambda_L)\right).\end{aligned} \tag{17.37}$$

The last inequality is by Lemma 17.9. The right side converges to zero if the inner box Λ_L^i is chosen to have a distance to $\partial\Lambda_L$ which is proportional to L^a with some $a\in(0,1)$. This shows (17.34).

The convergence result is extended to all $g\in L^1(\mathbb{R})$ using the same approximation argument as in the proof of Lemma 17.7. □

17.3.3. Verifying the assumptions of Proposition 17.5. In order to finish the proof of Theorem 17.6 it remains to show that if M_L diverges as L^α for some $\alpha > 0$ with $L \to \infty$, then for any $E \in \mathcal{L}_{\rm FM} \cap \mathcal{L}_{\rm DOS}$ and all $g \in C_c^+(\mathbb{R})$,

$$\lim_{L\to\infty} \mathbb{E}\left[e^{-\mu_L^E(g)}\right] = \exp\left(-\nu(E)\int_{\mathbb{R}}\left(1-e^{-g(u)}\right)du\right); \tag{17.38}$$

i.e., the superposition $\sum_{j=1}^{M_L}\mu_{L,j}^E$ converges to a Poisson process with an intensity measure given by $\overline{\mu}(I) = \nu(E)\,|I|$.

Proof of Theorem 17.6. In view of the elementary estimate $\left|e^a - e^b\right| \le |a-b|\,e^{\max\{\mathrm{Re}\,a,\mathrm{Re}\,b\}}$ valid for all $a,b \in \mathbb{C}$, Lemma 17.7 ensures that μ_L^E and the approximands $\sum_{j=1}^{M_L}\mu_{L,j}^E$ converge to the same limit if one of them converges; i.e., for any $g \in C_c^+(\mathbb{R})$,

$$\lim_{L\to\infty} \mathbb{E}\left[\left|e^{-\mu_L^E(g)} - e^{-\sum_{j=1}^{M_L}\mu_{L,j}^E(g)}\right|\right] = 0\,. \tag{17.39}$$

For a proof of (17.38) it therefore remains to establish the convergence of the approximands. This is done with the help of Proposition 17.5, whose assumptions need to be checked. In this context, it is useful to note that

$$\mu_{L,j}^E(I;\omega) = \operatorname{tr} P_{E+|\Lambda_L|^{-1}I}\left(H_{L,j}(\omega)\right) \tag{17.40}$$

for any bounded Borel set $I \subset \mathbb{R}$. Therefore:

1. The Wegner estimate, Theorem 4.1, guarantees that

$$\mathbb{P}\left(\mu_{L,j}^E(I) \ge 1\right) \le \mathbb{E}\left[\mu_{L,j}^E(I)\right] \le \|\varrho\|_\infty \frac{|I|}{M_L}\,.$$

2. Condition (17.13) can be deduced from the so-called Minami estimate, Theorem 17.11, which is presented below. It guarantees that

$$\sum_{m=2}^{\infty}\mathbb{P}\left(\mu_{L,j}^E(I) \ge m\right) \le \mathbb{E}\left[\mu_{L,j}^E(I)\left(\mu_{L,j}^E(I)-1\right)\right] \le \|\varrho\|_\infty^2 \frac{|I|^2}{M_L^2}\,.$$

If the number of subboxes tends to infinity, $M_L \to \infty$, as $L \to \infty$, then assumptions (17.11) and (17.13) are satisfied. In order to check (17.12), it is useful to observe that $\sum_j \mathbb{P}\left(\mu_{L,j}^E(I) \ge 1\right)$ asymptotically coincides with an average,

$$\begin{aligned}\lim_{L\to\infty}\sum_{j=1}^{M_L}&\left|\mathbb{E}\left[\mu_{L,j}^E(I)\right] - \mathbb{P}\left(\mu_{L,j}^E(I)\ge 1\right)\right| \\ &\le \lim_{L\to\infty}\sum_{j=1}^{M_L}\sum_{m=2}^{\infty}\mathbb{P}\left(\mu_{L,j}^E(I)\ge m\right) = 0\,.\end{aligned} \tag{17.41}$$

The emerging intensity measure hence coincides with the limit of the averaged finite-volume density of states. The latter is identified through a combination of Lemmas 17.7 and 17.10, which yield

$$\lim_{L\to\infty} \sum_{j=1}^{M_L} \mathbb{E}\left[\mu_{L,j}^E(I)\right] \;=\; \lim_{L\to\infty} \mathbb{E}\left[\mu_L^E(I)\right] \;=\; \nu(E)\,|I|\,. \tag{17.42}$$

This concludes the proof of (17.38), up to the proof of the Minami estimate to which we turn next. □

17.4. The Minami bound and its CGK generalization

As seen above, to learn about the eigenvalue gap statistics one needs to go beyond the Wegner bound on the density of states and ask about the probability of there being at least two eigenvalues in a given small interval I. A useful bound on this probability was produced through an ingenious argument by N. Minami [**301**]. The Minami bound (for $n = 2$) was eventually generalized by J.-M. Combes, F. Germinet and A. Klein (CGK) [**92**], through a different argument which yields the following generalization of Corollary 4.9.

Theorem 17.11 (Combes-Germinet-Klein). *Let $H(\omega) = T + \lambda V(\omega)$ with $\lambda > 0$ be an operator on $\ell^2(\mathbb{G})$ over a finite graph $\mathbb{G}$, with a random potential of joint probability distribution ρ and the corresponding concentration function S_ρ (defined in (4.11)). Then for any $n \in \mathbb{N}$ and any interval $I \subset \mathbb{R}$*

$$\boxed{\mathbb{P}\left(\operatorname{tr} P_I(H) \geq n\right) \;\leq\; \frac{(4+\pi/\lambda)^n}{n!}\,|\mathbb{G}|^n\, S_\rho(|I|)^n}\,. \tag{17.43}$$

Proof. The CGK argument starts with the following extension to $n > 1$ of the Chebyshev-type bound (4.23):

$$\mathbb{P}\left(\operatorname{tr} P_I(H) \geq n\right) \;\leq\; \frac{1}{n!}\,\mathbb{E}\left[\prod_{j=0}^{n-1}\left(\operatorname{tr} P_I(H) - j\right)\right]\,. \tag{17.44}$$

Next, using induction on $n \geq 1$ it will be shown that

$$\mathbb{E}\left[\prod_{j=0}^{n-1}\left(\operatorname{tr} P_I(H) - j\right)\right] \;\leq\; (4+\pi/\lambda)^n\,|\mathbb{G}|^n\, S_\rho(|I|)^n\,. \tag{17.45}$$

For $n = 1$ the assertion coincides with the Wegner estimate (4.21). For the induction step, $n \to n+1$, one may separate the first factor and write the expression on the right side in (17.45) by first conditioning on $V_{\neq x} = \{V(y)\}_{y \neq x}$, as

$$\sum_{x \in \mathbb{G}} \mathbb{E}\left[\mathbb{E}\left[\langle \delta_x, P_I(H)\, \delta_x\rangle \prod_{j=1}^{n} (\operatorname{tr} P_I(H) - j) \,\middle|\, V_{\neq x}\right]\right]. \tag{17.46}$$

The conditional expectation amounts to an integration over the random variable $V(x)$ which enters the operator $H(\omega)$ as the coefficient of a rank-one perturbation term. When its value is changed to $\widehat{V}(x)$ which is not tied to the original $V(x)$, the number of eigenvalues of the resulting operator in the interval I can change by at most one, and hence for all $\widehat{V}(x)$: $\operatorname{tr} P_I(H_{L,\widehat{V}(x)}) \leq \operatorname{tr} P_I(H_{L,V(x)}) + 1$ (cf. Figure 5.1). Therefore the conditional expectation of the product in (17.46) can be estimated as follows:

$$\begin{aligned} &\mathbb{E}\left[\langle \delta_x, P_I(H(\omega))\, \delta_x\rangle \prod_{j=1}^{n} (\operatorname{tr} P_I(H(\omega)) - j) \,\middle|\, V_{\neq x}\right] \\ &\quad \leq \prod_{j=0}^{n-1} (\operatorname{tr} P_I(H(\omega)) - j) \;\; \mathbb{E}\left[\langle \delta_x, P_I(H(\omega))\, \delta_x\rangle \middle|\, V_{\neq x}\right], \end{aligned} \tag{17.47}$$

where the value of $V(x)$ in the last product could be arbitrary, but for convenience was reset to its original one. Through the argument of (4.22), the last factor is bounded by $(4 + \pi/\lambda)$. Upon the insertion of the resulting expression in (17.46) one may apply the induction hypothesis. Using it to bound the average of the remaining product yields the bound (17.45) for $n+1$. □

17.5. Level statistics on finite tree graphs

Tree graphs provide a natural testing ground for Conjecture 17.1 on the relation of the spectral gap statistics of operators on finite graphs to the spectral characteristics of the corresponding operator on the graph's infinite limit.

However, the infinite graph which captures the limit of finite trees, of fixed branching ratio (K) except at the leaves, is not the infinite regular tree! Instead, it is the **canopy graph**, using the terminology of [**16**], which is depicted in Figure17.3. The relevant observation is that if a site is selected uniformly at random on the finite tree, it will be at a distance of a few steps to the graph's canopy, i.e., the set of its leaves. From the perspective of such a site, in the infinite-volume limit the graph has a unique path to infinity and takes the form depicted in Figure 17.3(b).

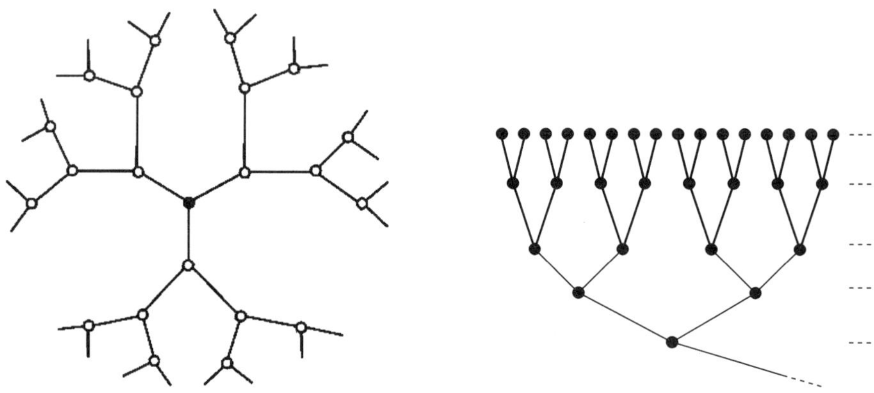

(a) the Bethe lattice $\mathbb{B}$ (b) the canopy graph

Figure 17.3. A large finite tree graph as it appears from the perspectives of (a) the point at its center, or points deeply below the tree's top layer, and (b) one of the leaves. The graph limits depend on the perspective: In case (a) the limit is a regular tree $\mathbb{T}$, of uncountably infinite boundary (i.e., directions towards infinity), and in case (b) the limit is the canopy graph which has only one boundary point.

On the canopy graph any finite set can be disconnected from infinity by the removal of a single edge. As stated in Chapter 12, under this condition the absolutely continuous spectrum of operators such as $H(\omega)$ is almost surely empty [**113**]. Thus it would be in line with the general conjecture to expect that the eigenvalue statistics of the finite tree graphs $\mathbb{T}_L$ would scale to the Poisson process at all energies, i.e., even for energies E within the *absolutely continuous* spectrum of $\mathbb{T}$. Under an additional assumption, that was indeed proven to be the case [**16**].

Omitting some of the proof details let us note that the argument which was presented for the Poisson limit of spectra on $\Lambda_L \subset \mathbb{Z}^d$ does not apply to $\mathbb{T}_L$ since on tree graphs the surface\volume ratio for sets of intermediate size (and the related Cheeger constant) are uniformly bounded away from 0. Instead, on trees the divisibility of the eigenvalue distribution of $\mathbb{T}_L$ can be approached by noting that upon the removal of a finite neighborhood of the root the graphs split into a number of independent similar trees. One still needs to show that the interaction which occurs when the subtrees are coupled through the root does not produce significant level repulsion.

It is also true that, unlike the spectrum of the Bethe lattice, the spectrum of the operator on the canopy graph is essentially localized. Since this graph is the relevant reference for $\mathbb{T}_L$, one may regard the above observation to be in line with Conjecture 17.1.

17.6. Regular trees as the large N limit of d-regular graphs

Once it is understood that the absolute continuity of the spectrum of the Schrödinger operator on the regular tree does not relate to the spectral statistics of its finite subtrees, one may enquire whether there are finite graphs for which the regular tree serves as the natural limit. A case can be made that such a characterization does apply to the ensemble of random d-regular graphs, of the fixed degree $d = K + 1$ and N vertices, in the limit $N \to \infty$. The term "random" refers in this context to graphs selected with uniform distribution among all graphs within the specified ensemble. A relevant combinatorial statement is the following:

Theorem 17.12 ([**60**]). *For the ensemble of random d-regular simple graphs (i.e., without single-point loops and double edges) of N vertices, except for probability which tends to 0 as $N \to \infty$, for all but a fraction $o(1)$ of sites the graph locally coincides with a regular tree up to distances at least $C_d \log N$ with some $C_d < \infty$.*

A random Schrödinger operator on a random d-regular (simple) graph has two sources of randomness—one is the way the local tree is wired "at infinity", i.e., at distances of the order of $\log N$, and the other is the potential. From Theorem 17.12 it can be deduced that the mean density of states of the Schrödinger operator on random d-regular graphs tends to that of the corresponding operator on the regular tree graph (Exercise 17.2). The corresponding statement for the graph Laplacian was proven by B. D. McKay [**294**].

A natural expression of Conjecture 17.1 in reference to the different spectra which are known for the infinite trees could therefore be sought in the spectral statistics of the random Schrödinger operator on random d-regular graphs. Consistent with that, in numerical studies it was found that the level statistics of the Laplacian alone on such graphs does exhibit GOE-like level repulsion [**201**]. On the other hand, for the Schrödinger operator on the d-regular graphs, at sufficiently high disorder Poisson statistics could be established, with a technique different from those presented in the previous sections [**167**]. However, the full proof of Conjecture 17.1 in either of these contexts still remains an open problem.

Notes

In relation to the dichotomy presented by the spectral statistics conjecture it should be mentioned that there is a large body of interesting works which link and interpolate between random matrices and random Schrödinger operators. Included in it are recent results and concepts which are still being developed.

The original random matrix ensembles were classified through the symmetry (GOE/GUE/GSE) of the probability distribution on the space of $N \times N$ matrices [**28, 296**]. The corresponding eigenvalue point processes form particular cases ($\beta = 1, 2, 4$) of the one-parameter family of point processes, with the law

$$\mathbb{P}_\beta = \frac{1}{Z(N,\beta)} \prod_{1 \le j < k \le N} |E_j - E_k|^\beta \exp\left(-\frac{1}{2} \sum_{j=1}^{N} E_j^2 \right) \prod_{j=1}^{N} dE_j \,. \tag{17.48}$$

Only relatively recently was it recognized that also for the other values of β this point process describes the spectrum of random matrices, which have a natural description in terms of random tridiagonal matrices (called the GβE ensembles) with some resemblance to random Schrödinger operators [**131, 225, 71**].

J. Breuer, P. Forrester, and U. Smilansky [**71**] pointed out that the spectra of the limiting Jacobi operators associated with the GβE ensemble exhibit a transition from pure-point spectrum at $\beta \le 2$, with power-law localized eigenfunctions, to singular continuous spectrum for $\beta > 2$ (the localization being stated here with respect to the original basis in which the matrix is tridiagonal). The GβE eigenvalue statistics also feature in random CMV matrices, which for a certain class of distribution of their coefficients exhibits a phase transition from Poisson to clock statistics via random matrix ensembles [**226**]. The limiting process of the rescaled eigenvalues, sometimes called Sine(β), was investigated in by B. Valkó and B. Virág [**388**].

Also related to the subject of this chapter are studies of one-dimensional operators $H_L(\omega) = -\Delta + \lambda V(\omega)$ on $\ell^2([0,L] \cap \mathbb{Z})$ in the following scaling regimes:

1. decaying coupling: $\lambda = 1/\sqrt{L}$ with iid random potential or
2. decaying randomness: $\lambda = 1$ but the iid potential is scaled by $1/\sqrt{|x|}$.

In both cases the rescaled process of eigenvalues converges, in the first case to the so-called Schrödinger process and in the second case to the Sine(β)

process [**271**]. These scaling limits are somewhat less universal since the parameters of the limiting processes such as β depend on the strength of disorder. See also [**312**] for the continuum version.

More in line with the universality conjecture were the results of [**389, 338**] for the above random Schrödinger operator in strips $([0,L] \times [0,W]) \cap \mathbb{Z}^2$. It was shown there that under suitable scaling of $\lambda = o(1/\sqrt{L})$ and the strip's width $1 \ll W \ll L$, the rescaled eigenvalues in the vicinity of energy $E \in (-2,2)$ (i.e., away from the band-edges ± 2) converge to the Sine(1) point process. The statement is closely related to the Dorokhov-Mello-Pereyra-Kumar (DMPK) theory of scaling limits of transfer operators; cf. [**47**] and the references therein.

In addition to Poisson limits for the process of eigenvalues in the localization regime, it is natural to consider the joint rescaled process of eigenvalues and localization centers. While the latter may also be shown to exhibit some repulsion [**302, 311**], that is ineffective in the localization regime $\mathcal{L}_{\mathrm{FM}} \cap \mathcal{L}_{\mathrm{DOS}}$ for which the joint Poisson limit was proven in [**224**].

The universality conjecture is also formulated in reference to the eigenvalue point processes of the continuum random Schrödinger operators. However, for that case, beyond certain one-dimensional operators [**302**] at present even the Poisson part is still an open problem. The stumbling block is the lack of a continuum analogue of Minami's estimate. More information on that can be found in [**180, 51, 174, 377**].

Throughout this chapter we chose to express the main results in the language of random point measures, which offer a convenient way of describing point processes. Further background on the terminology and basic results may be found in textbooks such as [**216**].

Exercises

17.1. a. What fraction of eigenvalues of the operator in Theorem 17.6 within an a priori fixed interval $[E, E+\Delta E] \subset \mathcal{L}_{\mathrm{DOS}} \cap \mathcal{L}_{\mathrm{FM}}$ of width $1/|\Lambda_L| \ll |\Delta E| \ll 1$ has a gap of size $|E_{n+1} - E_n| \leq t$?

b. Does the gap between the first eigenvalue above E and the first below E have the same probability distribution? (Explain the paradox.)

17.2. Let $(\mathbb{G}_n, H_n)$ be a sequence of graphs, rooted at $x_0^{(n)}$, paired with Schrödinger operators acting in $\ell^2(\mathbb{G}_n)$, of potentials V_n. Assume the following:
i) The graphs converge locally to a rooted graph $\mathbb{G}$, in the sense that for each $L < \infty$ at n large enough the subgraphs

$$\mathbb{G}_n^L \ := \ \{x \in \mathbb{G}_n \mid \text{dist}(x, x_0^{(n)}) \leq L\}$$

are isomorphic to the similarly defined L-neighborhood of $x_0 \in \mathbb{G}$.
ii) For $n \to \infty$ the potentials also converge pointwise, under the corresponding isomorphism, to the potential of a Schrödinger operator H acting in $\ell^2(\mathbb{G})$. Prove that the following limits exist and satisfy the following:
(a) For each $k < \infty$,

$$\lim_{n\to\infty} \langle \delta_{x_0^{(n)}}, H_n^k \, \delta_{x_0^{(n)}} \rangle \ = \ \langle \delta_{x_0}, H^k \delta_{x_0} \rangle \, .$$

(b) For each $z \in \mathbb{C}^+$,

$$\lim_{n\to\infty} \left\langle \delta_{x_0^{(n)}}, \frac{1}{H_n - z} \delta_{x_0^{(n)}} \right\rangle \ = \ \left\langle \delta_{x_0}, \frac{1}{H_n - z} \delta_{x_0} \right\rangle \, .$$

(c) The density of state measures of H_n, associated with $\delta_{x_0^{(n)}}$, converge to the corresponding DOS of H associated with δ_{x_0}.

(In the above limits, (a) is rather elementary, (b) is less immediate but more useful, and (c) is equivalent to (b). The next exercise concerns the adaptation of the last statement to a random setting.)

17.3. Using the principle stated in Exercise 17.2(c) and the combinatorial statement of Theorem 17.12, prove that the density of states of the d-regular graph of n vertices converges, as $n \to \infty$, to the DOS of the d-regular tree (which is given by the Kesten-McKay distribution).

Appendix A

Elements of Spectral Theory

Since observables in quantum mechanics are described by self-adjoint operators on a Hilbert space, we briefly summarize here, in a somewhat informal manner, selected elements of the theory of such operators. For proofs the reader is referred to the extended existent literature, in particular [**380, 118, 331, 398, 207**].

A.1. Hilbert spaces, self-adjoint linear operators, and their resolvents

The state space of an isolated quantum systems is a separable complex Hilbert space $\mathcal{H}$. Its inner product is denoted here by $\langle\cdot,\cdot\rangle : \mathcal{H}\times\mathcal{H}\to\mathbb{C}$ and is taken to be linear in its second entry. The induced norm of $\psi\in\mathcal{H}$ is given by $\|\psi\| := \sqrt{\langle\psi,\psi\rangle}$.

An example is the space of square-summable, complex-valued sequences over the vertices $\mathbb{G}$ of a graph

$$\ell^2(\mathbb{G}) \;:=\; \Big\{\psi:\mathbb{G}\to\mathbb{C} \,\Big|\, \sum_{x\in\mathbb{G}} |\psi(x)|^2 \;<\; \infty\Big\}, \tag{A.1}$$

which is rendered a Hilbert space together with the scalar product

$$\langle\varphi,\psi\rangle \;:=\; \sum_{x\in\mathbb{G}} \overline{\varphi(x)}\,\psi(x)\,. \tag{A.2}$$

Here $\overline{(\cdot)}$ denotes complex conjugation. An example for $\mathbb{G}$ is the d-dimensional lattice $\mathbb{Z}$.

A linear operator $A : \mathcal{D}(A) \to \mathcal{H}$ defined on a domain, i.e., a dense linear subspace $\mathcal{D}(A) \subset \mathcal{H}$, is bounded if and only if there exists some $a < \infty$ such that $\|A\psi\| \leq a\|\psi\|$ for all $\psi \in \mathcal{D}(A)$. The smallest possible constant in this inequality defines the **operator norm**

$$\|A\| := \sup_{\substack{\psi \in \mathcal{D}(A) \\ \|\psi\|=1}} \|A\psi\| .$$

Following are some examples of linear operators which are frequently encountered throughout the book.

The **graph Laplacian**, with the initial domain $\mathcal{D}(\Delta)$ consisting of functions of finite support, is the closure of the operator:

(A.3)

$$\begin{aligned} &\Delta : \mathcal{D}(\Delta) \to \ell^2(\mathbb{G}) , \\ &\quad \Delta\psi(x) := \sum_{\substack{y \in \mathbb{G} \\ [x,y] \in E_{\mathbb{G}}}} (\psi(y) - \psi(x)) = \sum_{\substack{y \in \mathbb{G} \\ [x,y] \in E_{\mathbb{G}}}} \psi(y) - \deg_{\mathbb{G}}(x)\, \psi(x) , \end{aligned}$$

where $\deg_{\mathbb{G}}(x) := |\{y \in \mathbb{G} \,|\, [x,y] \in E_{\mathbb{G}}\}|$ (the degree function). The operator is bounded if and only if $\deg_{\mathbb{G}} := \sup_{x \in \mathbb{G}} \deg_{\mathbb{G}}(x) < \infty$. Applying the triangle inequality,

(A.4)

$$\begin{aligned} \|\Delta\psi\| &= \Big(\sum_{x \in \mathbb{G}} \Big| \sum_{y\colon [x,y] \in E_{\mathbb{G}}} (\psi(y) - \psi(x)) \Big|^2 \Big)^{1/2} \\ &\leq \Big[\sum_{x \in \mathbb{G}} \Big| \sum_{y\colon [x,y] \in E_{\mathbb{G}}} \psi(y) \Big|^2 \Big]^{1/2} + \Big[\sum_{x \in \mathbb{G}} \deg_{\mathbb{G}}(x)^2 \, |\psi(x)|^2 \Big]^{1/2} \\ &\leq 2 \deg_{\mathbb{G}} \|\psi\| , \end{aligned}$$

and consequently $\|\Delta\| \leq 2 \deg_{\mathbb{G}}$. This bound is saturated (i.e., holds as equality) on finite graphs which are regular and bipartite, and also on $\mathbb{G} = \mathbb{Z}^d$ where $\|\Delta\| = 4d$.

A variant is the **magnetic Laplacian** in two-space dimension. It is defined on $\ell^2(\mathbb{Z}^2)$ by

$$\left(\Delta^{(A)} \psi\right)(x) := \sum_{|y-x|=1} \left(e^{iA(x,y)} \psi(y) - \psi(x) \right) \tag{A.5}$$

in terms of an antisymmetric function $A : \mathbb{Z}^2 \times \mathbb{Z}^2 \to \mathbb{R}$, $(x, y) \mapsto A(x, y) = A(x, y)\, \delta_{|x-y|,1}$, which is supported on the edges of the lattice $\mathbb{Z}^2$. Similarly as above one may prove $\|\Delta^{(A)}\| \leq 8$ independently of A.

We will mainly be concerned with self-adjoint operators. Recall that the **adjoint** of $A : \mathcal{D}(A) \to \mathcal{H}$ is the unique operator $A^\dagger : \mathcal{D}(A^\dagger) \to \mathcal{H}$ acting as

$$\langle \varphi, A\psi \rangle = \langle A^\dagger \varphi, \psi \rangle$$

for all $\psi \in \mathcal{D}(A)$ and $\varphi \in \mathcal{D}(A^\dagger) := \{\varphi \in \mathcal{H} \,|\, \text{there is } \varrho \in \mathcal{H} \text{ such that for all } \psi \in \mathcal{D}(A) : \ \langle \varphi, A\psi \rangle = \langle \varrho, \psi \rangle\}$. The operator A is **self-adjoint** if and only if $A = A^\dagger$, which in particular requires $\mathcal{D}(A) = \mathcal{D}(A^\dagger)$.

Any bounded, real-valued sequence $V : \mathbb{G} \to \mathbb{R}$ on any graph with a bounded degree gives rise to a bounded and self-adjoint **Schrödinger operator** $H : \ell^2(\mathbb{G}) \to \ell^2(\mathbb{G})$ defined by

$$(H\psi)(x) := -(\Delta\psi)(x) + V(x)\,\psi(x) \tag{A.6}$$

on $\psi \in \ell^2(\mathbb{G})$. In fact, each of the terms individually define bounded self-adjoint operators and hence their sum is bounded and self-adjoint. The self-adjointness of the **multiplication operator** $V : \ell^2(\mathbb{G}) \to \ell^2(\mathbb{G})$ corresponding to V is straightforward. The self-adjointness of the graph Laplacian follows from its representation as a **quadratic form**,

$$\langle \varphi, -\Delta\psi \rangle = \sum_{[x,y] \in E_{\mathbb{G}}} \overline{(\varphi(y) - \varphi(x))}\, (\psi(y) - \psi(x)) \,, \tag{A.7}$$

which implies $\langle \varphi, -\Delta\psi \rangle = \langle -\Delta\varphi, \psi \rangle$ for all $\varphi, \psi \in \ell^2(\mathbb{G})$.

The **resolvent set** of an operator $A : \mathcal{D}(A) \to \mathcal{H}$ is given by

$$\varrho(A) := \{z \in \mathbb{C} \,|\, (A - z) : \mathcal{D}(A) \to \mathcal{H} \text{ is bijective}\} \,.$$

For $z \in \varrho(A)$ the operator $(A-z)$ is hence invertible and its inverse $(A-z)^{-1} : \mathcal{H} \to \mathcal{D}(A)$ is called the resolvent. The **spectrum** of A is the set

$$\sigma(A) := \mathbb{C} \setminus \varrho(A) \,,$$

which is always closed. If A is bounded, $\sigma(A)$ is compact. In case A is self-adjoint, $\sigma(A) \subset \mathbb{R}$ and for all $z \in \varrho(A)$,

$$\|(A - z)^{-1}\| \leq \frac{1}{\operatorname{dist}(z, \sigma(A))} \leq \frac{1}{|\operatorname{Im} z|} \,, \tag{A.8}$$

where the last inequality requires $z \in \mathbb{C} \setminus \mathbb{R}$.

As an example, let us determine the spectrum of the Laplacian for the lattice $\mathbb{Z}^d$. The Fourier transformation

$$\mathcal{F} : \ell^2(\mathbb{Z}^d) \to L^2([0, 2\pi]^d)$$

$$(\mathcal{F}\psi)(k) := \frac{1}{(2\pi)^{d/2}} \sum_{x \in \mathbb{Z}^d} e^{-ik\cdot x}\, \psi(x) \tag{A.9}$$

defines a unitary operator. By an explicit calculation, one checks that the discrete Laplacian is unitarily equivalent to a multiplication operator; i.e.,

$$\left(\mathcal{F}\Delta\mathcal{F}^{-1}\varphi\right)(k) \;=\; 2\sum_{\nu=1}^{d} (\cos(k_\nu) - 1)\; \varphi(k) \;\;=: h(k)\, \varphi(k)\,, \tag{A.10}$$

for all $\phi \in L^2([0, 2\pi]^d)$, where $k = (k_1, \ldots, k_d) \in [0, 2\pi]^d$. The spectrum of any multiplication operator is the range of the corresponding function, which in the case of h is the interval $[-4d, 0]$. Since the spectrum is invariant under unitary transformation, we therefore conclude

$$\sigma(-\Delta) \;=\; [0, 4d]\,. \tag{A.11}$$

This implies $\|\Delta\| = 4d$ since the operator norm coincides with the maximal modulus of points in the spectrum.

Important tools for spectral analysis are the first and second resolvent equations.

Proposition A.1 (Resolvent equations). *For a linear operator $A : \mathcal{D}(A) \to \mathcal{H}$ and $z_1, z_2 \in \varrho(A)$,*

$$(A - z_1)^{-1} - (A - z_2)^{-1} \;=\; (z_1 - z_2)(A - z_1)^{-1}(A - z_2)^{-1}\,. \tag{A.12}$$

Moreover, if $B : \mathcal{D}(B) \to \mathcal{H}$ is defined on a common domain, $\mathcal{D}(A) = \mathcal{D}(B)$,

$$\begin{aligned} (A - z)^{-1} - (B - z)^{-1} \;&=\; (A - z)^{-1}(B - A)\,(B - z)^{-1} \\ &=\; (B - z)^{-1}(B - A)\,(A - z)^{-1} \end{aligned} \tag{A.13}$$

for any $z \in \varrho(A) \cap \varrho(B)$.

The proof is elementary and can be found in [**398**].

An immediate corollary is the fact that the distance of the spectra of two self-adjoint operators $A, B : \mathcal{D}(A) \to \mathcal{H}$ whose difference is a bounded operator is bounded by

$$\operatorname{dist}\left(\sigma(A), \sigma(B)\right) \;\leq\; \|A - B\|\,. \tag{A.14}$$

A.2. Spectral calculus and spectral types

The spectral calculus provides an extension to self-adjoint operators of the benefits which the eigenfunction expansion provides for self-adjoint matrices. The subject can be approached in a variety of ways. Since the spectral representation of Herglotz-Pick functions has been of independent interest here, we shall use that also as a door to this subject.

Recall that for any $\psi \in \mathcal{H}$, the function given by

$$F_\psi(z) := \langle \psi \,, (A-z)^{-1}\, \psi \rangle \tag{A.15}$$

is holomorphic for $z \in \varrho(A) \supset \mathbb{C}^+ := \{z \in \mathbb{C} \mid \operatorname{Im} z > 0\}$. Using (A.8) we conclude that

$$\begin{aligned} |F_\psi(z)| \;&\leq\; \left\|(A-z)^{-1}\right\| \, \|\psi\|^2 \leq \frac{\|\psi\|^2}{\operatorname{Im} z} \quad \text{for any } z \in \mathbb{C}^+, \\ \overline{F_\psi(z)} &= \langle (A-z)^{-1}\, \psi \,,\, \psi \rangle = \langle \psi \,, (A-\overline{z})^{-1}\, \psi \rangle = F_\psi(\overline{z})\,. \end{aligned}$$

Moreover, the first resolvent equation (A.12) implies that

$$\operatorname{Im} F_\psi(z) \;=\; \frac{1}{2i}\left(F_\psi(z) - \overline{F_\psi(z)}\right) = \operatorname{Im} z \,\left\|(A-z)^{-1}\,\psi\right\|^2 \;>0$$

for any $\psi \neq 0$; hence F_ψ is a **Herglotz-Pick function**, i.e., a holomorphic function $F_\psi : \mathbb{C}^+ \to \mathbb{C}^+$ mapping the upper half-plane into itself.

The representation theorem for Herglotz functions, which is summarized in Appendix B, entails that there exists a unique finite Borel measure μ_ψ on $\mathbb{R}$, called the **spectral measure** of A associated with ψ, such that

$$F_\psi(z) \;=\; \int \frac{\mu_\psi(d\lambda)}{\lambda - z}\,, \quad z \in \mathbb{C}^+\,. \tag{A.16}$$

By polarization, for any pair $\varphi, \psi \in \mathcal{H}$ one may identify complex finite Borel measures $\mu_{\varphi,\psi}$ such that $\langle \varphi, (A-z)^{-1}\psi\rangle \;=\; \int (\lambda - z)^{-1} \mu_{\varphi,\psi}(d\lambda)$.

The spectral measures $\mu_{\varphi,\psi}(d\lambda)$ enable the definition of functions $f(A)$ of a self-adjoint operators; i.e., for any $f \in L^\infty(\mathbb{R})$,

$$\langle \varphi, f(A)\, \psi\rangle \;:=\; \int f(\lambda)\, \mu_{\varphi,\psi}(d\lambda)\,. \tag{A.17}$$

This yields a natural extensions of $f, g \in L^\infty(\mathbb{R})$ to operator-valued functions of self-adjoint operators satisfying the basic rules:

$$(f+g)(A) \;=\; f(A) + g(A)\,, \quad (fg)(A) \;=\; f(A)\, g(A)\,, \quad \overline{f}(A) \;=\; f(A)^\dagger\,.$$

If $f \geq 0$, then $\langle \psi, f(A)\psi\rangle \geq 0$ for all $\psi \in \mathcal{H}$.

Moreover, $\|f(A)\| = \sup_{\lambda \in \sigma(A)} |f(\lambda)|$.

Of particular importance are indicator functions,

$$1_J(x) := \begin{cases} 1 & \text{if } x \in J, \\ 0 & \text{otherwise,} \end{cases} \qquad J \subset \mathbb{R}\,.$$

They define orthogonal projections $P_J(A) := 1_J(A)$. In fact, the map $J \mapsto P_J(A)$ on the Borel sets $J \subset \mathbb{R}$ defines a **projection-valued measure**, i.e., as follows:

1. $P_{\mathbb{R}}(A) = 1$ and $\mathbb{P}_{\emptyset}(A) = 0$.
2. If $J = \bigcup_n J_n$ with $J_n \cap J_m = \emptyset$ for $n \neq m$, then $P_J(A) = \sum_n P_{J_n}(A)$.

The support of this measure coincides with the spectrum:

$$\sigma(A) \;=\; \big\{\lambda \in \mathbb{R} \,\big|\, P_{(\lambda-\varepsilon,\lambda+\varepsilon)}(A) \neq 0 \text{ for all } \varepsilon > 0\big\}\,. \tag{A.18}$$

The **spectral theorem** ensures that there is one-to-one correspondence between self-adjoint operators and projection-valued measures. This is summarized by (A.17) through the relation $\langle \varphi\,, P_J(A)\,\psi\rangle = \mu_{\varphi,\psi}(J)$.

In its **multiplication operator** version [**331, 398**], the spectral theorem provides the following representation of the action of any self-adjoint operator A within the cyclic subspace $\mathcal{H}_{A,\phi} := \overline{\operatorname{span}\{(A-z)^{-1}\phi \,|\, z \in \mathbb{C}\backslash\mathbb{R}\}}$ associated with a vector $\phi \in \mathcal{H}$. For each ϕ the corresponding subspace is invariant under the action of A. In case $\mathcal{H}_{A,\phi} = \mathcal{H}$ one calls ϕ a cyclic vector for A.

Proposition A.2 (Spectral theorem; multiplication operator version). *For any self-adjoint operator A on a Hilbert space $\mathcal{H}$ and any $\phi \in \mathcal{H}$, there is a unitary equivalence between the cyclic subspace $\mathcal{H}_{A,\phi}$ and $L^2(\mathbb{R}, \mu_\phi)$ under which ϕ is mapped to the constant function 1 and the action of A within $\mathcal{H}_{A,\phi}$ corresponds to the operator of multiplication by x in $L^2(\mathbb{R}, \mu_\phi)$.*

According to the Radon-Nikodym theorem every Borel measure μ can be uniquely decomposed with respect to the Lebesgue measure into three mutually singular parts:

$$\mu = \mu^{\mathrm{pp}} + \mu^{\mathrm{sc}} + \mu^{\mathrm{ac}}\,.$$

Whereas μ^{ac} is absolutely continuous with respect to Lebesgue measure, the sum of the first two terms is singular. It consists of a pure-point component, μ^{pp}, and a singular continuous remainder, μ^{sc}.

Accordingly, for a self-adjoint operator $A : \mathcal{D}(A) \to \mathcal{H}$ one decomposes the Hilbert space into closed subspaces:

$$\mathcal{H}^{\#} \;:=\; \left\{\psi \in \mathcal{H} \,|\, \mu_\psi = \mu_\psi^{\#}\right\}, \quad \# = \mathrm{pp,\ sc,\ ac}.$$

This decomposition turns out to be orthogonal, $\mathcal{H} = \mathcal{H}^{\mathrm{pp}} \oplus \mathcal{H}^{\mathrm{sc}} \oplus \mathcal{H}^{\mathrm{ac}}$ and the above subspaces are left-invariant under the action of A.

In the case of a bounded self-adjoint operator $A : \mathcal{H} \to \mathcal{H}$, the restriction to these subspaces defines the components of the spectrum:

$$\sigma^{\#}(A) \; := \; \sigma\big(A\,\big|_{\mathcal{H}^{\#}}\big)\,, \quad \# = \text{ pp, sc, ac}\,, \tag{A.19}$$

which are called the **pure-point**, **singular continuous**, and **absolutely continuous spectrum**.

The pure-point spectrum coincides with the closure of the set of eigenvalues:

$$\sigma^{\mathrm{pp}}(A) = \overline{\{\lambda \in \mathbb{R} \,|\, \lambda \text{ is an eigenvalue of } A\}}\,.$$

The multiplicity of an eigenvalue is the dimension of the eigenspace $\{\varphi \in D(A) \,|\, A\varphi = \lambda\varphi\}$ associated to λ. An eigenvalue λ is called **isolated** if there is an $\varepsilon > 0$ such that $\sigma(A) \cap (\lambda - \varepsilon, \lambda + \varepsilon) = \{\lambda\}$. Any isolated point in the spectrum is always an eigenvalue. The **discrete spectrum** $\sigma_{\mathrm{dis}}(A)$ is the set of all isolated eigenvalues of finite multiplicity. The **essential spectrum** is defined by $\sigma_{\mathrm{ess}}(A) := \sigma(A) \setminus \sigma_{\mathrm{dis}}(A)$.

Let us conclude this section with two examples.

Example A.3. For a multiplication operator corresponding to a real-valued sequence $V \in \ell^\infty(\mathbb{G})$, one identifies the spectral measure corresponding to $\psi \in \ell^2(\mathbb{G})$ as a weighted sum of Dirac measures:

$$\mu_\psi \; = \sum_{x \in \mathbb{G}} |\psi(x)|^2 \; \delta_{V(x)}\,;$$

i.e., $\mu_\psi \; = \mu_\psi^{\mathrm{pp}}$. The eigenvectors are given by the localized vectors $\delta_x \in \ell^2(\mathbb{G})$;

$$\delta_x(y) \; := \; \begin{cases} 1 & \text{if } x = y, \\ 0 & \text{otherwise.} \end{cases} \tag{A.20}$$

The corresponding eigenvalues are $\{V(x) \,|\, x \in \mathbb{G}\}$, whose closure is the pure-point spectrum $\sigma^{\mathrm{pp}}(V)$. The essential spectrum $\sigma_{\mathrm{ess}}(V)$ coincides with all infinitely degenerate eigenvalues and the set of accumulation points of $\{V(x) \,|\, x \in \mathbb{G}\}$.

Example A.4. For the Laplacian on $\ell^2(\mathbb{Z}^d)$ we use the unitarily equivalent representation as multiplication operator on $L^2([0, 2\pi]^d)$ given in (A.10) to determine the spectral measure:

$$\begin{aligned} \langle \psi, f(\Delta)\psi \rangle \; &= \; \langle \mathcal{F}\psi\,, \mathcal{F}\, f(\Delta)\, \mathcal{F}^{-1}\, \mathcal{F}\psi \rangle = \langle \mathcal{F}\psi\,,\, f\big(\mathcal{F}\Delta\mathcal{F}^{-1}\big)\, \mathcal{F}\psi \rangle \\ &= \; \int_{[0,2\pi]^d} |(\mathcal{F}\psi)\,(k)|^2\, f(h(k))\, dk \; = \; \int f(\lambda)\, \mu_\psi(d\lambda)\,. \end{aligned} \tag{A.21}$$

Hence, $\mu_\psi = \mu_\psi^{\mathrm{ac}}$ for all $\psi \in \ell^2(\mathbb{Z}^d)$ and the spectrum is only absolutely continuous; i.e., $\sigma^{\mathrm{ac}}(-\Delta) = [0, 4d]$. The whole spectrum is essential.

A.3. Relevant notions of convergence

The space of linear operators on $\mathcal{H}$ may be equipped with various topologies. For sequences of bounded operators the most important concepts are those of **weak**, **strong**, and **norm convergence**; i.e.,

$$\begin{aligned}
A_n &\to A \text{ weakly} && \text{iff } \langle \varphi, (A_n - A)\,\psi\rangle \to 0 \text{ for all } \varphi, \psi \in \mathcal{H},\\
A_n &\to A \text{ strongly} && \text{iff } \|(A_n - A)\,\psi\| \to 0 \text{ for all } \psi \in \mathcal{H},\\
A_n &\to A \text{ in norm} && \text{iff } \|A_n - A\| \to 0\,.
\end{aligned}$$

Clearly, norm convergence implies strong convergence which in turn implies weak convergence.

Often one deals with sequences of unbounded self-adjoint operators (A_n) which approximate a self-adjoint operator A. Two natural notions of proximity are the following:

(i) $A_n \to A$ in the **strong-resolvent sense**

$$\text{if } (A_n - z)^{-1} \to (A - z)^{-1} \text{ strongly for all } z \in \mathbb{C}\backslash\mathbb{R}.$$

(ii) $A_n \to A$ in the **norm-resolvent sense**

$$\text{if } (A_n - z)^{-1} \to (A - z)^{-1} \text{ in norm for all } z \in \mathbb{C}\backslash\mathbb{R}.$$

As is evident, the second notion is stronger than the first. Also there is no natural notion of weak-resolvent convergence for self-adjoint operators since it is equivalent to strong-resolvent convergence ([**331**]).

Among the useful implications of the above notions is the convergence of continuous functions of the operator:

$$f(A_n) \to f(A) \tag{A.22}$$

$$\begin{aligned}
&\text{strongly and for all } f \in C_b(\mathbb{R}) \text{ in case (i)},\\
&\text{in norm and for all } f \in C_0(\mathbb{R}) \text{ in case (ii)},
\end{aligned}$$

where $C_b(\mathbb{R})$ and $C_0(\mathbb{R})$ denote the continuous functions which are bounded or decay to zero at infinity, respectively. A proof can be found in [**331**]. It is in fact based on the Stone-Weierstrass theorem which deals with subsets of the space $C_0(\mathbb{R})$. A subset $\mathcal{D}$ is called an involutive subalgebra of $C_0(\mathbb{R})$ if it is a linear subspace and if for $f, g \in \mathcal{D}$ both the product $f \cdot g$ and the complex conjugate $\overline{f}$ belong to $\mathcal{D}$. We say that $\mathcal{D}$ separates points if for $x, y \in \mathbb{R}$ there is a function $f \in \mathcal{D}$ such that $f(x) \neq f(y)$ and both $f(x)$ and $f(y)$ are non-zero.

Proposition A.5 (Stone-Weierstrass). *If $\mathcal{D}$ is an involutive subalgebra of $C_0(\mathbb{R})$ which separates points, then $\mathcal{D}$ is dense in $C_0(\mathbb{R})$ with respect to the topology of uniform convergence.*

A proof of this proposition can be found in [**331, 380**]. As indicated above, Proposition A.5 is especially useful in connection with resolvents. Since by elementary algebra the set of linear combinations of the functions $f_z(x) = \frac{1}{x-z}$ for all $z \in \mathbb{C}\setminus\mathbb{R}$ forms an involutive subalgebra of $C_0(\mathbb{R})$ which separates points, it is dense.

Returning to the notion of strong-resolvent convergence of $A_n \to A$, another consequence, which is immediate from (A.22), is the vague convergence of all spectral measures associated to these operators and some vectors. The general (and standard) definition of this convergence is [**46**]

Definition A.6. A sequence of Borel measures μ_n on $\mathbb{R}$ is said to **converge weakly** to a Borel measure μ if and only if for all $f \in C_b(\mathbb{R})$,

$$\lim_{n\to\infty} \int f(x)\, \mu_n(dx) = \int f(x)\, \mu(dx). \tag{A.23}$$

If (A.23) holds for all $f \in C_c(\mathbb{R})$, the sequence is said to **converge vaguely**.

Clearly, weak convergence requires the measures to be finite. Moreover, weak convergence is a stronger notion than vague convergence. Only if no mass is lost at infinity does vague convergence imply weak convergence ([**46**]). The Portmanteau theorem states that vague convergence implies the convergence (A.23) for all compactly supported bounded functions f which are μ-almost everywhere continuous.

The following general criterion for weak convergence allows us to apply some of the useful properties of the resolvent.

Proposition A.7. *Within the class of probability measures on $\mathbb{R}$, a sufficient condition for the weak convergence of sequence (μ_n) to μ is that for a dense collection of $z \in \mathbb{C}\setminus\mathbb{R}$,*

$$\lim_{n\to\infty} \int \frac{\mu_n(dx)}{x-z} = \int \frac{\mu(dx)}{x-z}. \tag{A.24}$$

Since it is assumed that μ is a probability measure, it is enough to check (A.23) for all $\varphi \in C_0(\mathbb{R})$; cf. [**46**, Thm. 30.6 and Thm. 30.8]. A proof can then be based on the Stone-Weierstrass theorem.

Notes

The material in this appendix including its proofs can be found in standard textbooks [**331, 398**]. The construction of the spectral measure using the representation theorem for Herglotz-Pick functions is discussed in detail in [**380, 118**].

Appendix B

Herglotz-Pick Functions and Their Spectra

A natural arena for the discussion of the resolvents of self-adjoint operators is the class of holomorphic mappings of the upper half-plane into itself, $F : \mathbb{C}^+ \to \mathbb{C}^+$. Combining the mixed terminology which can be found in the literature, we refer to it as the **Herglotz-Pick** (HP) class [**128, 118**]. Following are some useful facts about functions in this class. A more thorough discussion and the omitted proofs can be found in [**118, 380, 207, 360**].

B.1. Herglotz representation theorems

Proposition B.1 (Representation I). *Any function $F : \mathbb{C}^+ \to \mathbb{C}^+$ which is analytic in $\mathbb{C}^+$ and satisfies $\operatorname{Im} F(z) > 0$ for all $\operatorname{Im} z > 0$ admits a unique representation in the form*

$$F(z) = az + b + \int \left(\frac{1}{u-z} - \frac{u}{1+u^2} \right) \mu(du) \tag{B.1}$$

with some $a \geq 0$, $b \in \mathbb{R}$ and a Borel measure μ satisfying $\int (1+u^2)^{-1} \mu(du) < \infty$. The measure μ can be recovered from F using

$$\frac{1}{2} \left(\mu\left((u_1, u_2)\right) + \mu\left([u_1, u_2]\right) \right) \; = \; \lim_{\varepsilon \downarrow 0} \frac{1}{\pi} \int_{u_1}^{u_2} \operatorname{Im} F(u + i\varepsilon) \, du \, . \tag{B.2}$$

A subclass of HP functions arises in spectral theory as the **Borel-Stieltjes transformation** of finite Borel measures μ on $\mathbb{R}$:

$$F(z) = \int \frac{\mu(du)}{u - z}, \quad z \in \mathbb{C}^+ . \tag{B.3}$$

The relation of the functions to the measures is summarized in the correspondingly restricted representation theorem.

Proposition B.2 (Representation II). *The Borel-Stieltjes transform of any finite Borel measure μ is a Herglotz function satisfying*

$$|F(z)| \leq \frac{\mu(\mathbb{R})}{\operatorname{Im} z} . \tag{B.4}$$

Conversely, if F is a Herglotz function satisfying

$$|F(z)| \leq \frac{C}{\operatorname{Im} z}, \quad z \in \mathbb{C}^+ , \tag{B.5}$$

then there exists a unique Borel measure μ with $\mu(\mathbb{R}) \leq C$ and F its Borel-Stieltjes transformation.

The statement is an elementary consequence of Proposition B.1.

B.2. Boundary function and its relation to the spectral measure

The following basic result underlies our frequent reference to the boundary value of Herglotz functions.

Proposition B.3 (de la Vallée-Poussin). *For any Herglotz function $F : \mathbb{C}^+ \to \mathbb{C}^+$, the limit (or boundary value)*

$$F(u + i0) := \lim_{\varepsilon \downarrow 0} F(u + i\varepsilon) \tag{B.6}$$

exists and is finite for Lebesgue-almost every $u \in \mathbb{R}$.

More can be said on the relation of the boundary value to the underlying spectral measure of the Herglotz function.

Proposition B.4. *For any Herglotz function $F : \mathbb{C}^+ \to \mathbb{C}^+$, the spectral measure μ associated to it in the representation* (B.1) *bears the following relations to the boundary value function:*

1. *For Lebesgue-almost every $u \in \mathbb{R}$,*

$$\frac{\mu^{\mathrm{ac}}(du)}{du} = \frac{1}{\pi} \operatorname{Im} F(u + i0) . \tag{B.7}$$

2. *The singular part,* $\mu^{\mathrm{s}} := \mu^{\mathrm{sc}} + \mu^{\mathrm{pp}}$*, is supported on the set*

$$\{u \in \mathbb{R} \,|\, \operatorname{Im} F(u+i0) = \infty\} \,.$$

3. *A point mass of* μ *at* u *is obtained by* $\mu(\{u\}) = \lim_{\varepsilon \downarrow 0} \varepsilon \operatorname{Im} F(u+i\varepsilon)$.

Thus the boundary value of a Herglotz function determines the function uniquely. In fact, the function is already uniquely determined from its boundary value on a Borel set of positive Lebesgue measure (see, e.g., [**118**, Cor. 1.4.9]).

Proposition B.5. *If the boundary values of two Herglotz functions* $F, G : \mathbb{C}^+ \to \mathbb{C}^+$ *agree Lebesgue-almost everywhere on some Borel set* $I \subset \mathbb{R}$ *of positive Lebesgue measure, then* $F = G$.

The proofs of Propositions B.1–B.4 can be based on classic properties of holomorphic functions $f : \mathbb{D} \to \mathbb{C}$ on the open unit disc $\mathbb{D} := \{z \in \mathbb{C} \,|\, |z| < 1\}$. Via the standard Möbius transformation $m : \mathbb{D} \to \mathbb{C}^+$ given by

$$m(w) := \frac{w+1}{i(w-1)}, \tag{B.8}$$

Herglotz-Pick functions F are related to such functions by $f = F \circ m$ where the property $\operatorname{Im} F > 0$ translates to $\operatorname{Im} f > 0$.

B.3. Fractional moments of HP functions

The following observation is of relevance for fractional moments.

Theorem B.6. *Let* (X, κ) *be a measure space and let* $F : X \to \mathbb{C}^+$. *Then for* F^s *defined using the principal value of* $\arg F$,

$$\int |F(x)|^s \kappa(dx) \;\leq\; \frac{1}{\cos(\pi s/2)} \left| \int F(x)^s \kappa(dx) \right| \tag{B.9}$$

as long as $\operatorname{Re}[e^{-is\pi/2} F^s] (\geq 0)$ *is* κ*-integrable.*

Proof. The bound readily follows from the observation that $\operatorname{Im} F \geq 0$ implies that $|F^s| \leq \operatorname{Re}[e^{-is\pi/2} F^s] / \cos(\pi s/2)$ for any $0 < s < 1$. □

For Herglotz-Pick functions $z \mapsto F(z;\omega)$ which also depend (measurably) on a parameter ω which takes values in a probability space, the above bound is useful in two different situations:

1. At specified z, a probability average is taken over ω.
2. An integral of such a function is taken over $E = \operatorname{Re} z$, with ω either fixed or also averaged over.

In the latter case, integrals of the form $\int_I F(E+i\eta;\omega)^s dE$ over intervals can be estimated taking advantage of the analytic structure of the function. This yields yet another proof of the finiteness of (local) fractional moments (8.23) for the boundary value of functions defined through the Borel-Stieltjes transform and a related *Kolmogorov inequality* for fractional moments of functions in the H^p space [**128**].

B.4. Relation to operator monotonicity

A function $F:(0,\infty)\mapsto\mathbb{R}$ is said to be operator monotone if for self-adjoint matrices of arbitrary order $F(A)\geq F(B)$ whenever $A\geq B\geq 0$.

A simple example is provided by the functions $F(x) = -(x+u)^{-1}$, with fixed $u\geq 0$. By an easy implication of the spectral representation, any positive linear combination of such functions and hence functions of the form

$$F(x) = ax+b+\int_{-\infty}^{0}\left(\frac{1}{u-x}-\frac{u}{1+u^2}\right)\mu(du) \tag{B.10}$$

with $a\geq 0$, $b\in\mathbb{R}$, and μ any non-negative Borel measure with $\int(1+u^2)^{-1}\,\mu(du)<\infty$ are operator monotone. An important result of K. Löwner [**291**] states that the converse is also true:

Theorem B.7 (Löwner). *A function $F:(0,\infty)\mapsto\mathbb{R}$ is operator monotone if and only if F is the boundary value of an HP function whose spectral measure is supported on $(-\infty,0]$.*

More on operator monotone functions can be found in [**124**].

B.5. Universality in the distribution of the values of random HP functions

A noteworthy property of Herglotz-Pick functions with pure-point spectrum is the identity of G. Boole [**63**]. The original proof is presented here in the Notes of Chapter 8, and a more general statement is given here in Proposition 8.2. This result offers another perspective on the finiteness of the fractional moments of HP functions. It also underlies the general observation [**297, 22**] that for any random Herglotz-Pick function with a shift-invariant distribution and only singular spectral measure, the random variables $F(u+i0)$ are each distributed as $(a+bX)$ with X a **Cauchy random variable** and $b>0, a\in\mathbb{R}$.

Bibliography

[1] E. Abrahams, P. W. Anderson, D. C. Licciardello, T. V. Ramakrishnan: Scaling theory of localization: Absence of quantum diffusion in two dimensions. Phys. Rev. Lett. 42, 673–676 (1979).

[2] R. Abou-Chacra, P. W. Anderson, D. J. Thouless: A selfconsistent theory of localization. J. Phys. C: Solid State Phys. 6, 1734–1752 (1973).

[3] R. Abou-Chacra, D. J. Thouless: Self-consistent theory of localization. II. localization near the band edges. J. Phys. C: Solid State Phys. 7, 65–75 (1974).

[4] M. Aizenman: Geometric analysis of Φ^4 fields and Ising models. Comm. Math. Phys. 86, 1–48 (1982).

[5] M. Aizenman: Localization at weak disorder: Some elementary bounds. Rev. Math. Phys. 6, 1163–1182 (1994).

[6] M. Aizenman, C. M. Newman: Tree graph inequalities and critical behavior in percolation models. J. Stat. Phys. 36, 107–143 (1984).

[7] M. Aizenman, D. Barsky: Sharpness of the phase transition in percolation models. Comm. Math. Phys. 108, 489–526 (1987).

[8] M. Aizenman, S. Molchanov: Localization at large disorder and at extreme energies: An elementary derivation. Comm. Math. Phys. 157, 245–278 (1993).

[9] M. Aizenman, A. Elgart, S. Naboko, J. Schenker, G. Stolz: Moment analysis for localization in random Schrödinger operators. Invent. Math. 163, 343–413 (2006).

[10] M. Aizenman, G. M. Graf: Localization bounds for an electron gas. J. Phys. A, Math. Gen. 31, No. 32, 6783–6806 (1998).

[11] M. Aizenman, J. H. Schenker, R. M. Friedrich, D. Hundertmark: Finite-volume fractional-moment criteria for Anderson localization. Comm. Math. Phys. 224, No. 1, 219–253 (2001).

[12] M. Aizenman, M. Shamis, S. Warzel: Resonances and partial delocalization on the complete graph. Ann. Henri Poincaré 16, 1969–2003 (2015).

[13] M. Aizenman, B. Simon: Local Ward identities and the decay of correlations in ferromagnets. Comm. Math. Phys. 77, 137–143 (1980).

[14] M. Aizenman, R. Sims, S. Warzel: Stability of the absolutely continuous spectrum of random Schrödinger operators on tree graphs. Probab. Theory Relat. Fields 136, 363–394 (2006).

[15] M. Aizenman, R. Sims, S. Warzel: Absolutely continuous spectra of quantum tree graphs with weak disorder. Comm. Math. Phys. 264, 371–389 (2006).

[16] M. Aizenman, S. Warzel: The canopy graph and level statistics for random operators on trees. Math. Phys. Anal. Geom. 9, 291–333 (2007).

[17] M. Aizenman, S. Warzel: Localization bounds for multiparticle systems. Comm. Math. Phys. 290, 903–934 (2009).

[18] M. Aizenman, S. Warzel: Extended states in a Lifshitz tail regime for random Schrödinger operators on trees. Phys. Rev. Lett. 106, 136804 (2011).

[19] M. Aizenman, S. Warzel: Absence of mobility edge for the Anderson random potential on tree graphs at weak disorder. Europhys. Lett. 96, 37004 (2011).

[20] M. Aizenman, S. Warzel: Absolutely continuous spectrum implies ballistic transport for quantum particles in a random potential on tree graphs. J. Math. Phys. 53, 095205 (2012).

[21] M. Aizenman, S. Warzel: Resonant delocalization for random Schrödinger operators on tree graphs. J. Eur. Math. Soc. 15, 1167–1222 (2013).

[22] M. Aizenman, S. Warzel: On the ubiquity of the Cauchy distribution in spectral problems. Probab. Theor. Relat. Fields 163, 61–87 (2015).

[23] M. Aizenman, S. Warzel: A boosted Simon-Wolff spectral criterion and resonant delocalization. To appear in Commun. Pure Appl. Math.

[24] O. Ajanki, F. Huveneers: Rigorous scaling law for the heat current in disordered harmonic chain. Comm. Math. Phys. 301, 841–883 (2011).

[25] B. L. Altshuler, Y. Gefen, A. Kamenev, L. S. Levitov: Quasiparticle lifetime in a finite system: A nonperturbative approach. Phys. Rev. Lett. 78, 2803 (1997).

[26] W. Amrein, V. Georgescu: On the characterization of bound states and scattering states. Helv. Phys. Acta 46, 635–658 (1973).

[27] P. W. Anderson: Absence of diffusion in certain random lattices. Phys. Rev. 109, 1492–1505 (1958).

[28] G. W. Anderson, A. Guionnet, O. Zeitouni: *An introduction to random matrices.* Cambridge (2010).

[29] T. Ando, Y. Matsumoto, Y. Uemura: Theory of Hall effect in a two-dimensional electron system. J. Phys. Soc. Jpn. 39, 279–288 (1975).

[30] N. Aronszajn: On a problem of Weyl in the theory of Sturm-Liouville equations. Am. J. Math. 79, 597–610 (1957).

[31] J. Asch, O. Bourget, A. Joye: Localization properties of the Chalker-Coddington model. Ann. Henri Poincaré 11, 1341–1373 (2010).

[32] J. Asch, O. Bourget, A. Joye: Dynamical localization of the Chalker-Coddington model far from transition. J. Stat. Phys. 147, 194–205 (2012).

[33] W. Aschbacher, V. Jaksic, Y. Pautrat, C.-A. Pillet: Transport properties of quasi-free fermions. J. Math. Phys. 48, 032101 (2007)

[34] S. Aubry, G. André: Analyticity breaking and Anderson localization in incommensurate lattices. Ann. Israel Phys. Soc. 3, 133–164 (1980).

[35] A. Avila: The absolutely continuous spectrum of the almost Mathieu operator. Preprint. arXiv:0810.2965 (2008).

[36] A. Avila, S. Jitomirskaya: The ten martini problem. Annals of Math. 170, 303–342 (2009).

[37] J. Avron, B. Simon: Almost periodic Schrödinger operators. II: The integrated density of states. Duke Math. J. 50, 369–391 (1983).

[38] J. Avron, R. Seiler, B. Simon: Homotopy and quantization in condensed matter physics. Phys. Rev. Lett. 51, 51–53 (1983).

[39] J. Avron, R. Seiler, B. Simon: Quantization of the Hall conductance for general, multiparticle Schrödinger operators. Phys. Rev. Lett. 54, 259–262 (1985).

[40] J. E. Avron: Adiabatic quantum transport. In: E. Akkermans et. al. *Les Houches* LXI, 1994. North-Holland, Amsterdam (1995).

[41] J. Avron, R. Seiler, B. Simon: Charge deficiency, charge transport and comparison of dimensions. Comm. Math. Phys. 159, 399–422 (1994).

[42] J. E. Avron, D. Osadchy, R. Seiler: A topological look at the quantum Hall effect. Physics Today, 28–42, August 2003.

[43] J. M. Barbaroux, J. M. Combes, P. D. Hislop: Localization near band edges for random Schrödinger operators. Helv. Phys. Acta 70, No. 1-2, 16–43 (1997).

[44] D. M. Basko, I. L. Aleiner, B. L. Altshuler: Metal-insulator transition in a weakly interacting many-electron system with localized single-particle states. Annals of Physics 321, 1126–1205 (2006).

[45] V. Bapst: The large connectivity limit of the Anderson model on tree graphs. J. Math. Phys. 55, 092101 (2014).

[46] H. Bauer: *Measure and integration theory.* de Gruyter (2001).

[47] S. Bachmann, W. De Roeck: From the Anderson model on a strip to the DMPK equation and random matrix theory. J. Stat. Phys. 139, 541–564 (2010).

[48] J. Bellissard: Ordinary quantum Hall effect and non-commutative cohomology. In: *Localization in disordered systems*, edited by W. Weller, P. Zieche. Leipzig: Teubner (1988).

[49] J. Bellissard, A. van Elst, H. Schulz-Baldes: The noncommutative geometry of the quantum Hall effect. J. Math. Phys. 35, 5373–5451 (1994).

[50] J. Bellissard, P. Hislop: Smoothness of correlations in the Anderson model at strong disorder. Ann. Henri Poincaré 8, 1–28 (2007).

[51] J. Bellissard, P. Hislop, G. Stolz: Correlations estimates in the lattice Anderson model. J. Stat. Phys. 129, 649–662 (2007).

[52] R. Ben Saad, C.-A. Pillet: A geometric approach to the Landauer-Büttiker formula. J. Math. Phys. 55, 075202 (2014).

[53] Yu. M. Berezanskij: On an eigenfunction expansion for selfadjoint operators. Am. Math. Soc. Transl. Ser. II. 93, 227–238 (1970).

[54] J. van den Berg, H. Kesten: Inequalities with applications to percolation and reliability. J. Appl. Prob. 22, 556–569 (1985).

[55] M. V. Berry, M. Tabor: Level clustering in the regular spectrum, Proc. Roy. Soc. A 356, 375–394 (1977).

[56] S. de Bièvre, F. Germinet: Dynamical localization for discrete and continuous random Schrödinger operators. Comm. Math. Phys. 194, 323–341 (1998).

[57] M. Birman, M. Solomyak: Remarks on the spectral shift function, J. Soviet Math. 3, 408–419 (1975).

[58] G. Biroli, A. C. Ribeiro-Teixeira, M. Tarzia: Difference between level statistics, ergodicity and localization transitions on the Bethe lattice. arXiv:1211.7334.

[59] M. Biskup, W. König: Long-time tails in the parabolic Anderson model. Ann. of Prob. 29, No. 2, 636–682 (2001).

[60] B. Bollobás: *Random graphs.* Cambridge Studies in Advanced Mathematics 73, Cambridge Univ. Press (2001).

[61] V. L. Bonch-Bruevich, R. Enderlein, B. Esser, R. Keiper, A. G. Mironov, I. P. Zvyagin: *Elektronentheorie ungeordneter Halbleiter.* Berlin: VEB Deutscher Verlag der Wissenschaften (1984).

[62] O. Bohigas, M.-J. Giannoni, C. Schmit: Characterization of chaotic quantum spectra and universality of level fluctuation laws. Phys. Rev. Lett. 52, 1–4 (1984).

[63] G. Boole: On the comparison of transcendents, with certain applications to the theory of definite integrals. Phil. Trans. Royal Soc. 147, 745–803 (1857).

[64] R. E. Borland: The nature of the electronic states in disordered one-dimensional systems. Proc. R. Soc. Lond. A 274, 529–545 (1963).

[65] J.-M. Bouclet, F. Germinet, A. Klein, J. Schenker: Linear response theory for magnetic Schrödinger operators in disordered media. J. Funct. Anal. 226, 301–372 (2005).

[66] P. Bougerol, J. Lacroix: *Products of random matrices with applications to Schrödinger operators.* Progress in Probability and Statistics 8, Springer (1985).

[67] J. Bourgain: On random Schrödinger operators on $\mathbb{Z}^d$. Discrete Contin. Dyn. Syst. 8, 1–15 (2002).

[68] J. Bourgain: *Green's function estimates for lattice Schrödinger operators and applications.* Princeton Univ. Press (2004).

[69] J. Bourgain, C. Kenig: On localization in the continuous Anderson-Bernoulli model in higher dimension. Invent. Math. 161, 389–426 (2005).

[70] J. Bourgain, A. Klein: Bounds on the density of states for Schrödinger operators. Invent. Math. 194, 41–72 (2013).

[71] J. Breuer, P. Forrester, U. Smilansky: Random Schrödinger operators from random matrix theory. J. Phys. A: Math. Theor. 40, F1–F8 (2007).

[72] J. Breuer, E. Ryckman, B. Simon: Equality of the spectral and dynamical definitions of reflection. Comm. Math. Phys. 295, 531–550 (2010).

[73] K. Broderix, D. Hundertmark, W. Kirsch, H. Leschke: The fate of Lifshitz tails in magnetic fields. J. Stat. Phys. 80, 1–22 (1995).

[74] J.-B. Bru, W. de Siqueira Pedra, C. Kurig: Heat production of free fermions subjected to electric fields. Commun. Pure Appl. Math. 68, 964–1013 (2014).

[75] J.-B. Bru, W. de Siqueira Pedra, C. Kurig: Macroscopic conductivity of free fermions in disordered media. Rev. Math. Phys. 26, 1450008 (2014).

[76] J.-B. Bru, W. de Siqueira Pedra: Microscopic conductivity of lattice fermions at equilibrium — Part II: Interacting particles. To appear in Lett. Math. Phys.

[77] J.-B. Bru, W. de Siqueira Pedra: From the 2nd law of thermodynamics to ac-conductivity measures of interacting fermions in disordered media. Math. Models Methods Appl. Sci. (DOI: 10.1142/S0218202515500566).

[78] L. Bruneau, V. Jaksic, Y. Last, C.-A. Pillet: Landauer-Büttiker and Thouless conductance. Commun. Math. Phys. 338, 347–366 (2015).

[79] A. Bufetov: Markov averaging and ergodic theorems for several operators. In: *Topology, ergodic theory, real algebraic geometry.* Amer. Math. Soc. Transl. 202, Amer. Math. Soc., Providence, RI, pp. 39–50 (2001).

[80] R. Carmona: Exponential localization in one dimensional disordered systems. Duke Math. J. 49, 191–213 (1982).

[81] R. Carmona, A. Klein, F. Martinelli: Anderson localization for Bernoulli and other singular potentials. Comm. Math. Phys. 108, 41–66 (1987).

[82] R. Carmona, J. Lacroix: *Spectral theory of random Schrödinger operators*. Probability and Its Applications. Birkhäuser, Basel (1990).

[83] G. Casati, L. Molinari, F. Izrailev: Scaling properties of band random matrices. Phys. Rev. Lett. 64, 1851–1854 (1990).

[84] G. Casati, B. V. Chirikov, I. Guarneri, F. M. Izrailev: Band-random-matrix model for quantum localization in conservative systems. Phys. Rev. E 48, R1613–R1616 (1993).

[85] A. Casher, J. L. Lebowitz: Heat flow in regular and disordered harmonic chains. J. Math. Phys. 12, 1701–1711 (1971).

[86] J. T. Chalker, P. D. Coddington: Percolation, quantum tunnelling and the integer Hall effect. J. Phys. C: Solid State Phys. 21, 2665–2679 (1988).

[87] D. C. Champeney: *A handbook of Fourier theorems*. Cambridge Univ. Press (1987).

[88] V. Chulaevsky, Y. Suhov: Eigenfunctions in a two-particle Anderson tight binding model. Comm. Math. Phys. 289, 701–723 (2009).

[89] V. Chulaevsky, Y. Suhov: Multi-particle Anderson localisation: Induction on the number of particles. Math. Phys. Anal. Geom. 12, 117–139 (2009).

[90] V. Chulaevsky, Y. Suhov: *Multi-scale analysis for random quantum systems with interaction*. Progress in Mathematical Physics 65, Birkhäuser/Springer, New York (2014).

[91] J.-M. Combes: Connections between quantum dynamics and spectral properties of time-evolution operators. In: *Differential equations with applications to mathematical physics*, edited by W. F. Ames, E. M. Harrel, J. V. Herod. Academic Press, Boston, MA, pp. 59–68 (1993).

[92] J.-M. Combes, F. Germinet, A. Klein: Generalized eigenvalue-counting for the Anderson model, J. Stat. Phys. 135, 201–216 (2009).

[93] J. M. Combes, F. Germinet, A. Klein: Conductivity and the current-current correlation measure. J. Phys. A 43, 474010 (2010).

[94] J. M. Combes, P. D. Hislop: Localization for some continuous, random Hamiltonians in d-dimensions. J. Funct. Anal. 124, 149–180 (1994).

[95] J. M. Combes, P. D. Hislop: Landau Hamiltonians with random potentials: Localization and the density of states. Comm. Math. Phys. 177, 603–629 (1996).

[96] J. M. Combes, P. D. Hislop, F. Klopp: An optimal Wegner estimate and its application to the global continuity of the integrated density of states for random Schrödinger operators. Duke Math. J. 140, 469–498 (2007).

[97] J. M. Combes, P. D. Hislop, E. Mourre: Spectral averaging, perturbation of singular spectra, and localization. Trans. Am. Math. Soc. 348, 4883–4894 (1996).

[98] J. M. Combes, L. Thomas: Asymptotic behaviour of eigenfunctions for multiparticle Schrödinger operators. Comm. Math. Phys. 34, 251–270 (1973).

[99] A. Connes: Non-commutative differential geometry. Publ. IHES 62, 257–360 (1986).

[100] H. D. Cornean, A. Jensen, V. Moldoveanu: A rigorous proof of the Landauer-Büttiker formula. J. Math. Phys. 46, 042106 (2005).

[101] H. D. Cornean, P. Duclos, G. Nenciu, R. Purice: Adiabatically switched-on electrical bias in continuous systems, and the Landauer-Büttiker formula. J. Math. Phys. 49, 102106 (2008).

[102] R. Courant: Über die Eigenwerte bei den Differentialgleichungen der mathematischen. Physik. Math. Z. 7, 1–57 (1920).

[103] W. Craig, B. Simon: Log Hölder continuity of the integrated density of states for stochastic Jacobi matrices. Comm. Math. Phys. 90, 207–218 (1983).

[104] W. Craig, B. Simon: Subharmonicity of the Lyapunov exponent. Duke Math. J. 50, 551–560 (1983).

[105] H. L. Cycon, R. G. Froese, W. Kirsch, B. Simon: *Schrödinger operators, with application to quantum mechanics and global geometry.* Texts and Monographs in Physics. Springer, Berlin (1987).

[106] D. J. Daley, D. Vere-Jones: *An introduction to the theory of point processes.* Vols. 1 and 2, Springer (2008).

[107] D. Damanik, R. Sims, G. Stolz: Localization for one-dimensional, continuum, Bernoulli-Anderson models. Duke Math. J. 114, No. 1, 59–100 (2002).

[108] D. Damanik, P. Stollmann: Multi-scale analysis implies strong dynamical localization. Geom. Funct. Anal. 11, 11–29 (2001).

[109] E. B. Davies: *Heat kernels and spectral theory.* Cambridge Tracts in Mathematics, 92. Cambridge University Press, Cambridge (1989).

[110] E. B. Davies: *Spectral theory and differential operators.* Cambridge Studiesl in Advanced Mathematics, 42. Cambridge: Cambridge University Press (1995).

[111] R. del Rio, S. Jitomirskaya, Y. Last, B. Simon: Operators with singular continuous spectrum IV: Hausdorff dimensions, rank one perturbations and localization. J. d'Analyse Math. 69, 153–200 (1996).

[112] R. del Rio, S. Jitomirskaya, Y. Last, B. Simon: What is localization? Phys. Rev. Lett. 75, 117–119 (1995).

[113] F. Delyon, H. Kunz, B. Souillard: One-dimensional wave equations in disordered media. J. Phys. A: Math. Gen. 16, 25–42 (1983).

[114] F. Delyon, B. Souillard: Remark on the continuity of the density of states of ergodic finite difference operators. Comm. Math. Phys. 94, 289–291 (1984).

[115] F. Delyon, Y. Lèvy, B. Souillard: Anderson localization for multi-dimensional systems at large disorder or large energy. Comm. Math. Phys. 100, 463–470 (1985).

[116] A. Dembo, O. Zeitouni: *Large deviations techniques and applications.* Springer, New York (1998).

[117] A. B. De Monvel, S. Naboko, P. Stollmann, G. Stolz: Localization near fluctuation boundaries via fractional moments and applications. Journal d'Analyse Mathematique 100, 83–116. (2006).

[118] M. Demuth, M. Krishna: *Determining spectra in quantum theory.* Progress in Mathematical Physics, 44, Birkhäuser, Boston (2005).

[119] M. Disertori, F. Merkl, S. W. W. Rolles: Localization for a nonlinear sigma model in a strip related to vertex reinforced jump processes. Comm. Math. Phys. 332, 783–825 (2014).

[120] M. Disertori, T. Spencer, M. Zirnbauer: Quasi-diffusion in a 3D supersymmetric hyperbolic sigma model. Comm. Math. Phys. 300, 435–486 (2010).

[121] M. Disertori, T. Spencer: Anderson localization for a supersymmetric sigma model. Comm. Math. Phys. 300, 659–671 (2010).

[122] R. L. Dobrushin, S. B. Shlosman: Completely analytical interactions: Constructive description. J. Stat. Phys. 46, 983–1014 (1986).

[123] N. Dombrowski, F. Germinet: Linear response theory for random Schrödinger operators and noncommutative integration. Markov Process. Related Fields 14, 403–426 (2008).

[124] W. F. Donoghue Jr.: *Monotone matrix functions and analytic continuation.* Springer, Berlin (1976).

[125] M. D. Donsker, S. R. S. Varadhan: Asymptotic for the Wiener sausage. Comm. Pure Appl. Math. 28, 525–565 (1975).

[126] H. von Dreifus: On the effects of randomness in ferromagnetic models and Schrödinger operators. NYU PhD thesis (1987).

[127] H. von Dreifus, A. Klein: A new proof of localization in the Anderson tight binding model. Comm. Math. Phys. 124, No. 2, 285–299 (1989).

[128] P. L. Duren: *Theory of H^p spaces.* Dover (2000).

[129] F. J. Dyson: A Brownian motion model for the eigenvalues of a random matrix. J. Math. Phys. 3, 1191–1198 (1962).

[130] M. S. P. Eastham: *The spectral theory of periodic differential equations.* Texts in Mathematics. Scottish Academic Press, Edinburgh-London (1973).

[131] I. Dumitriu, A. Edelman: Matrix models for beta ensembles. J. Math. Phys. 43, 5830–5847 (2002).

[132] J. T. Edwards, D. J. Thouless: Numerical studies of localization in disordered systems. J. Phys. C: Solid State Phys. 5, 807–820 (1972).

[133] K. B. Efetov: *Supersymmetry in disorder and chaos.* Cambridge University Press (1987).

[134] P. Elbau, G. M. Graf: Equality of bulk and edge Hall conductance revisited. Comm. Math. Phys. 229, 415–432 (2002).

[135] A. Elgart: Lifshitz tails and localization in the three-dimensional Anderson model. Duke Math. J. 146, 331–360 (2009).

[136] A. Elgart, B. Schlein: Adiabatic charge transport and the Kubo formula for Landau-type Hamiltonians. Comm. Pure Appl. Math. 57, 590–615 (2004).

[137] A. Elgart, M. Shamis, S. Sodin: Localisation for non-monotone Schrödinger operators. J. Eur. Math. Soc. 16, 909–924 (2014).

[138] V. Enss: Asymptotic completeness for quantum-mechanical potential scattering, I. Short range potentials. Comm. Math. Phys. 61, 285–291 (1978).

[139] L. Erdös: Lifschitz tail in a magnetic field: The nonclassical regime. Probab. Theory Relat. Fields 112, No. 3, 321–371 (1998).

[140] L. Erdös, M. Salmhofer, H.-T. Yau: Quantum diffusion of the random Schrödinger evolution in the scaling limit I. The non-recollision diagrams. Acta Mathematica 200, 211–277 (2008).

[141] L. Erdös, M. Salmhofer, H.-T. Yau: Quantum diffusion of the random Schrödinger evolution in the scaling limit II. The recollision diagrams. Comm. Math. Phys. 271, 1–53 (2007).

[142] L. Erdös, M. Salmhofer, H.-T. Yau: Quantum diffusion for the Anderson model in the scaling limit. Ann. Henri Poincaré 8, 621–685 (2007).

[143] L. Erdös, S. Péché, J. A. Ramirez, B. Schlein, H.-T. Yau: Bulk universality for Wigner matrices. Comm. Pure and Appl. Math. 63, 895–925 (2010).

[144] L. Erdös, H.-T. Yau: A comment on the Wigner-Dyson-Mehta bulk universality conjecture for Wigner matrices. Electron. J. Probab. 17, 1–5 (2012).

[145] L. Erdös, H.-T. Yau, J. Yin: Bulk universality for generalized Wigner matrices. Probab. Theor. Relat. Fields 154, 341–407 (2012).

[146] P. Exner, M. Helm, P. Stollmann: Localization on a quantum graph with a random potential on the edges. Rev. Math. Phys. 19, 923–939 (2007).

[147] M. Fauser, S. Warzel: Multiparticle localization for disordered systems on continuous space via the fractional moment method. Rev. Math. Phys. 27, 1550010 (2015).

[148] E. Feenberg: A note on perturbation theory. Phys. Rev. 74, 206–208 (1948).

[149] E. Feenberg: Theory of scattering processes. Phys. Rev. 74, 664–669 (1948).

[150] M. Fekete: Über die Verteilung der Wurzeln bei gewissen algebraischen Gleichungen mit ganzzahligen Koeffizienten. Mathematische Zeitschrift 17, 228–249 (1923).

[151] R. P. Feynman, A. R. Hibbs: *Quantum mechanics and path integrals.* McGraw-Hill (1965).

[152] A. Figotin, A. Klein: Localization of classical waves I: Acoustic waves. Comm. Math. Phys. 180, 439–482 (1996).

[153] A. Figotin, A. Klein: Localization of classical waves II. Electromagnetic waves. Comm. Math. Phys. 184, 411–441 (1997).

[154] E. Fischer: Über quadratische Formen mit reellen Koeffizienten. Monatsheft Math. Phys. 16, 234–249 (1905).

[155] W. Fischer, H. Leschke, P. Müller: Spectral localization by Gaussian random potentials in multi-dimensional continuous space J. Stat. Phys. 101, 935–985 (2000).

[156] M. E. Fisher: Critical temperatures of anisotropic Ising lattices, II. General upper bounds. Phys. Rev. 162, 480–485 (1967).

[157] S. Fishman, D. R. Grempel, R. E. Prange: Chaos, quantum recurrences and Anderson localization. Phys. Rev. Lett. 49, 509–512 (1982).

[158] V. Fock: Bemerkung zur Quantelung des harmonischen Oszillators im Magnetfeld. Z. Physik 47, 446–448 (1928).

[159] R. Froese, F. Halasan, D. Hasler: Absolutely continuous spectrum for the Anderson model on a product of a tree with a finite graph. J. Funct. Anal. 262, 1011–1042 (2012).

[160] R. Froese, D. Hasler, W. Spitzer: Transfer matrices, hyperbolic geometry and absolutely continuous spectrum for some discrete Schrödinger operators on graphs. J. Funct. Anal. 230, 184–221 (2006).

[161] R. Froese, D. Hasler, W. Spitzer: Absolutely continuous spectrum for the Anderson model on a tree: A geometric proof of Klein's theorem. Comm. Math. Phys. 269, 239–257 (2007).

[162] R. Froese, D. Hasler, W. Spitzer: Absolutely continuous spectrum for a random potential on a tree with strong transverse correlations and large weighted loops. Rev. Math. Phys. 21, 709–733 (2009).

[163] R. Froese, D. Lee, C. Sadel, W. Spitzer, G. Stolz: Localization for transversally periodic random potentials on binary trees. To appear in J. Spectr. Theory.

[164] J. Fröhlich, F. Martinelli, E. Scoppola, T. Spencer: Constructive proof of localization in the Anderson tight binding model. Comm. Math. Phys. 101, 21–46 (1985).

[165] J. Fröhlich, T. Spencer: Absence of diffusion in the Anderson tight binding model for large disorder or low energy. Comm. Math. Phys. 88, 151–184 (1983).

[166] H. Furstenberg: Non commuting random products. Trans. Amer. Math. Soc. 108, 377–428 (1963).

[167] L. Geisinger: Poisson eigenvalue statistics for random Schrödinger operators on regular graphs. Ann. Henri Poincaré 16, 1779–1806 (2015).

[168] H.-O. Georgii: *Gibbs measures and phase transitions.* de Gruyter (1988).

[169] F. Germinet, A. Klein: Bootstrap multiscale analysis and localization in random media. Comm. Math. Phys. 222, 415–448 (2001).

[170] F. Germinet, A. Klein: A characterization of the Anderson metal-insulator transport transition. Duke Math. J. 124, 309–351 (2004).

[171] F. Germinet, A. Klein, J. Schenker: Dynamical delocalization in random Landau Hamiltonians. Annals Math. 166, 215–244 (2007).

[172] F. Germinet, A. Klein, J. Schenker: Quanization of the Hall conductance and delocalisation in ergodic Landau Hamiltonians. Rev. Math. Phys. 21, 1045–1080 (2009).

[173] F. Germinet A. Klein, B. Mandy: Dynamical delocalization in random Landau Hamiltonians with unbounded random couplings. In: *Spectral and scattering theory for quantum magnetic systems*, Contemp. Math. 500, Amer. Math. Soc., pp. 87–100 (2009).

[174] F. Germinet, F. Klopp: Enhanced Wegner and Minami estimates and eigenvalue statistics of random Anderson models at spectral edges. Ann. Henri Poincaré 5, 1263–1285 (2013).

[175] F. Gesztesy, E. Tsekanovskii: On matrix-valued Herglotz functions. Mathematische Nachrichten 218, 61–138 (2000).

[176] I. Goldsheid, S. Molchanov, L. Pastur: A pure point spectrum of the stochastic one-dimensional Schrödinger equation. Funct. Anal. Appl. 11, 1–10 (1977).

[177] A. Y. Gordon, S. Jitomirskaya, Y. Last, B. Simon: Duality and singular continuous spectrum in the almost Mathieu equation. Acta Math. 178, 169–183 (1997).

[178] G. M. Graf: Anderson localization and the space-time characteristic of continuum states. J. Stat. Phys. 75, 337–346 (1994).

[179] G. M. Graf: Aspects of the integer quantum Hall effect. In: *Spectral theory and mathematical physics: A festschrift in honor of Barry Simon's 60th birthday.* Proc. Sympos. Pure Math. 76, Part 1, Amer. Math. Soc., pp. 429–442 (2007).

[180] G. M. Graf, A. Vaghi: A remark on the estimate of a determinant by Minami. Lett. Math. Phys. 79, 17–22 (2007).

[181] G. Grimmett: *Percolation.* Springer-Verlag (1999).

[182] I. Guarneri: On an estimate concerning quantum diffusion in the presence of a fractal spectrum. Europhys. Lett. 21, 729–733 (1993).

[183] F. Halasan: Absolutely continuous spectrum for the Anderson model on some tree-like graphs. Ann. Henri Poincaré 13, 789–811 (2012).

[184] P. Hall, C. C. Heyde: *Martingale limit theory and its application.* Academic Press (1980).

[185] B. I. Halperin: Quantized Hall conductance, current-carrying edges states, and the existence of extended states in a two-dimensional disordered potential. Phys. Rev. B 25, 2185–2190 (1982).

[186] J. M. Hammersley: Percolation processes: Lower bounds for the critical probability. Ann. Math. Statist. 28, 790–795 (1957).

[187] E. Hamza, A. Joye, G. Stolz: Dynamical localization for unitary Anderson models. Math. Phys. Anal. Geom. 12, 381–444 (2009).

[188] E. Hamza, A. Joye: Correlated Markov quantum walks. Ann. Henri Poincaré 13, 1767–1805 (2012).

[189] E. Hamza, A. Joye: Spectral transition for random quantum walks on trees. Comm. Math. Phys. 326, 415–439 (2014).

[190] E. Hamza, R. Sims, G. Stolz: A note on fractional moments for the one-dimensional continuum Anderson model. Journal of Mathematical Analysis and Applications 365, 435–446 (2010).

[191] M. B. Hastings, S. Michalakis: Quantization of Hall conductance for interacting electrons on a torus. Commun. Math. Phys. 334, 433–471 (2015).

[192] D. C. Herbert, R. Jones: Localized states in disordered systems. J. Phys. C: Solid St. Phys. 4, 1145–1161 (1971).

[193] T. Hirota, K. Ishii: Exactly soluble models of one-dimensional disordered systems. Prog. Theor. Phys. 45, 1713–1715 (1971).

[194] P. D. Hislop, O. Lenoble: Basic properties of the current-current correlation measure for random Schrödinger operators. J. Math. Phys. 47, 112106 (2006).

[195] D. Hofstädter: Energy levels and wave functions of Bloch electrons in rational and irrational magnetic field. Phys. Rev. B 14, 2240–2249 (1976).

[196] F. Hoecker-Escuti: Localization for random Schrödinger operators with low density potentials. Preprint (2012).

[197] D. Hundertmark: A short introduction to Anderson localization. Chapter 9 in: *Analysis and stochastics of growth processes and interface models*, edited by P. Mörters, R. Moser, M. Penrose, H. Schwetlick, J. Zimmer. Oxford (2008).

[198] J. Z. Imbrie: On many-body localization for quantum spin chains. arXiv:1403.7837.

[199] Y. Imry: *Introduction to mesoscopic physics.* 2nd ed., Oxford Univ. Press (2008).

[200] K. Ishii: Localization of eigenstates and transport phenomena in the one-dimensional disordered system. Progress of Theoretical Physics Supplement 53, 77–138 (1973).

[201] D. Jakobson, S. Miller, I. Rivin, Z. Rudnick: Level spacings for regular graphs. In *Emerging applications of number theory*, edited by D. Hejhal, J. Friedman, M. Gutzwiller, A. Odlyzko, IMA Vol. Math. Appl. 109, Springer (1999).

[202] V. Jaksic, S. Molchanov, L. Pastur: On the propagation properties of surface waves. In *Wave propagation in complex media.* IMA Vol. Math. Appl. 96, 143 (1998).

[203] V. Jaksic, S. Molchanov: Localization of surface spectra. Comm. Math. Phys. 208, 153–172 (1999).

[204] V. Jaksic, Y. Last: Corrugated surfaces and a.c. spectrum, Rev. Math. Phys. 12, 1465–1503 (2000).

[205] V. Jaksic, Y. Last: Spectral structure of Anderson type Hamiltonians. Inv. Math. 141, 561–577 (2000).

[206] V. Jaksic, Y. Last: Simplicity of singular spectrum in Anderson-type Hamiltonians. Duke Math. J. 133, 185–204 (2006).

[207] V. Jaksic: Topics in spectral theory. Open Quantum Systems I. The Hamiltonian Approach. Lecture Notes in Mathematics, 1880, 235–312, Springer (2006).

[208] V. Jaksic, B. Landon, A. Panati: Note on reflectionless Jacobi matrices. Comm. Math. Phys. 332, 827–838 (2014).

[209] S. Jitomirskaya: Metal-insulator transition for the almost Mathieu operator. Ann. of Math. 150 (3), 1159–1175 (1999).

[210] S. Jitomirskaya: Ergodic Schrödinger operators (on one foot). In: *Spectral theory and mathematical physics: A festschrift in honor of Barry Simon's 60th birthday*, edited by F. Gesztesy, P. Deift, C. Galvez, P. Perry, W. Schlag, pp. 613–647 (2007).

[211] S. Jitomirskaya, H. Schulz-Baldes, G. Stolz: Delocalization in random polymer models. Comm. Math. Phys. 233, 27–48 (2003).

[212] A. Joye: Fractional moment estimates for random unitary operators. Lett. Math. Phys. 72, 51–64 (2005).

[213] A. Joye: Random time-dependent quantum walks. Commun. Math. Phys. 307, 65–100 (2011).

[214] A. Joye: Dynamical localization for d-dimensional random quantum walks. Quant. Inf. Process. 11, 1251–1269 (2012).

[215] J. Karamata: Neuer Beweis und Verallgemeinerung der Tauberschen Sätze, welche die Laplacesche und Stieltjes Transformation betreffen. J. Reine Angew. Math. 164, 27–39 (1931).

[216] O. Kallenberg: *Foundations of modern probability.* 2nd ed., Springer (2002).

[217] T. Kato: *Perturbation theory for linear operators.* 2nd corr. print. of the 2nd ed. Grundlehren der Mathematischen Wissenschaften, 132. Berlin, etc.: Springer-Verlag (1984).

[218] M. Keller: Absolutely continuous spectrum for multi-type Galton Watson trees. Ann. Henri Poincaré 13, 1745–1766 (2012).

[219] M. Keller, D. Lenz, S. Warzel: On the spectral theory of trees with finite cone type. Israel Journal of Mathematics 194, 107–135 (2013).

[220] M. Keller, D. Lenz, S. Warzel: Absolutely continuous spectrum for random operators on trees of finite cone type. Journal d'Analyse Mathematique 118, 363–396 (2012).

[221] M. Keller, D. Lenz, S. Warzel: An invitation to trees of finite cone type: Random and deterministic operators. To appear in Markov Proc. Relat. Fields.

[222] H. Kesten: Symmetric random walks on groups. Trans. Amer. Math. Soc. 92, 336–354 (1959).

[223] R. Ketzmerick, G. Petschel, T. Geisel: Slow decay of temporal correlations in quantum systems with Cantor spectra. Phys. Rev. Lett. 69, 695–698 (1992).

[224] R. Killip, F. Nakano: Eigenfunction statistics in the localized Anderson model. Ann. Henri Poincaré 8, 27–36 (2007).

[225] R. Killip, I. Nenciu: Matrix models for circular ensembles. Int. Math. Res. Not. 50, 2665–2701 (2004).

[226] R. Killip, M. Stoiciu: Eigenvalue statistics for CMV matrices: From Poisson to clock via random matrix ensembles. Duke Math. J. 146, 361–399 (2009).

[227] J. F. C. Kingman: *Poisson processes.* Oxford Univ. Press (1993).

[228] W. Kirsch: Random Schrödinger operators. In: *Schrödinger operators.* Proc. Nord. Summer Sch. Math., Sandbjerg Slot, Sonderborg/Denmark 1988, Lect. Notes Phys. 345, 264–370 (1989).

[229] W. Kirsch: Wegner estimates and Anderson localization for alloy-type potentials. Math. Z. 221, No. 3, 507–512 (1996).

[230] W. Kirsch: An invitation to random-Schrödinger operators. Panoramas et Syntheses 25, 1–119 (2008).

[231] W. Kirsch, O. Lenoble, L. Pastur: On the Mott formula for the ac conductivity and binary correlators in the strong localization regime of disordered systems. J. Phys. A, Math. Gen. 36, 12157–12180 (2003).

[232] W. Kirsch, F. Martinelli: On the density of states of Schrödinger operators with a random potential. J. Phys. A 15, 2139–2156 (1982).

[233] W. Kirsch, F. Martinelli: On the ergodic properties of the spectrum of general random operators. J. Reine Angew. Math. 334, 141–156 (1982).

[234] W. Kirsch, F. Martinelli: On the spectrum of Schrödinger operators with a random potential. Comm. Math. Phys. 85, 329–350 (1982).

[235] W. Kirsch, F. Martinelli: Large deviations and Lifshitz singularity of the integrated density of states of random Hamiltonians. Comm. Math. Phys. 89, 27–40 (1983).

[236] W. Kirsch, S. Kotani, B. Simon: Absence of absolutely continuous spectrum for some one dimensional random but deterministic Schrödinger operators. Ann. Henri Poincaré 42, 383–406 (1985).

[237] W. Kirsch, B. Metzger: The integrated density of states for random Schrödinger operators. In: *Spectral theory and mathematical physics: A festschrift in honor of Barry Simon's 60th birthday*, edited by F. Gesztesy, P. Deift, C. Galvez, P. Perry, W. Schlag, pp. 649–696 (2007).

[238] W. Kirsch, B. Simon: Lifshitz tails for periodic plus random potential. J. Stat. Phys. 42, 799–808 (1986).

[239] W. Kirsch, P. Stollmann, G. Stolz: Localization for random perturbations of periodic Schrödinger operators. Random Oper. Stoch. Equ. 6, 241–268 (1998).

[240] W. Kirsch, P. Stollmann, G. Stolz: Anderson localization for random Schrödinger operators with long range interactions. Comm. Math. Phys. 195, 495–507 (1998).

[241] W. Kirsch, S. Warzel: Lifshits tails caused by anisotropic decay: The emergence of a quantum-classical regime. Math. Phys. Anal. Geometry 8, 257–285 (2005).

[242] W. Kirsch, S. Warzel: Anderson localization and Lifshits tails for random surface potentials. J. Func. Anal. 230, 222–250 (2006).

[243] A. Kiselev, Y. Last: Solutions, spectrum, and dynamics for Schrödinger operators on infinite domains. Duke Math. J. 102, 125–150 (2000).

[244] A. Klein: Absolutely continuous spectrum in the Anderson model on the Bethe lattice. Math. Res. Lett. 1, No. 4, 399–407 (1994).

[245] A. Klein: The Anderson metal-insulator transition on the Bethe lattice. In *Proceedings of the XIth International Congress on Mathematical Physics, Paris, France, July 18–23, 1994*, edited by Daniel Iagolnitzer, Cambridge, MA: International Press, pp. 383–391 (1995).

[246] A. Klein: Extended states in the Anderson model on the Bethe lattice. Adv. Math. 133, No. 1, 163–184 (1998).

[247] A. Klein: Multiscale analysis and localization of random operators, Panoramas et Syntheses 25, 121–159 (2008).

[248] A. Klein, O. Lenoble, P. Müller: On Mott's formula for the ac-conductivity in the Anderson model. Ann. Math.166, 549–577 (2007).

[249] A. Klein, S. Molchanov: Simplicity of eigenvalues in the Anderson model. J. Stat. Phys. 122, 95–99 (2006).

[250] A. Klein, P. Müller: The conductivity measure for the Anderson model. Zh. Mat. Fiz. Anal. Geom. 4, 128–150 (2008).

[251] A. Klein, P. Müller: Ac-conductivity and electromagnetic energy absorption for the Anderson model in linear response theory. To appear in Markov Proc. Relat. Fields.

[252] A. Klein, C. Sadel: Absolutely continuous spectrum for random Schrödinger operators on the Bethe strip. Math. Nachr. 285, 5–26 (2012).

[253] K. von Klitzing, G. Dorda, M. Pepper: New method for high-accuracy determination of the fine-structure constant based on quantized Hall resistance. Physical Review Letters 45, 494–497 (1980).

[254] F. Klopp: Localization for some continuous random Schrödinger operators. Comm. Math. Phys. 167, 553–569 (1995).

[255] F. Klopp: Lifshitz tails for random perturbations of periodic Schrödinger operators. Proc. Indian Acad. Sci., Math. Sci. 112, 147–162 (2002).

[256] F. Klopp: Precise high energy asymptotics for the integrated density of states of an unbounded random Jacobi matrix. Rev. Math. Phys. 12, 575–620 (2000).

[257] F. Klopp: Lifshitz tails for random perturbations of periodic Schrödinger operators. Proc. Indian Acad. Sci., Math. Sci. 112, 147–162 (2002).

[258] F. Klopp: Weak disorder localization and Lifshitz tails. Comm. Math. Phys. 232, 125–155 (2002).

[259] F. Klopp: Weak disorder localization and Lifshitz tails: Continuous Hamiltonians. Ann. Henri Poincaré 3, No. 4, 711–737 (2002).

[260] F. Klopp: Internal Lifshitz tails for Schrödinger operators with random potentials. J. Math. Phys. 43, No. 6, 2948–2958 (2002).

[261] F. Klopp, T. Wolff: Lifshitz tails for 2-dimensional random Schrödinger operators. J. Anal. Math. 88, 63–147 (2002).

[262] F. Klopp, K. Pankrashkin: Localization on quantum graphs with random edge lengths. Lett. Math. Phys. 87, 99–114 (2009).

[263] S. Kotani: Ljapunov indices determine absolutely continuous spectra of stationary random one-dimensional Schrödinger operators. North-Holland Math. Libr. 32, 225–247 (1984).

[264] S. Kotani: Lyapunov exponents and spectra for one-dimensional random Schrödinger operators. Contemp. Math. 50, 277–286 (1986).

[265] S. Kotani, B. Simon: Localization in general one-dimensional random systems. II. Continuum Schrödinger operators. Comm. Math. Phys. 112, 103–119 (1987).

[266] T. Kottos, U. Smilansky: Periodic orbit theory and spectral statistics for quantum graphs. Ann. Phys. 274, 76–124 (1999).

[267] M. G. Krein: On the trace formula in perturbation theory (Russian), Mat. Sbornik N.S. 33, 597–626 (1953).

[268] U. Krengel: *Ergodic theorems.* With a supplement by Antoine Brunel. Walter de Gruyter, Berlin-New York (1985).

[269] M. Krishna: Anderson models with decaying randomness: Existence of extended states. Proc. Indian Acad. Sci. Math. 100, 285–294 (1990).

[270] M. Krishna: Absolutely continuous spectrum for sparse potentials. Proc. Indian Acad. Sci. Math. 103, 333–339 (1993).

[271] E. Kritchevski, B. Valkó, B. Virág: The scaling limit of the critical one-dimensional random Schrödinger operator. Comm. Math. Phys. 314, 775–806 (2012).

[272] R. Kubo: *Statistical mechanics.* North Holland, Amsterdam (1965).

[273] R. Kubo, M. Toda, N. Hashitsume: *Nonequilibrium statistical mechanics.* Springer, New York (1985).

[274] H. Kunz, B. Souillard: Sur le spectre des opérateurs aux diffrénces finies aléatoires. Comm. Math. Phys. 78, 201–246 (1980).

[275] H. Kunz: The quantum Hall effect for electrons in a random potential. Communications in Mathematical Physics 112, 121–145 (1987).

[276] A. Lagendijk, B. van Tiggelen, D. S. Wiersma: Fifty years of Anderson localization. Phys. Today 62, 24–29 (2009).

[277] J. W. Lamperti: *Probability. A survey of the mathematical theory.* 2nd ed. Wiley Series in Probability and Mathematical Statistics. Wiley, New York (1996).

[278] J. W. Lamperti: *Stochastic processes. A survey of the mathematical theory.* Applied Mathematical Sciences. 23. New York-Heidelberg-Berlin: Springer-Verlag (1977).

[279] L. Landau: Diamagnetismus der Metalle. Z. Physik 64, 629–637 (1930).

[280] R. Landauer: Spatial variation of currents and fields due to localized scatterers in metallic conduction. IBM Journal of Research and Development 1, 223–231 (1957).

[281] R. Lang: *Spectral theory of random Schrödinger operators. A genetic introduction.* Lecture Notes in Mathematics. 1498. Berlin: Springer-Verlag (1991).

[282] R. B. Laughlin: Quantized Hall conductivity in two dimensions. Phys. Rev. B 23, 5632–5633 (1981).

[283] Y. Last: Quantum dynamics and decompositions of singular continuous spectra. J. Funct. Anal. 142, 406–445 (1996).

[284] P. Lax: *Functional analysis.* Wiley, New York (2002).

[285] H. Leschke, P. Müller, S. Warzel: A survey of rigorous results on random Schrödinger operators for amorphous solids, Markov Proc. Relat. Fields 9, 729–760 (2003).

[286] I. M. Lifshitz: Energy spectrum structure and quantum states of disordered condensed systems. Sov. Phy. Usp. 7, 549 (1965).

[287] E. H. Lieb: A refinement of Simon's inequality. Comm. Math. Phys. 77, 127–135 (1980).

[288] E. H. Lieb, D. W. Robinson: The finite group velocity of quantum spin systems. Com. Math. Phys. 28, 251–257 (1972).

[289] I. M. Lifshits, S. A. Gredeskul, L. A. Pastur: *Introduction to the theory of disordered systems.* Wiley, New York (1988).

[290] P. Lloyd: Exactly solvable model of electronic states in a three-dimensional disordered Hamiltonian: Non-existence of localized states. J. Phys. C 2, 1717–1725 (1969).

[291] K. Löwner: Über monotone Matrixfunktionen, Math. Z. 38, 177–216 (1934).

[292] F. Martinelli, H. Holden: On absence of diffusion near the bottom of the spectrum for a random Schrödinger operator on $L^2(\mathbb{R}^\nu)$. Comm. Math. Phys. 93, 197–217 (1984).

[293] F. Martinelli, E. Scoppola: Remark on the absence of absolutely continuous spectrum for d-dimensional Schrödinger operators with random potential for large disorder or low energy. Comm. Math. Phys. 97, 465–471 (1985).

[294] B. D. McKay: The expected eigenvalue distribution of a random labelled regular graph. Linear Algebra and its Applications 40, 203–216 (1981).

[295] A. J. McKean, M. Stone: Localization as an alternative to Goldstone's theorem. Ann. Phys. 131, 36–55 (1981).

[296] M. L. Mehta: *Randon matrices*, 3rd ed., Elsevier (2004).

[297] P. A. Mello: Theory of random matrices: Spectral statistics and scattering problems. In: *Mesoscopic quantum physics*, edited by E. Akkermans, G. Montambaux, J.-L.Pichard, J. Zinn-Justin. Elsevier Amsterdam, pp. 435–491 (1995).

[298] M. Menshikov: Coincidence of critical points in percolation problems. Soviet Math. Dokl. 33, 856–859 (1986).

[299] G. A. Mezincescu: Internal Lifschitz singularities of disordered finite-difference Schrödinger operators. Comm. Math. Phys. 103, 167–176 (1986).

[300] J. D. Miller, B. Derrida: Weak disorder expansion for the Anderson model on a tree. J. Stat. Phys, 75, 357–388 (1993).

[301] N. Minami: Local fluctuation of the spectrum of a multidimensional Anderson tight binding model. Comm. Math. Phys. 177, 709–725 (1996).

[302] S. A. Molchanov: The structure of eigenfunctions of one-dimensional unordered structures. Math. USSR Izv. 12, 69–101 (1978).

[303] S. A. Molchanov: The local structure of the spectrum of the one-dimensional Schrödinger operator. Comm. Math. Phys. 78, 429–446 (1981).

[304] S. A. Molchanov: Ideas in the theory of random media. Acta Appl. Math. 22, 139–282 (1991).

[305] S. A. Molchanov: Lectures on random media. In: *Lectures on probability theory*, edited by Dominique Bakry et al., Ecole d'Ete de Probabilites de Saint-Flour XXII-1992. Summer School, 9th–25th July, 1992, Saint-Flour, France. Berlin: Springer-Verlag. Lect. Notes Math. 1581, 242–411 (1994).

[306] N. F. Mott, W. D. Twose: The theory of impurity conduction. Advances in Physics 10, 107–163 (1961).

[307] N. F. Mott: Electrons in disordered structures. Advances Phys. 16, 49–144 (1967).

[308] S. Naboko, R. Nichols, G. Stolz: Simplicity of eigenvalues in Anderson-type models, Ark. Mat. 51, 157-183 (2011).

[309] H. Najar: Lifshitz tails for random acoustic operators. J. Math. Phys. 44, 1842–1867 (2003).

[310] F. Nakano: Absence of transport in Anderson localization. Rev. Math. Phys. 14, 375–407 (2002).

[311] F. Nakano: The repulsion between localization centers in the Anderson model. J. Stat. Phys. 123, 803–810 (2006).

[312] F. Nakano: Level statistics for one-dimensional Schrödinger operators and Gaussian beta ensemble. J. Stat. Phys. 156, 66–93 (2014).

[313] S. Nakao: On the spectral distribution of the Schrödinger operator with random potential. Jap. J. Math., New Ser. 3, 111–139 (1977).

[314] S. Nakamura, J. Bellissard: Low energy bands do not contribute to quantum Hall effect. Comm. Math. Phys. 131, 283–305 (1990).

[315] G. Nenciu: Independent electrons model for open quantum systems: Landauer-Büttiker formula and strict positivity of the entropy production. J. Math. Phys. 48, 033302 (2007).

[316] K. S. Novoselov, Z. Jiang, Y. Zhang, S. V. Morozovl, H. L. Stormer, U. Zeitler, J. C. Maan, G. S. Boebinger, P. Kim, A. K. Geim: Room-temperature quantum Hall effect in graphene. Science 315, 1379 (2007).

[317] V. I. Oseledec: A multiplicative ergodic theorem. Ljapunov characteristic numbers for dynamical systems. Trans. Moscow Math. Soc. 19, 197–231 (1968).

[318] A. Pal, D. A. Huse: Many-body localization phase transition. Phys. Rev. B 82, 174411 (2010).

[319] L. A. Pastur: On the Schrödinger equation with a random potential. Theor. Math. Phys. 6, 299–306 (1971). [Russian original: Teor. Mat. Fiz. 6, 415–424 (1971)].

[320] L. A. Pastur: Spectra of random self adjoint operators. Russ. Math. Surv. 28, No. 1, 1–67 (1973).

[321] L. A. Pastur: Behaviour of some Wiener integrals for $t \to \infty$ and the density of states of the Schrödinger equation with random potential. Teor. Mat. Fiz. 32, 88–95 (1977).

[322] L. A. Pastur: Spectral properties of disordered systems in the one-body approximation. Comm. Math. Phys. 75, 179–196 (1980).

[323] L. A. Pastur: On a heuristic expansion method in the strong localization regime of the theory of disordered systems. In: *Applied and industrial mathematics*, Venice-2, 1998, Kluwer, Dordrecht, pp. 173–185 (2000).

[324] L. Pastur, A. Figotin: *Spectra of random and almost-periodic operators*. Grundlehren der Mathematischen Wissenschaften. 297. Springer, Berlin (1992).

[325] D. Pearson: Singular continuous measures in scattering theory. Comm. Math. Phys. 60, 13–36 (1978).

[326] T. Poerschke, G. Stolz, J. Weidmann: Expansions in generalized eigenfunctions of selfadjoint operators. Math. Z. 202, No. 3, 397–408 (1989).

[327] E. Prodan: The non-commutative geometry of the complex classes of topological insulators. Topological Quantum Matter 1, 1–16 (2014).

[328] M. Reed, B. Simon: *Methods of modern mathematical physics. II: Fourier analysis, self-adjointness*. New York-San Francisco-London: Academic Press, a subsidiary of Harcourt Brace Jovanovich, Publishers (1975).

[329] M. Reed, B. Simon: *Methods of modern mathematical physics. IV: Analysis of operators*. New York-San Francisco-London: Academic Press (1978).

[330] M. Reed, B. Simon: *Methods of modern mathematical physics. III: Scattering theory*. New York, San Francisco, London: Academic Press (1979).

[331] M. Reed, B. Simon: *Methods of modern mathematical physics. I: Functional analysis*. Rev. and enl. ed. New York etc.: Academic Press, a subsidiary of Harcourt Brace Jovanovich, Publishers (1980).

[332] C. Remling: The absolutely continuous spectrum of Jacobi matrices. Annals of Math. 174, 125–171 (2011).

[333] A. Rényi: Remarks on the Poisson process. Studia Scientiarum Math. Hungarica 2, 119–123 (1967).

[334] V. Rivasseau: Lieb's correlation inequality for plane rotors. Comm. Math. Phys. 77, 145–147 (1980).

[335] I. Rodnianski, W. Schlag: Classical and quantum scattering for a class of long range random potentials. Int. Math. Res. Not. 5, 243–300 (2003).

[336] W. Rudin: *Functional analysis*. McGraw-Hill (1973).

[337] D. Ruelle: A remark on bound states in potential scattering theory. Rivista Nuevo Cimento 61A, 655–662 (1969).

[338] C. Sadel, B. Virág: A central limit theorem for products of random matrices and GOE statistics for the Anderson model on long boxes. Preprint. arXiv:1411.1040.

[339] H. Schanz, U. Smilansky: Periodic-orbit theory of Anderson-localization on graphs. Phys. Rev. Lett. 84, 1427–1430 (2000).

[340] J. Schenker: Eigenvector localization for random band matrices with power law band width. Comm. Math. Phys. 290, 1065–1097 (2009).

[341] H. Schulz-Baldes, J. Bellissard: Anomalous transport: A mathematical framework. Rev. Math. Phys. 10, 1–46 (1998).

[342] H. Schulz-Baldes, J. Bellissard: A kinetic theory for quantum transport in aperiodic media. J. Stat. Phys. 91, 991–1026 (1998).

[343] H. Schulz-Baldes, J. Kellendonk, T. Richter: Simultaneous quantization of edge and bulk Hall conductivity. J. Phys. A33, L27–L32 (2000).

[344] M. Segev, Y. Silberberg, D. N. Christodoulides: Anderson localization of light. Nature Photonics 7, 197–208 (2013).

[345] M. Shamis: Resonant delocalization on the Bethe strip. Ann. Henri Poincaré 15, 1549–1567 (2014).

[346] S. Shlosman: From the seminar on mathematical statistical physics in Moscow State University, 1962–1994. Constructive criteria. The European Physical Journal H 37, 595–603 (2012).

[347] B. Simon: Correlation inequalities and the decay of correlations in ferromagnets. Comm. Math. Phys. 77, 137–143 (1980).

[348] B. Simon: Schrödinger semigroups. Bull. Amer. Math. Soc., New Ser. 7, 447–526 (1982).

[349] B. Simon: Kotani theory for one dimensional stochastic Jacobi matrices. Comm. Math. Phys. 89, 227–234 (1983).

[350] B. Simon: Lifshitz tails for the Anderson model. J. Stat. Phys. 38, 65–76 (1985).

[351] B. Simon: Internal Lifshitz tails. J. Stat. Phys. 46, 911–918 (1987).

[352] B. Simon: Cyclic vectors in the Anderson model. Rev. Math. Phys. 6, 1183–1185 (1994).

[353] B. Simon: *The statistical mechanics of lattice gases.* Princeton Univ. Press, Princeton (1993).

[354] B. Simon: Spectral analysis of rank one perturbations and applications. In: *Mathematical quantum theory II: Schrödinger operators*, edited by J. Feldman et al., Providence, RI: American Mathematical Society. CRM Proc. Lect. Notes. 8, 109–149 (1995).

[355] B. Simon: Operators with singular continuous spectrum: I. General operators. Ann. of Math. 141, 131–145 (1995).

[356] B. Simon: Spectral averaging and the Krein spectral shift, Proc. Amer. Math. Soc. 126, 1409–1413 (1998).

[357] B. Simon: *Functional integration and quantum physics.* 2nd ed., Amer. Math. Soc. (2005).

[358] B. Simon: *Orthogonal polynomials on the unit circle, Part 2: Spectral theory,* AMS Colloquium Series, American Mathematical Society, Providence, RI (2005).

[359] B. Simon: Equilibrium measures and capacities in spectral theory. Inverse Problems and Imaging 1, 713–772 (2007).

[360] B. Simon: *A comprehensive course in analysis Part 3: Harmonic analysis.* Amer. Math. Soc. (2015).

[361] B. Simon, T. Spencer: Trace class perturbations and the absence of absolutely continuous spectra. Comm. Math. Phys. 125, 113–125 (1989).

[362] B. Simon, M. Taylor: Harmonic analysis on SL(2,(R) and smoothness of the density of states in the one-dimensional Anderson model. Comm. Math. Phys. 101, 1–19 (1985).

[363] B. Simon, T. Wolff: Singular continuous spectrum under rank one perturbations and localization for random Hamiltonians. Comm. Pure Appl. Math. 39, 75–90 (1986).

[364] A. Sommerfeld: Zur Elektronentheorie der Metalle auf Grund der Fermischen Statistik. I. Teil: Allgemeines, Strömungs und Austrittsvorgänge. Z. Physik 47, 1–42 (1928).

[365] T. Spencer: The Schrödinger equation with a random potential. A mathematical review. In: *Critical phenomena, random systems, gauge theories.* Proc. Summer Sch. Theor. Phys., Sess. 43, Les Houches/France 1984, Pt. 2, 895–942 (1986).

[366] T. Spencer: Lifshitz tails and localization. Preprint (1993).

[367] P. Stollmann: *Caught by disorder. Bound states in random media.* Progress in Mathematical Physics 20. Boston: Birkhäuser (2001).

[368] P. Stollmann: Lifshitz asymptotics via linear coupling of disorder. Math. Phys. Anal. Geom. 2, No. 3, 279–289 (1999).

[369] P. Stollmann: Wegner estimates and localization for continuum Anderson models with some singular distributions. Arch. Math. 75, 307–311 (2000).

[370] G. Stolz: Localization for random Schrödinger operators with Poisson potential. Ann. Inst. Henri Poincaré, Phys. Theor. 63, 297–314 (1995).

[371] G. Stolz: An introduction to the mathematics of Anderson localization. In: *Entropy and Quantum II.* Proceedings of the Arizona School of Analysis and Applications. Contemporary Mathematics 552 (2010).

[372] R. Strichartz: Fourier asymptotics of fractal measures. J. Funct. Anal. 89, 154–187 (1990).

[373] A. S. Sznitman: *Brownian motion, obstacles and random media.* Springer Monographs in Mathematics. Berlin: Springer (1998).

[374] T. Tao, V. Vu: Random matrices: Universality of local eigenvalue statistics up to the edge. Comm. Math. Phys. 298, 549–572 (2010).

[375] T. Tao, V. Vu: The Wigner-Dyson-Mehta bulk universality conjecture for Wigner matrices. Electronic J. Probab. 16, 2104–2121 (2011).

[376] M. Tautenhahn: Localization criteria for Anderson models on locally finite graphs. J. Stat. Phys. 144, 60–75 (2011).

[377] M. Tautenhahn, I. Veselic: Minami's estimate: Beyond rank one perturbation and monotonicity. Ann. Henri Poincaré 15, 737–754 (2014).

[378] G. Temple: The theory of Rayleigh's principle as applied to continuous systems. Proc. Roy. Soc. London. Ser. A 119, 276–293 (1928).

[379] G. Teschl: *Jacobi operators and completely integrable nonlinear lattices.* Amer. Math. Soc. (1999).

[380] G. Teschl: *Mathematical methods in quantum mechanics.* Amer. Math. Soc. (2009).

[381] D. J. Thouless: A relation between the density of states and range of localization for one dimensional random systems. J. Phys. C: Solid St. Phys. 5, 77–81 (1971).

[382] D. J. Thouless: Electrons in disordered systems and the theory of localization. Phys. Repts. 13, 93–142 (1974).

[383] D. J. Thouless: Maximum metallic resistance in thin wires. Phys. Rev. Lett. 39, 1167–1169 (1977).

[384] D. J. Thouless: Introduction to disordered systems. In: *Critical phenomena, random systems, gauge theories.* Proc. Summer Sch. Theor. Phys., Sess. 43, Les Houches/France 1984, Pt. 2, 685–722 (1986).

[385] D. J. Thouless: Quantization of particle transport. Phys. Rev. B 27, 6083–6087 (1983).

[386] D. J. Thouless, M. Kohmoto, M. P. Nightingale, M. den Nijs: Quantized Hall conductance in a two-dimensional periodic potential. Phys. Rev. Lett. 49, 405–408 (1982).

[387] D. C. Tsui, H. L. Stormer, A. C. Gossard: Two-dimensional magnetotransport in the extreme quantum limit. Phys. Rev. Lett. 48, 1559–1562 (1982).

[388] B. Valkó, B. Virág: Continuum limits of random matrices and the Brownian carousel. Inventiones Math. 177, 463–508 (2009).

[389] B. Valkó, B. Virág: Random Schrödinger operators on long boxes, noise explosion and the GOE, Trans. Amer. Math. Soc. 366, 3709–3728 (2014).

[390] I. Veselic: Localization for random perturbations of periodic Schrödinger operators with regular Floquet eigenvalues. Ann. Henri Poincaré 3, 389–409 (2002).

[391] I. Veselic: Integrated density and Wegner estimates for random Schrödinger operators. In: *Spectral theory of Schrödinger operators*, edited by Rafael del Rio, Rafael et al. Lecture notes from a workshop, Mexico, December 3–7, 2001. Providence, RI: American Mathematical Society. Contemp. Math. 340, 97–183 (2004).

[392] I. Veselic: *Existence and regularity properties of the integrated density of states of random Schrödinger operators.* Lecture Notes in Mathematics, Vol. 1917. Springer, Berlin (2008).

[393] J. von Neumann, E. Wigner: Über das Verhalten von Eigenwerten bei adiabatischen Prozessen, Physikalische Zeitschrift 30, 467–470 (1929).

[394] S. Warzel: On Lifshitz tails in magnetic fields. Dissertation. Universität Erlangen-Nürnberg, Logos (2001).

[395] S. Warzel: Surprises in the phase diagram of the Anderson model on the Bethe lattice. In: *XVIIth International Congress on Mathematical Physics*, edited by A. Jensen. World Scientific, pp. 239–253 (2013).

[396] F. Wegner: Electrons in disordered systems. Scaling near the mobility edge. Z. Physik B 25, 327–337 (1976).

[397] F. Wegner: Bounds on the density of states in disordered systems. Z. Physik B 44, 9–15 (1981).

[398] J. Weidmann: *Linear operators in Hilbert spaces.* Graduate Texts in Mathematics, Vol. 68. Springer, New York-Heidelberg-Berlin (1980).

[399] J. M. Ziman: *Models of disorder.* Cambridge (1979).

Index

Selected Published Titles in This Series

168 **Michael Aizenman and Simone Warzel,** Random Operators, 2015
164 **Terence Tao,** Expansion in Finite Simple Groups of Lie Type, 2015
163 **Gérald Tenenbaum,** Introduction to Analytic and Probabilistic Number Theory, Third Edition, 2015
162 **Firas Rassoul-Agha and Timo Seppäläinen,** A Course on Large Deviations with an Introduction to Gibbs Measures, 2015
161 **Diane Maclagan and Bernd Sturmfels,** Introduction to Tropical Geometry, 2015
160 **Marius Overholt,** A Course in Analytic Number Theory, 2014
159 **John R. Faulkner,** The Role of Nonassociative Algebra in Projective Geometry, 2014
158 **Fritz Colonius and Wolfgang Kliemann,** Dynamical Systems and Linear Algebra, 2014
157 **Gerald Teschl,** Mathematical Methods in Quantum Mechanics: With Applications to Schrödinger Operators, Second Edition, 2014
156 **Markus Haase,** Functional Analysis, 2014
155 **Emmanuel Kowalski,** An Introduction to the Representation Theory of Groups, 2014
154 **Wilhelm Schlag,** A Course in Complex Analysis and Riemann Surfaces, 2014
153 **Terence Tao,** Hilbert's Fifth Problem and Related Topics, 2014
152 **Gábor Székelyhidi,** An Introduction to Extremal Kähler Metrics, 2014
151 **Jennifer Schultens,** Introduction to 3-Manifolds, 2014
150 **Joe Diestel and Angela Spalsbury,** The Joys of Haar Measure, 2013
149 **Daniel W. Stroock,** Mathematics of Probability, 2013
148 **Luis Barreira and Yakov Pesin,** Introduction to Smooth Ergodic Theory, 2013
147 **Xingzhi Zhan,** Matrix Theory, 2013
146 **Aaron N. Siegel,** Combinatorial Game Theory, 2013
145 **Charles A. Weibel,** The K-book, 2013
144 **Shun-Jen Cheng and Weiqiang Wang,** Dualities and Representations of Lie Superalgebras, 2012
143 **Alberto Bressan,** Lecture Notes on Functional Analysis, 2013
142 **Terence Tao,** Higher Order Fourier Analysis, 2012
141 **John B. Conway,** A Course in Abstract Analysis, 2012
140 **Gerald Teschl,** Ordinary Differential Equations and Dynamical Systems, 2012
139 **John B. Walsh,** Knowing the Odds, 2012
138 **Maciej Zworski,** Semiclassical Analysis, 2012
137 **Luis Barreira and Claudia Valls,** Ordinary Differential Equations, 2012
136 **Arshak Petrosyan, Henrik Shahgholian, and Nina Uraltseva,** Regularity of Free Boundaries in Obstacle-Type Problems, 2012
135 **Pascal Cherrier and Albert Milani,** Linear and Quasi-linear Evolution Equations in Hilbert Spaces, 2012
134 **Jean-Marie De Koninck and Florian Luca,** Analytic Number Theory, 2012
133 **Jeffrey Rauch,** Hyperbolic Partial Differential Equations and Geometric Optics, 2012
132 **Terence Tao,** Topics in Random Matrix Theory, 2012
131 **Ian M. Musson,** Lie Superalgebras and Enveloping Algebras, 2012
130 **Viviana Ene and Jürgen Herzog,** Gröbner Bases in Commutative Algebra, 2011
129 **Stuart P. Hastings and J. Bryce McLeod,** Classical Methods in Ordinary Differential Equations, 2012
128 **J. M. Landsberg,** Tensors: Geometry and Applications, 2012